1220 18⁵⁰

Musculoskeletal Pain: Diagnosis and Physical Treatment

Musculoskeletal Pain: Diagnosis and Physical Treatment

David A. Zohn, M.D.

Chief, Department of Physical Medicine and Rehabilitation, Northern Virginia Doctors Hospital; Chief, Department of Physical Medicine and Rehabilitation, National Orthopaedic and Rehabilitation Hospital, Arlington, Virginia

John McM. Mennell, M.D.

Associate Clinical Professor, Physical Medicine and Rehabilitation, University of California, Davis, School of Medicine; Chief, Rehabilitation Medicine Service, Veterans Administration Hospital, Martinez, California

Little, Brown and Company
Boston

First Edition

Library of Congress catalog card No. 75-22601

ISBN 0-316-98893-6

Printed in the United States of America

Contents

Preface

Frequently we hear the recommendation to use conservative treatment in the management of musculoskeletal pain. What is meant by the term *conservative treatment* is often nebulous, though the inference is that some empirical form of physical therapy should be tried. Yet we have not found any concise reference covering the proper use and the potential benefits of physical methods of treatment in pain problems arising from the musculoskeletal system. That is the reason for this book. We are deliberately omitting reference to the use of such therapies in chronic diseases, as, in contrast to the treatment of pain, there is voluminous literature on the treatment of such conditions.

The appropriate therapeutic use of a physical modality presumes that a proper diagnosis has been established. Part I, therefore, is devoted to the principles of clinical and laboratory examination necessary to establish a diagnosis of the condition causing pain. Since there traditionally musculoskeletal pain problems have been diagnosed in generalities, emphasis here has been placed on establishing the specific pathological process within the proper anatomical entity.

In Part II, the wide range of modalities of physical treatment available are described first and then illustrated in the application of physical treatment to a specific topographical area, the shoulder. Principles are stressed governing the choice of modalities in the healing phase of a pathological condition that gives rise to pain, and in the restorative phase when healing is accomplished. The avoidance of morbility in structures of the musculoskeletal system not primarily affected is the basis of the proper use of physical therapy in both phases.

Although in Parts I and II emphasis is on principles, the text is designed to be a practical help to the practicing physician, whether family physician or specialist, who is involved in the diagnosis and treatment of musculoskeletal pain. The book is also meant to remind physical therapists of the many modalities of physical therapy and their proper usage. Therefore, Part II of the text is devoted to a discussion of various pain problems that present special difficulty to the physician or that are of particular interest. The treatment programs discussed differ, some more than others, from those customarily prescribed. We hope these departures will stimulate rather than disconcert the reader.

The concept that pain may have mechanical causes and be treated by mechanical means is stressed by drawing attention to a normal anatomical range of motion—joint play—as a prerequisite to normal function. Movements of joint play cannot be produced by the use of voluntary muscles. The loss of play produces mechanical pathology, called joint dysfunction. Its treatment is also mechanical—joint manipulation.

In discussing the concept of mechanical causes of pain we draw attention to the irritable trigger point as a common cause of pain from muscle. This is perhaps especially timely in view of the attention now being directed to acupuncture. The treatment of the irritable trigger point by use of vapocoolant spray and stretching or by needling has been described as "Western acupuncture."

We express our profound gratitude to many people for their assistance in producing this book. Particularly, we thank Mr. Fred Belliveau, vice-president of Little, Brown for his patience in allowing us great latitude in meeting a deadline. We specially thank Mrs. Lin Richter, our editor, who was more directly involved in the time lag and who showed remarkable tolerance and understanding. Little, Brown has also allowed us to reproduce illustrations used in previous publications, and for this we are most grateful. Many of the new line drawings were prepared by Miss Terri Blattspielen and Stephen Blair. Others are the work of Roger Shannon, M.D., and Walker Thompson; to these people we express our thanks. The excellent new photographic illustrations are largely the work of Messrs. Hal Strong and Stan Smith and of the Martinez Veterans Administration Hospital. We are most grateful to them for their untiring work. Eugene Jacobs, M.D., Chief of Radiology at National Orthopaedic and Rehabilitation Hospital, and his staff have been extremely helpful in opening their extensive files to us to reproduce illustrative radiographs.

Thanks must be given to Robert W. Downie, M.D. He has read the manuscript and offered valuable criticism, assisted in the preparation of the section on laboratory studies, and taken many photographs for us. His help has been warmly appreciated. The manuscript was also carefully reviewed by Fred Gill, M.D., and Stephen Levin, M.D. Their numerous helpful suggestions have been incorporated into the text.

Mrs. Edna Johnson and Mrs. Joan Francis, our secretaries, deserve special thanks for a job well done.

There are many others without whose contributions we would never have completed our work. We ask that they accept our tribute anonymously to save us the embarrassment of leaving anyone out.

D. A. Z.
J. McM. M.

Preamble

1
Introduction to Principles of Diagnosis and Treatment

Dazzling medical advances of increasing complexity are heralded in the professional and lay press on a regular basis. Their dramatic announcements are in sharp contrast to improvement in diagnosis and treatment of the humdrum problems of musculoskeletal pain that we offer in this text. We place emphasis on treatment by means of the common, fairly inexpensive, readily available, and relatively harmless modalities of physical therapy, not by means of surgical procedures or sophisticated drugs.

The type of musculoskeletal pain problems we shall be dealing with has been a stepchild in medicine, neglected in the medical schools and treated somewhat cavalierly by practitioners in the expectation that these are self-limiting difficulties with a high rate of spontaneous cure. Treatment has therefore tended to be empirical. Yet reflection will reveal that such problems occupy a large percentage of the time of the practicing physician, are an enormous item in the nation's health budget, and produce large-scale economic disturbances in terms of time lost from work, workmen's compensation payments, and, of course, medicolegal liability litigation. Further, a great many patients with disease of the viscera or with systemic disease have either a presenting complaint of musculoskeletal pain or a significant component of their complaint relating to the musculoskeletal system, and accurate differentiation must be made early.

We emphasize principles rather than details in diagnosis and the prescription of treatment in the expectation that such details can then be applied appropriately.

Things not commonly discussed in medical literature—the trigger point and referred pain patterns, manipulation, fibrositis, the therapeutic use of cold, the use of epidural block in lumbar radiculitis, biofeedback techniques, and others—are incorporated into this text. Acupuncture, operant conditioning in chronic pain problems, and the use of sclerosing solutions are commented upon briefly. We hope thus to expand the horizons of clinicians so that they can see and use what is available for the benefit of their patients.

Before we discuss the principles of differential diagnosis and treatment, there are some basic concepts about physical treatment to be considered.

With few exceptions such as the use of ultraviolet light in the treatment of rickets (in conjunction with vitamin D), the use of iontophoresis using copper for the treatment of certain fungal conditions, and the use of manipulation for the treatment of joint dysfunction, no physical therapy modality by itself ever cures anyone of anything.

But lack of curative effect of physical therapy modalities does not vitiate our reasoning that properly prescribed therapy is essential to the resolution of many musculoskeletal problems. It is the "proper use" to which we shall address ourselves.

The proper choice of modality depends on which of two phases of disability the patient presents at the time of evaluation. Phase one is the healing phase; phase two is the restorative phase. The principles of prescription are different in each phase.

PHASE ONE: THE HEALING PHASE

Prevention of Morbidity

The first principle of the proper use of any physical therapy modality is that it should prevent morbidity. This principle has two subdivisions: First, *rest from function* of an injured or pathologically involved anatomical structure promotes healing; second, the *maintenance of as normal a physiological state as possible* in anatomical structures which may be secondarily affected, but in which there is no primary pathology, prevents morbidity. When dealing with joint pathology, for instance, one must not only keep muscle in as good a state as he can but also maintain the vascular tree, both arterial and venous, the neuromuscular mechanisms, and the lymphatic system.

Rest from Function

The concept of rest in the treatment of pain is not new. Even more important is the concept of rest *from function* in healing. Rest from function in no way means *do not move;* indeed, *movement is life.* But movement in the presence of pathology, if it is not to do harm, means *movement within the limit of pain.* One degree of movement is better than none; 5 degrees is better than 1; 10 degrees is better than 5; and so on. Yet if you move something 10 degrees which should be moved only 5 degrees, you are breaking a principle of treatment and increasing morbidity instead of preventing it. However, too little movement when more is indicated also increases morbidity. Movement in this context means passive movement carried out by the therapist within the range of normal voluntary movement. It is necessary to develop a sense of dosage of movement.

PHASE TWO: THE RESTORATIVE PHASE

In the restorative phase the proper use of any physical therapy modality is to restore function that has been lost during healing of the pathological state, and this is the guiding principle in the choice of therapy modality. Unfortunately, some loss of function is usually inevitable during the healing phase. But the restorative therapy must not create new traumatic pathology. In this context it is especially important to remember that muscles cannot move a joint that is not free to move. Muscle exercises alone, then, cannot effectively or safely restore lost function to a joint in which there is intrinsic dysfunction.

IMPROPER USE OF PHYSICAL THERAPY

The use of any physical therapy modality that contravenes these principles is improper. The proper use of physical therapy hastens healing and lessens disability and in terms of musculoskeletal pain problems should usually restore normalcy.

WHAT IS "NORMAL"?

Usual concepts of what is normal in a patient based on preconceived ideas of posture, gait, range of motion, and flexibility often prevent correct diagnosis. When a patient is first seen, there is no way of telling what his "normal" is. It may be far removed from one's idea of what it should be.

As is often pointed out, man has been subject through the millennia to phylogenetic changes, and these changes have caused difficulties. In the musculoskeletal system, adoption of the upright position as a cause of back pain and alteration of the arches and function of the feet as a cause of foot pain are examples. While such speculations are interesting from an anthropological point of view, they beg the clinical issues and do little to help the individual patient with his pain.

Further, most individuals are asymmetrical with regard to the two halves of the body. Although they have lived with this asymmetry in an asymptomatic fashion until musculoskeletal pain brings them to a physician, correction of anatomical asymmetries is often indicated as part of the overall treatment since rectification of any and all mechanical problems lessens the possibility of recurrence of symptoms.

TOPOGRAPHICAL MOVEMENT

Perhaps because medical students learn about anatomy on cadavers, they tend to develop some misconceptions about joints and their normal movement.

One is accustomed to thinking of wrists, elbows, shoulders, knees, ankles, hands, feet, and the topographical parts of the back in approaching a patient with pain in these areas. One is likely to differentiate weight-bearing joints of the lower extremities from non-weight-bearing joints of the upper extremities and to think little, if at all, of the synovial joints in the back.

Joints, ligaments, muscles, menisci, bursae, bones, articular cartilage, and synovium are the same wherever they occur in the musculoskeletal system; they do not change because of some topographical difference. Nor do they react differently to pathological change or traumatic stress because of their topographical differences. The frequency of injury may alter but not the nature of the change.

Even then, clinically, things do not happen to the wrist, the ankle, the back, or any other topographical area, unless the area has been subjected to gross external trauma or disease. Pathological conditions usually primarily involve an anatomical structure in the joint or a structure around it. Furthermore, every joint in a topographical area is concerned with a specific functional movement.

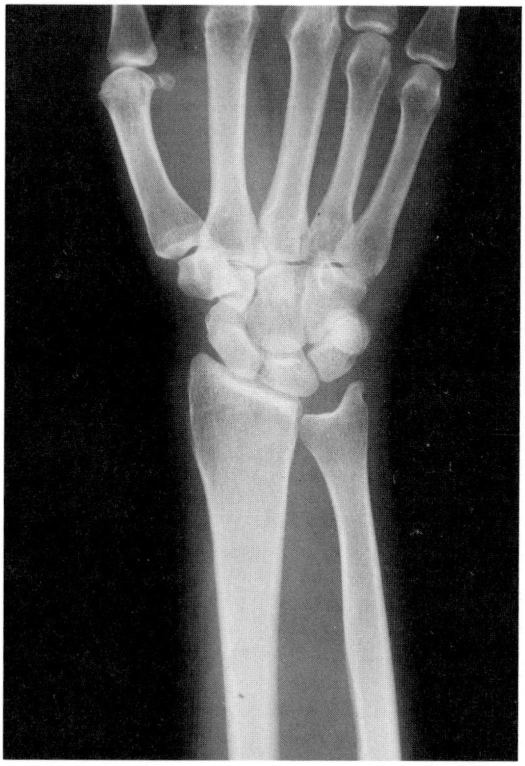

Figure 1-1. X-ray picture of a normal wrist to remind the reader of the intricacies of the carpus. From this are traced Figures 1-2 to 1-6.

The Wrist as an Example

Let us consider the wrist (Figure 1-1). There are four major joints at the wrist. These are the radiocarpal joint, the multiple-facet joint between the radius and the navicular and lunate, illustrated in Figure 1-2; the ulnomeniscotriquetral joint, illustrated in Figure 1-3; the distal radioulnar joint, illustrated in Figure 1-4; and the midcarpal joint, the joint between the

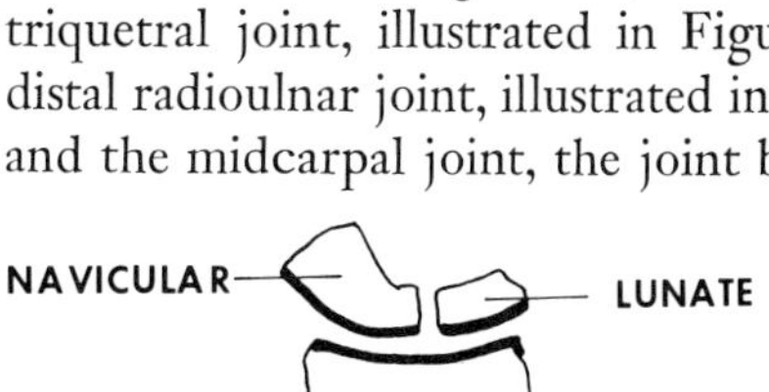

Figure 1-2. The radiocarpal joint—a multiple joint made up of the distal end of the radius and an articular surface of the navicular and the lunate bones—where most of the flexion at the wrist takes place.

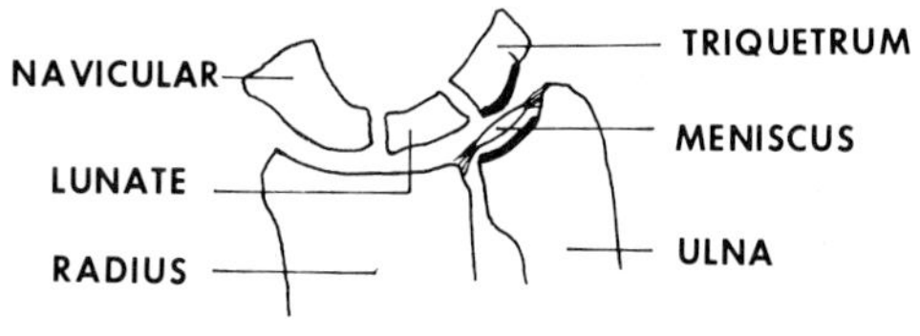

Figure 1-3. Ulnomeniscotriquetral joint with its intraarticular meniscus. The function of supination depends for the most part on the movement of this joint.

navicular and lunate proximally and the capitate and hamate distally, also a multiple-facet joint and illustrated in Figure 1-5. Figure 1-6 is a composite illustrating all four major joints. Of these, the ulnotriquetral joint is of special importance because it has an intraarticular meniscus incorporated in it.

RANGE OF MOVEMENT AT THE WRIST

Anatomists describe a voluntary range of movement of the wrist: flexion, extension (dorsiflexion), adduction, and abduction. The movements of supination and pronation of the hand and forearm are not usually associated with any movement in any joint in the wrist.

This is a correct description if movement is studied in the cadaver, in which all voluntary movement is lost; but it is an incomplete account of normal movement in the living wrist.

VOLUNTARY MOVEMENT AT THE WRIST

In fact, in the living wrist, flexion takes place almost exclusively at the radiocarpal joint (Figure 1-7), and extension takes place almost exclusively at the midcarpal joint (Figure 1-8). Although there is no voluntary movement at the ulnomeniscotriquetral joint in the anatomical sense of the word, abduction (radial deviation) depends almost entirely on freedom of movement of the ulnomeniscotriquetral joint. Similarly, adduction (ulnar deviation) depends almost entirely on freedom of movement at the radionavicular joint, at which there is no voluntary movement in the anatomical sense. However, supination of the hand and forearm is just as dependent on freedom of movement at the ulnomeniscotriquetral joint as it is on freedom of movement of the proximal radioulnar joint. Similarly, pronation of the hand and forearm is just as dependent on freedom of movement at the distal radioulnar joint as it is on freedom of movement of the proximal radioulnar joint.

Thus it is possible to break down functional movement in the other topographical areas of multiple joints throughout the musculoskeletal system. There is probably no single topographical area in which functional movement occurs where movement is solely concerned with one joint. For instance, the glenohumeral joint cannot move unless the sternoclavicular and acromioclavicular joints move freely, as

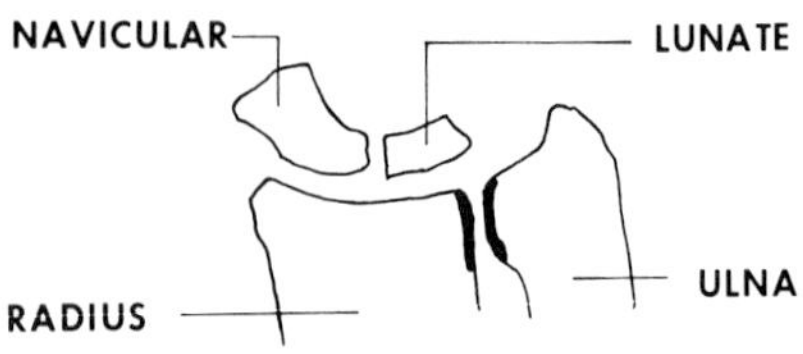

Figure 1-4. Distal radioulnar joint. The function of pronation of the hand depends for the most part on the movement of this joint.

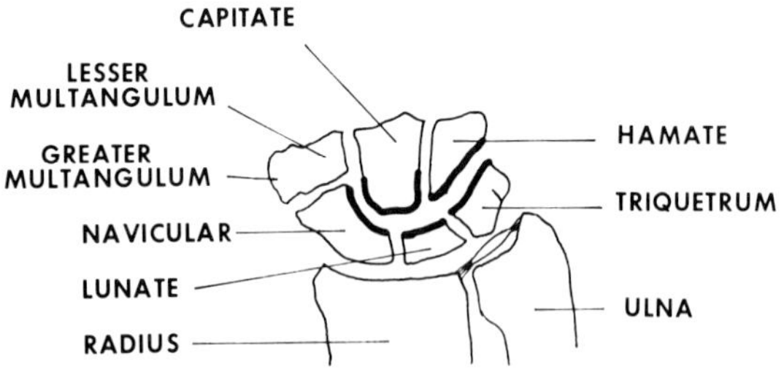

Figure 1-5. Midcarpal joint—a multiple joint made up of articular surfaces of the navicular and lunate bones proximally and of the capitate and hamate bones distally. Extension at the wrist for the most part takes place at this joint.

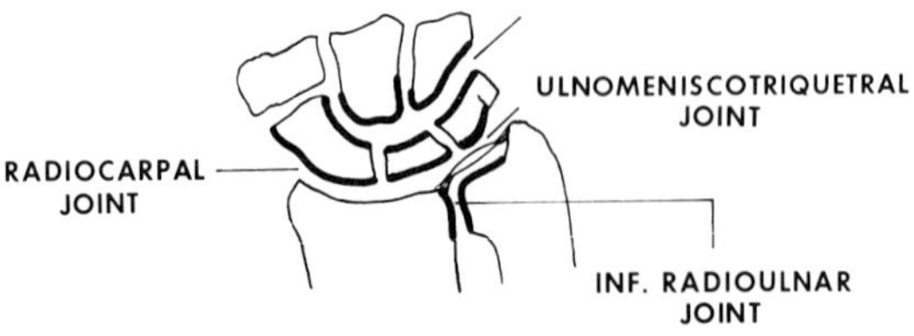

Figure 1-6. The four major joints that together make up the functioning topographical wrist.

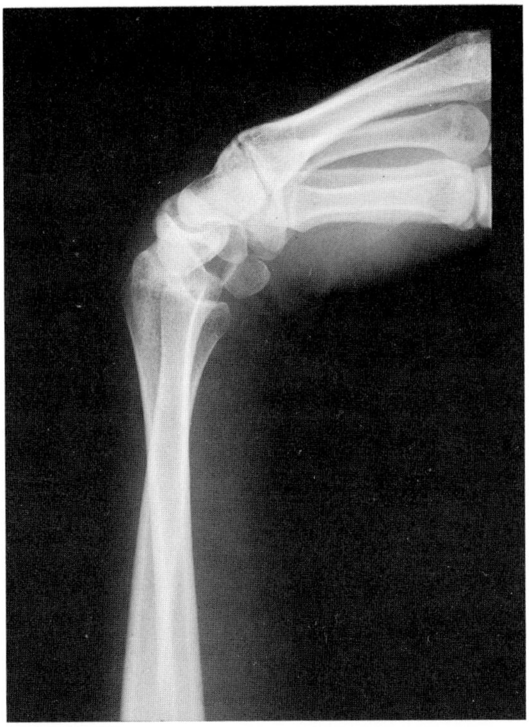

Figure 1-7. The wrist in flexion, illustrating that for the most part movement occurs at the radiocarpal joint. The relationship between the capitate and the lunate bones is relatively unchanged. This figure should be compared with Figure 1-8.

well as the scapula on the chest wall. Even the hip cannot move freely and normally without normal movement in the joints associated with the pelvis.

THE ULNOMENISCOTRIQUETRAL JOINT AND WRIST MOVEMENT

We have remarked above that there is no voluntary movement at the ulnomeniscotriquetral joint in the anatomical sense of the word, yet we claim that movement of this joint must be free to allow the voluntary movements of supination of the hand and forearm and abduction of the hand at the wrist.

There is a constant range of movement at this joint which can be demonstrated by an examiner. In the first of the movements the anatomical components of the joint can be distracted from each other in the long axis of the arm (Figure 1-9). A logical designation of this movement is long axis extension.

A further movement of the ulna on the triquetrum volarward and dorsalward can be demonstrated by placing the flexed proximal interphalangeal joint of the index finger over the pisiform immediately in front of the triquetrum, placing the thumb of the same hand behind the shaft of the ulna just proximal to its distal end, and pinching the thumb and index finger together while the other hand stabilizes the model's hand and the radius (Figure 1-10). A logical designation of this movement is anteroposterior glide (the posterior part of the movement being the guided rebound from the anterior part of it) of the triquetrum on the ulna through the meniscus, which lies between the articular surfaces of the two bones.

In a similar way, the triquetrum can be tilted

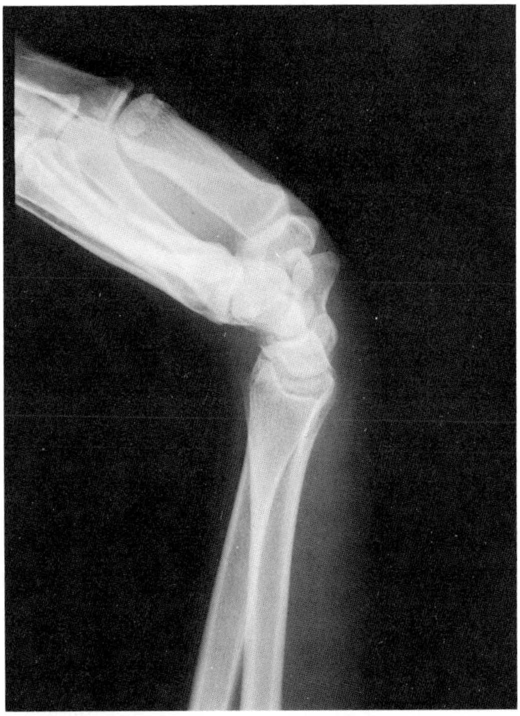

Figure 1-8. The wrist in extension (dorsiflexion), illustrating that for the most part movement occurs at the midcarpal joint. Comparison with Figure 1-7 should be made and the change of relationship between the lunate and capitate bones noted.

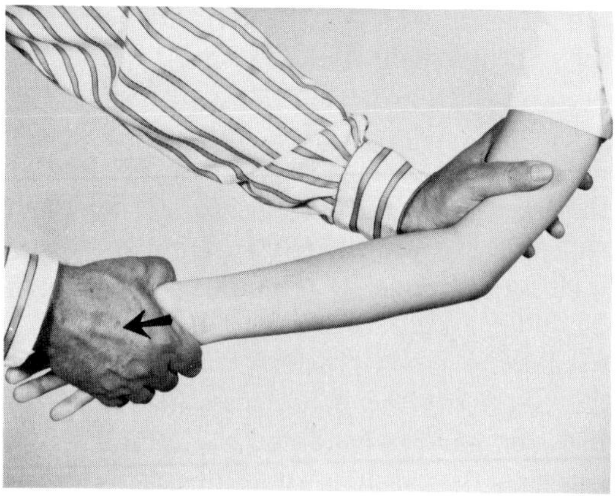

Figure 1-9. The position adopted to produce long axis extension at the ulnotriquetral joint. The arrow indicates the mobilizing pull by the examiner. It should be noted that long axis extension at the radiocarpal joint and the midcarpal joint has to be produced using this examining technique.

away from the ulna and the meniscus, and it is freedom of this movement that allows abduction to take place (Figure 1-11). A logical designation of this movement is lateral tilt.

THE DISTAL RADIOULNAR JOINT AND WRIST MOVEMENT

We also stated that there are no voluntary movements at the distal radioulnar joint in the anatomical sense of the word, yet the claim is made that movement of this joint must be free to allow the voluntary movement of pronation of the hand and forearm to take place.

The constant range of movement at this joint can be demonstrated by an examiner. The first of the movements is a gliding backward and forward of the ulna on the immobilized radius and a rotation clockwise and counterclockwise of the ulna in the long axis of its shaft on the immobilized radius at the joint. Figure 1-12 illustrates how the anteroposterior movement is performed, and Figure 1-13 illustrates how the rotation of the ulna is performed. These movements are logically designated anteroposterior glide and ulnar rotation.

JOINT PLAY

The movements that are not under control of the voluntary muscles are normal and demonstrable at every normal joint. If flexion and extension, for instance, are the formal designations for voluntary movements, some other terms must be found for these other movements.

It is useful to turn at this point to the science of mechanics. Everything that is made to move has a built-in factor of play to promote efficient functional movement. For example, a piston in a cylinder has play between the moving parts; a wheel on an axle has play to allow smooth rotation of the wheel; a hinge on a door has play between its components to al-

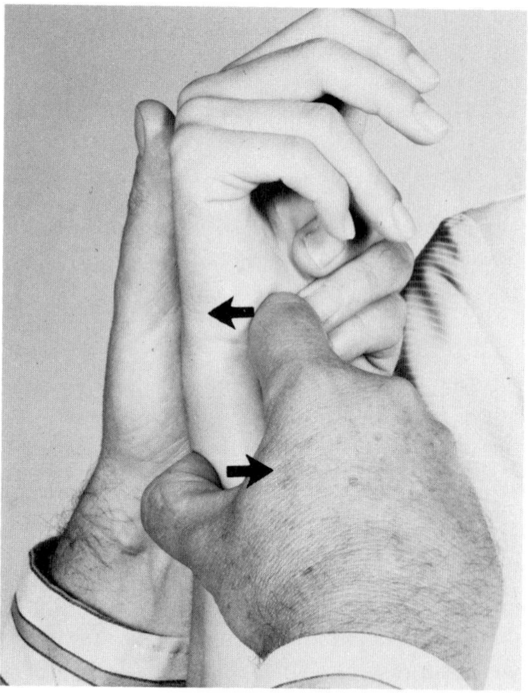

Figure 1-10. The position at the completion of the anterior phase of the anteroposterior glide of the ulna on the triquetrum, between which bones lies a meniscus. The posterior phase is achieved as a guided rebound from this position. Note that the examiner's left hand merely stabilizes the model's right hand and arm. The movement is achieved by the examiner's pinching his right thumb and forefinger together in parallels.

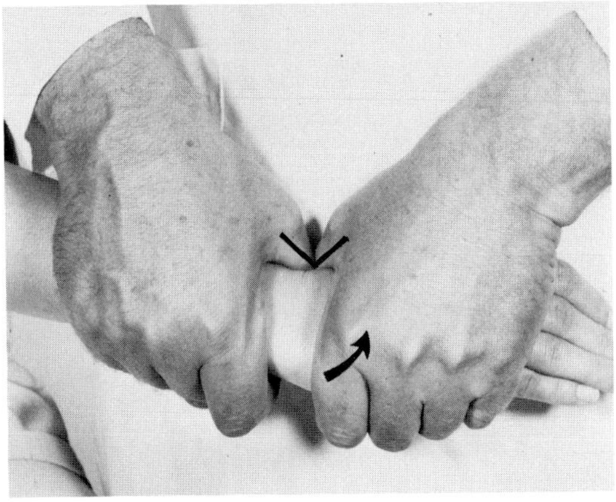

Figure 1-11. The position adopted to tilt the triquetrum away from the ulna and the meniscus on which it rests to perform the movement of side tilt at the ulnotriquetral joint. The examiner's right hand stabilizes the lower end of the ulna. Using both thumbs as a pivot, the examiner tilts the triquetrum away from the ulna when he swings his left hand into radial deviation.

low the door to open and close smoothly and easily. Why should there not be intrinsic play in human joints to promote efficiency of functional movement? It is our contention that there is, and the intrinsic movements which have been described are in fact this play; we designate them the movements of joint play.

A range of normal joint-play movements can be demonstrated in every synovial joint in the body in both the extremities and the spine. The movements at each joint are specific to each joint. They are described in further detail in Mennell's *Back Pain* [1960] and *Joint Pain* [1964].

JOINT DYSFUNCTION

Now what happens if by stress or strain or wear and tear the play is lost in anything that is made to move? The function of that moving part is impaired or lost, and impairment is associated with a squeak or some other abnormal noise—perhaps an indication of inanimate pain.

The same thing happens in a human joint when any joint-play movement is lost. Function of that joint is impaired or lost, and impairment is associated with the sensation of pain. This can be called joint dysfunction, which is a diagnosis.

Loss of function is common to all injuries or diseases of the musculoskeletal system; pain increases loss of function. These are symptoms and can be confirmed as physical signs. There may be additional signs associated with more serious pathological conditions: swelling, fluctuance, heat, skin discoloration, and atrophy of muscle. The degree of loss of function and pain varies with the seriousness of the underlying cause.

Joint Manipulation

Faced with a diagnosable cause of symptoms, one must try to find a specific treatment for it. The cause in this instance is mechanical, and it is logical that its treatment should be me-

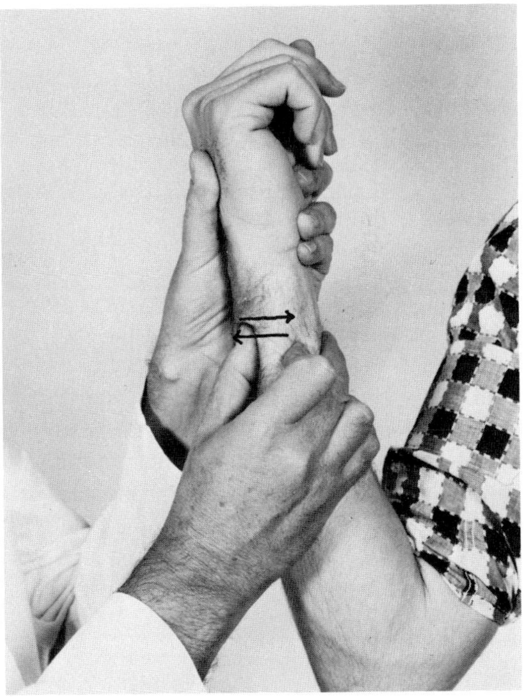

Figure 1-12. The position adopted to perform the movement of anteroposterior glide at the inferior (distal) radioulnar joint. The examiner's left hand immobilizes the model's hand and radius while his right hand moves the shaft of the ulna backward and forward. Note that these are two distinct movements.

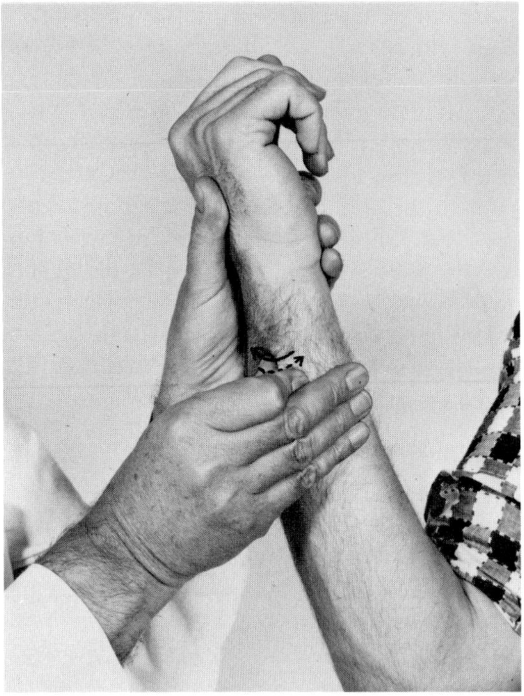

Figure 1-13. The position adopted to perform the movement of rotation clockwise and counterclockwise of the ulnar facet on the radial facet at the inferior (distal) radioulnar joint. The examiner's left hand stabilizes while the right hand mobilizes, as in Figure 1-12.

chanical. The cause is a lost joint-play movement. The treatment is to restore the lost movement. By definition the muscles cannot restore it, so restoration must be done for the patient. This and this alone is joint manipulation. The performance of a therapeutic manipulation is confined to the performance of a movement or movements of joint play. It has no relation in technique to the passive performance of any voluntary movement.

Etiology of Joint Dysfunction

There remains one other prerequisite to a hypothesis regarding a cause of joint pain. We have demonstrated a loss of normal movement—a diagnosis. We have indicated a specific method of restoring normalcy—a treatment. Still needed is etiology.

Joint dysfunction may be the result of:

1. Intrinsic trauma. To some this may be a new concept, but it can be equated etiologically to the spontaneous rupture of a tendon, as opposed to a traumatic rupture which presupposes the imparting of some extrinsic or external force.
2. Immobilization, including disuse and aging.
3. Resolution of some more serious pathological condition.

These observations require the acceptance of the concept of mechanical pathology as well as cellular and microbial pathology. They do not detract in any way from the concepts of the basic sciences of anatomy, physiology, kinesiology, and pathology. They add to them by recognizing the basic science of mechanics. They pertain to synovial joints wherever they may be and without exception—an unusual feature in any discussion involving musculoskeletal pain which should make the hypothesis especially attractive.

Though repetitious, it is necessary, we believe, once again to stress that therapeutic manipulative techniques must be concerned only with restoring lost joint-play movements, not voluntary functional movements. Before using manipulative techniques in therapy, one must learn what normal joint play is at every joint in the body before a loss of joint play can be detected on physical examination. Joint dysfunction is a common cause of patients' symptoms of pain and loss of function.

DIAGNOSIS WITHOUT PATHOLOGICAL BASIS

We wish to discourage the use of diagnostic terms such as *tennis elbow* and *bursitis* for any shoulder pain, *sinus tarsi syndrome*, *low back syndrome*, and *cervical spondylosis* for all causes of neck pain, and *flatfoot* and *dropped arches* for foot pain. *Arthritis*, which is meaningless without a prefix, is usually a spurious diagnosis when prefixed by *osteo*. If you successfully relieve pain in an osteoarthritic joint, the patient still has the osteoarthritic changes in the joint. He had them long before symptoms developed and will keep them long after symptoms are relieved. We discuss this further in detail on page 22.

BACK PAIN

The common reaction of many doctors when facing a patient who complains of symptoms in the back is, "Oh God, another back!" While realizing that the determination of the cause of back pain is not easy, it should become less of a problem if it is recognized that the back is just another part of the musculoskeletal system. It is no different anatomically, physiologically, or kinesiologically from any extremity with a problem for which a patient may seek advice. The diagnostic plan of attack should be the same. The principles guiding treatment should be the same.

Just because the mass of the back complicates breaking it down into its component parts for clinical diagnosis and because its component parts are not well defined radiographically as they are in the extremities, there is no reason to think that diagnosis cannot be made by logical inference and interpretation of symptoms and signs. For example, traumatic synovitis of a knee comes on slowly (maybe 24 hours after an injury), aches, and is warm whereas hemarthrosis comes on quickly, is very painful, and produces a hot joint. The same sequence of events may occur in any synovial joint in the back and can be logically inferred. In Chapter 8, p. 154, we expand on this sequence as it affects other joints.

PSYCHOSOMATIC CONSIDERATIONS

The question of whether musculoskeletal pain is a manifestation of psychoneurosis often arises. Certainly the musculoskeletal system, particularly the neck and low back, is a primary target area of psychosomatic symptoms, competing or even coexisting with headaches and head pains, peptic ulcers, and colitis. Sorting out organicity from predominantly psychic problems is often extremely difficult. It should be borne in mind, however, that whichever problem dominates, or whether problems coexist equally, a powerful role of physical therapy is the direct physical contact between therapist and patient during the treatment session. We are not in the least ashamed if it is the laying on of hands that produces relief of symptoms, although many tend to belittle physical therapy and hold that physical problems cannot be cured by psychological treatment.

It is commonplace for patients to tell us that they have never been examined when they really mean that their doctor never actually touched them during the examination. The mechanical application of treatment by machines is no substitute for the personal attention of the therapist. Because of this, patients tell therapists more about themselves than they do any other individual involved in health care.

SUMMARY

Certainly many problems remain to be faced. There are somatic components of visceral disease, symptom complexes from musculoskeletal causes that mimic visceral disease, and visceral disease that may simply mimic a musculoskeletal symptom. The problem of referred pain may confuse clinical thinking. There may be psychological components of physical problems and physical components of psychological problems. But these are the daily challenges of the practice of medicine. The physician cannot ignore occupational and environmental factors in assessing patients' problems. He must be open-minded about concepts that do not have universal acceptance, such as the possible effect of foci of infection, whether muscle spasm is a primary problem or (most often) a secondary effect, whether there is such an entity as fibrositis, and many other contentious subjects. This book attempts to bring all these problems and concepts into the proper perspective.

BIBLIOGRAPHY

Hilton, J. *On the Influence of Mechanical and Physiological Rest in the Treatment of Accidents and Surgical Diseases, and the Diagnostic Value of Pain.* London: Bell and Daldy, 1863.

Mennell, J. M. *Back Pain.* Boston: Little, Brown, 1960.

Mennell, J. M. *Joint Pain.* Boston: Little, Brown, 1964.

I
Diagnosis

2

Crossmatching Anatomy with Pathology as a Means to Differential Diagnosis

Our approach to differential diagnosis of musculoskeletal symptoms of pain local to the musculoskeletal system is very basic. We deliberately separate muscle and skeletal problems from problems that may arise from the whole of the locomotor system, as these would have to include problems of the nervous and vascular systems as well.

ANATOMICAL STRUCTURES OF THE MUSCULOSKELETAL SYSTEM

Seven anatomical structures make up the musculoskeletal system:

Bone and its periosteum
Hyaline cartilage
Joint capsule
Ligament
Muscle, tendon, and tendon sheath
Intraarticular meniscus
Bursa

Figure 2-1 diagramatically illustrates these seven structures.

Of these structures two, the hyaline cartilage and the meniscus, have no nerve supply, so it is unlikely that pain can arise from any primary pathological condition involving them. That they play a part in symptom complexes is certain, but pain does not arise directly from them. Impaired function may and does arise from them. Intraarticular menisci are present in only five joints in the body. Thus meniscal injury or anomaly need be considered in differential diagnosis only when there is pain in (1) the knee, (2) the ulnotriquetral joint, (3) the temporomandibular joint, (4) the sternoclavicular joint, or (5) the radiohumeral joint (Figure 2-2).*

Here we must comment on the intervertebral disc. It is not intraarticular in the sense that the other five joints are, but a damaged intervertebral disc acts like a damaged intraarticular meniscus. This point will be enlarged upon later (see Chapter 4, p. 53).

Before we leave generalizations and become specific, there are three structures—periosteum, bursae, and tendon sheaths—that merit special comment. Since sesamoid bones have no periosteum, when pain occurs in relation to sesamoid bones, especially the patella, it cannot be periosteal in origin.

With regard to bursae, symptoms cannot be due to bursitis if there is no bursa anatomically situated at a place where the symptoms are arising. When bursae are involved in a pathological process, the symptoms and signs are characteristic and clear. Here it should be re-

* Some authors draw attention to a meniscus in the acromioclavicular joint. Most surgeons with whom we have talked do not confirm its presence.

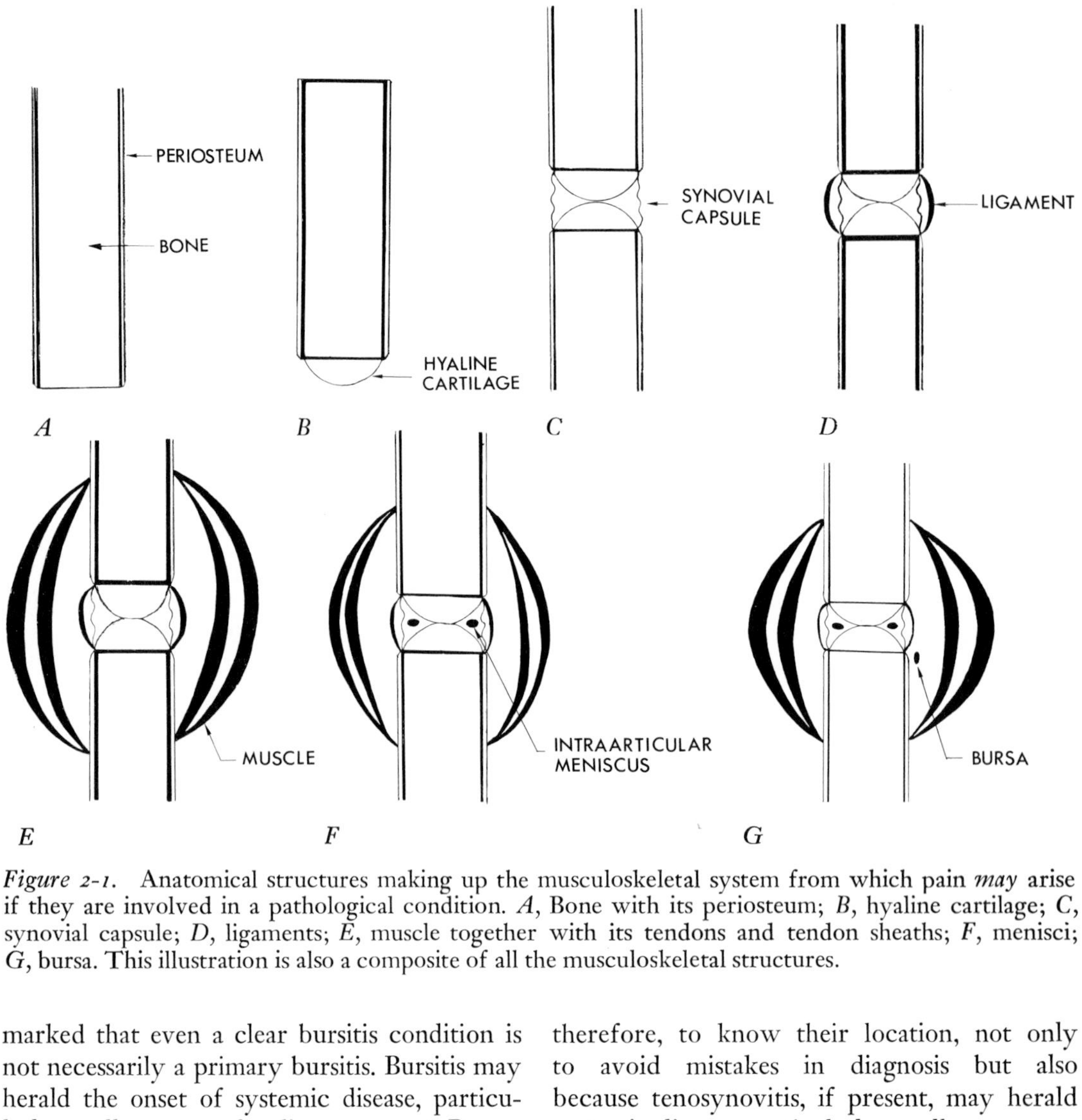

Figure 2-1. Anatomical structures making up the musculoskeletal system from which pain *may* arise if they are involved in a pathological condition. *A*, Bone with its periosteum; *B*, hyaline cartilage; *C*, synovial capsule; *D*, ligaments; *E*, muscle together with its tendons and tendon sheaths; *F*, menisci; *G*, bursa. This illustration is also a composite of all the musculoskeletal structures.

marked that even a clear bursitis condition is not necessarily a primary bursitis. Bursitis may herald the onset of systemic disease, particularly a collagen vascular disease or gout. Bursae may also become infected.

One should remember that there are no anatomical bursae in the back. However, adventitious bursae may form in the back but in only two anatomical situations: one over the angles of the ribs under the scapular border and the other between impinging spinous processes—"kissing spines"—of lumbar vertebrae.

The same comments also pertain to tendon sheaths. Symptoms cannot arise from tendon sheaths unless they are anatomically present where symptoms originate. It is important, therefore, to know their location, not only to avoid mistakes in diagnosis but also because tenosynovitis, if present, may herald systemic disease, particularly a collagen vascular disease or tuberculosis; pyogenic infection may also occur in tendon sheaths. Tendon sheaths are found only in the hands and wrist, at the shoulder, around the ankle, and in the foot.

PATHOLOGICAL CHANGES AFFECTING THE MUSCULOSKELETAL STRUCTURES

There are five pathological changes that may affect the seven anatomical structures and cause

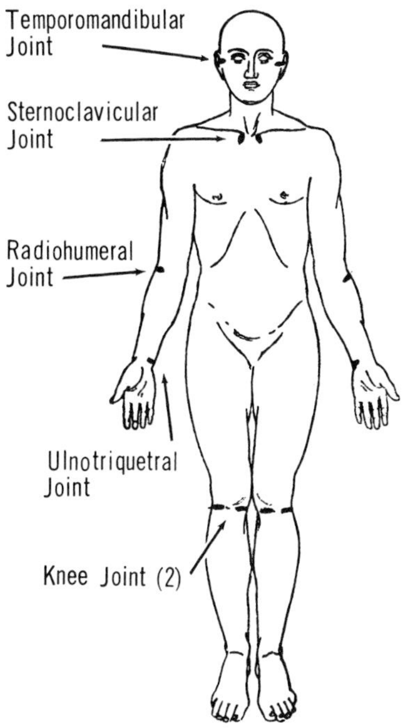

Figure 2-2. Joints in which intraarticular menisci are present.

musculoskeletal symptoms: (1) trauma, (2) inflammation, (3) metabolic disease, (4) neoplasm, and (5) congenital anomaly. Theoretically, there are 35 possibilities in the differential diagnosis of primary musculoskeletal pain. Not all the anatomical structures are affected by all the pathological possibilities, however, and the total comes to 23 in practice. Table 2-1 lists the anatomical structures and the pathological conditions with which they must be cross matched to arrive at a diagnosis.

DIFFERENTIAL DIAGNOSIS OF MUSCULOSKELETAL PAIN

To consolidate our thoughts on the differential diagnosis of musculoskeletal pain, we shall now crossmatch the seven anatomical structures with the five categories of pathological conditions. We recognize that our crossmatching is not all-embracing, but it is an exercise in applying principles to diagnostic thinking.

Table 2-1. Summary of Crossmatching Anatomical Structures and Pathological Conditions

Anatomical Structures That May Be the Seat of Pain-Producing Pathology	Pathological Changes or Conditions That May Affect These Anatomical Structures
1. Bone and its periosteum 2. Hyaline cartilage 3. Synovial capsule 4. Ligaments 5. Muscles, their tendons, and tendon sheaths 6. Intraarticular menisci 7. Bursae	1. Trauma (a) extrinsic (b) intrinsic 2. Inflammation 3. Metabolic disease 4. Neoplasms 5. Congenital anomalies

Bone and Periosteum

TRAUMA

Trauma over subcutaneous areas of bone may result in periosteal bruising, which, if not properly treated, may develop into a painful periosteal scar, a cause of chronic pain. Areas in which this condition is frequently found are around the ankle, knee, and elbow and over the sacrum. We repeat that sesamoid bones, having no periosteum, cannot be painful from this cause; sesamoid bones are affected by chondromalacia and osteochondritis, which *are* causes of pain. Trauma to bone results in fracture, and we remind the reader that fractures are not always initially detectable on x-ray examination of the injured part. Traumatic epiphyseal separations and greenstick fractures in children stem from trauma to bone. Avulsion fractures are considered later under ligamentous and muscular pathology (pp. 28, 29).

INFLAMMATION

Inflammation of bone, or osteomyelitis (Figure 2-3), may be acute, subacute, or chronic. Hematogenous osteomyelitis is acute and may be a lethal disease. Osteomyelitis caused by direct invasion of bone following open fractures or associated with pressure sores is usually sub-

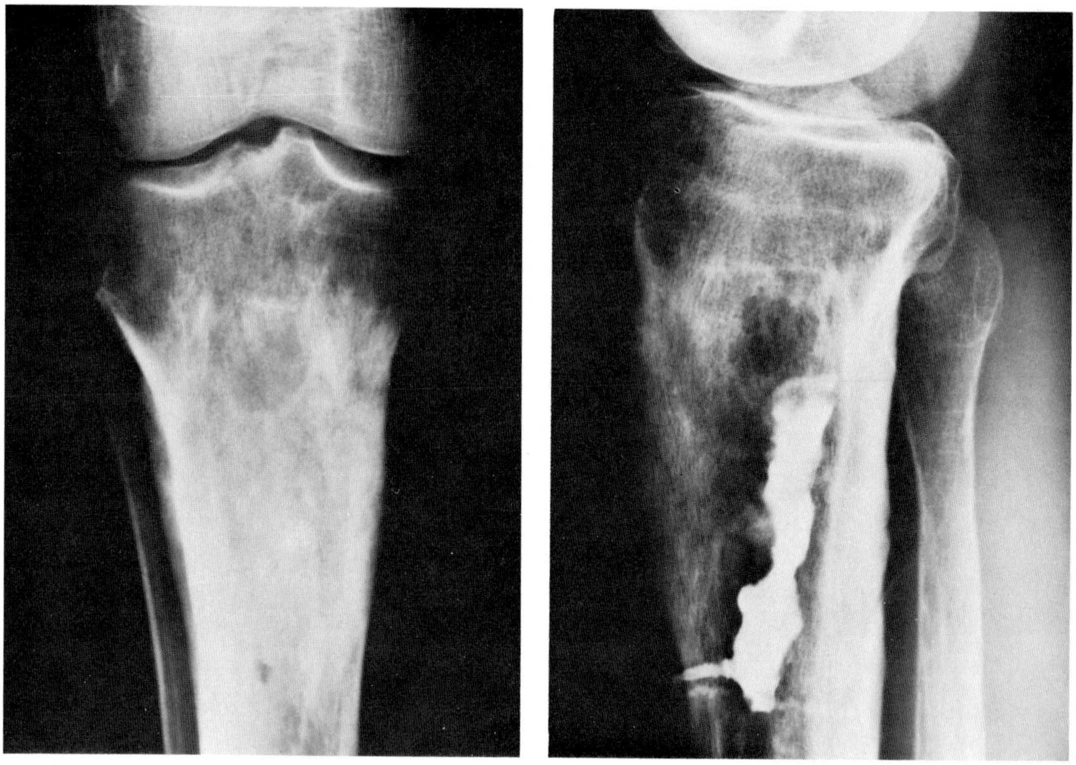

A *B*

Figure 2-3. *A,* Osteomyelitis of tibia with sequestrum and cloaca; *B,* injection of dye through the cloaca revealing (by sinography) the extent of sequestration. (Courtesy of David L. Berens, M.D.)

acute and less virulent. There are two chronic forms of osteomyelitis: the Brodie abscess illustrated in Figure 2-4 and the sclerosing osteomyelitis of Garré. The former is an abscess surrounded by a fibrous membrane and walled off by a dense ring of bone, while the latter is a sclerotic thickening of the cortex of bone.

Because it is potentially lethal, acute hematogenous osteomyelitis arouses the most concern, especially when it occurs in children. The diagnosis must be made on clinical grounds because radiographs usually do not reveal changes for about three weeks. The infection metastasizes to any organ in the body. If it is treated vigorously within the shortest possible time of the onset of diagnosable symptoms, it may be cured. If treatment is delayed pending confirmation of the diagnosis by radiography or even laboratory studies, the result of diagnostic indecision is likely to be a person crippled for life.

PERIOSTEUM AND INFLAMMATION

The periosteum is involved in inflammatory conditions. They may be pyogenic or granulomatous. Tuberculous or syphilitic periostitis may occur. Figure 2-5 shows two x-ray pictures of syphilitic periostitis of the femur in a child. Sarcoidosis may also produce a periosteal change.

METABOLIC DISEASE

Though there are many accepted classifications of metabolic bone disease we prefer the simplest, which recognizes three basic metabolic conditions: (1) osteoporosis senilis, (2) osteitis

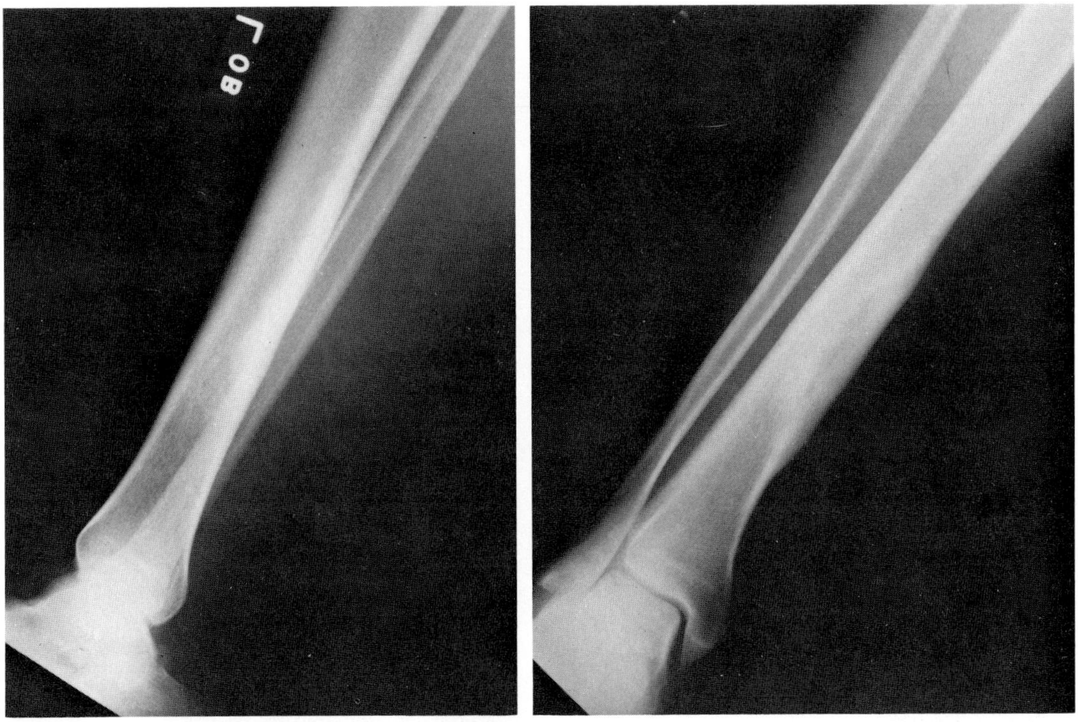

A

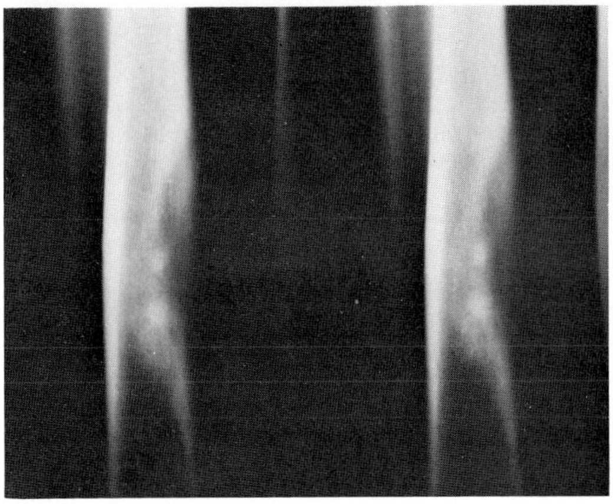

B

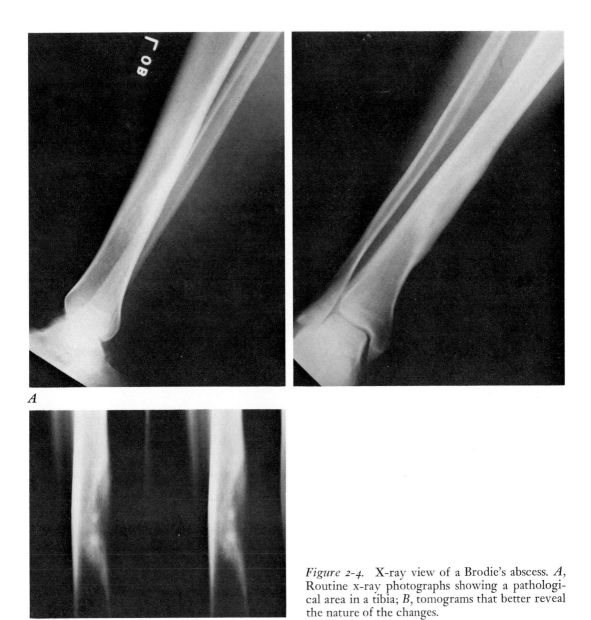

Figure 2-4. X-ray view of a Brodie's abscess. *A*, Routine x-ray photographs showing a pathological area in a tibia; *B*, tomograms that better reveal the nature of the changes.

fibrosa cystica, and (3) osteomalacia. Figure 2-6 illustrates the typical appearance of an osteoporotic spine with typical crush fractures and hypertrophic spurring, which is coincidental. All of these conditions have radiological manifestations of osteoporosis. Because osteoporotic changes are late in making their appearance on a radiograph, the diagnosis of pain from these diseases may be delayed. The radiological diagnosis of osteoporosis is made when the intensity of the shadow of bone is the same as that of the adjacent soft tissues (Figure 2-6). When osteoporosis is demonstrable, two "nevers" may aid in diagnosis. Osteoporosis senilis never affects the skull and never affects the lamina dura. When osteoporosis is due to one of the other two metabolic diseases, the lamina dura disappears. Figure 2-7*A* shows normal

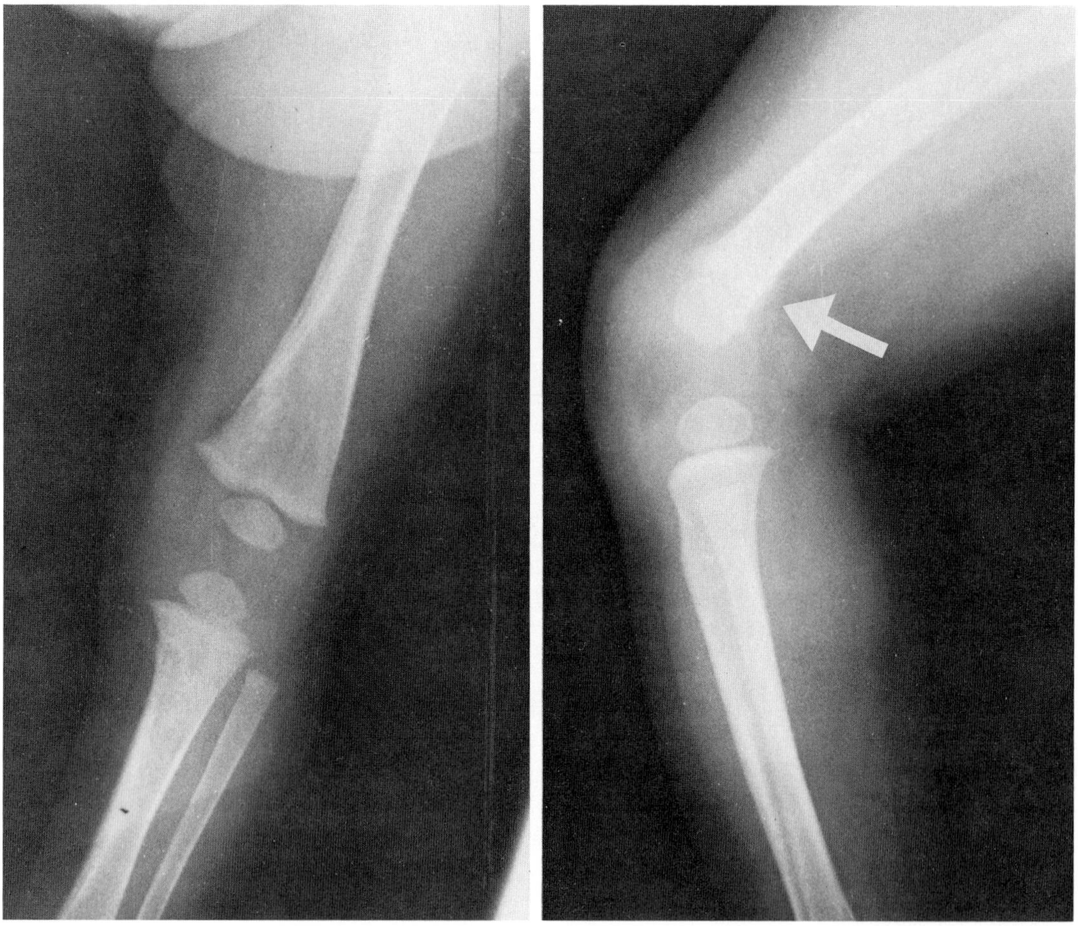

A *B*

Figure 2-5. *A*, Anteroposterior and *B*, lateral views of syphilitic periostitis. Note widening of the femoral metaphysis and thickening of its cortex medially. There is periosteal change, as shown by the arrow.

lamina dura around the roots of the teeth. In Figure 2-7*B* the lamina dura is absent. Of course, edentulous older people lack the lamina dura, and alveolar infection also destroys it.

There are, however, characteristic laboratory clues to the differential diagnosis of these metabolic diseases of bone, and they are discussed in Chapter 5, p. 84. In the common osteoporosis senilis the serum levels of calcium, phosphorus, and alkaline phosphatase are normal.

NEOPLASMS

It is beyond the scope of this work to enter into detail in the differential diagnosis of bone tumors, which may be primary or secondary, malignant or benign. However, there are a few features to which attention is drawn. Figure 2-8 illustrates some classic sites of bone tumors.

Malignant neoplasms of bone, especially in children, are lethal; their early diagnosis is vital if treatment is to be successful. Night pain of musculoskeletal origin is very suggestive of bone tumor. Night pain relieved by aspirin is characteristic of an osteoid osteoma.

CONGENITAL CONDITIONS

Musculoskeletal pain associated with congenital problems of bone should present no difficulty in diagnosis, providing one is well ac-

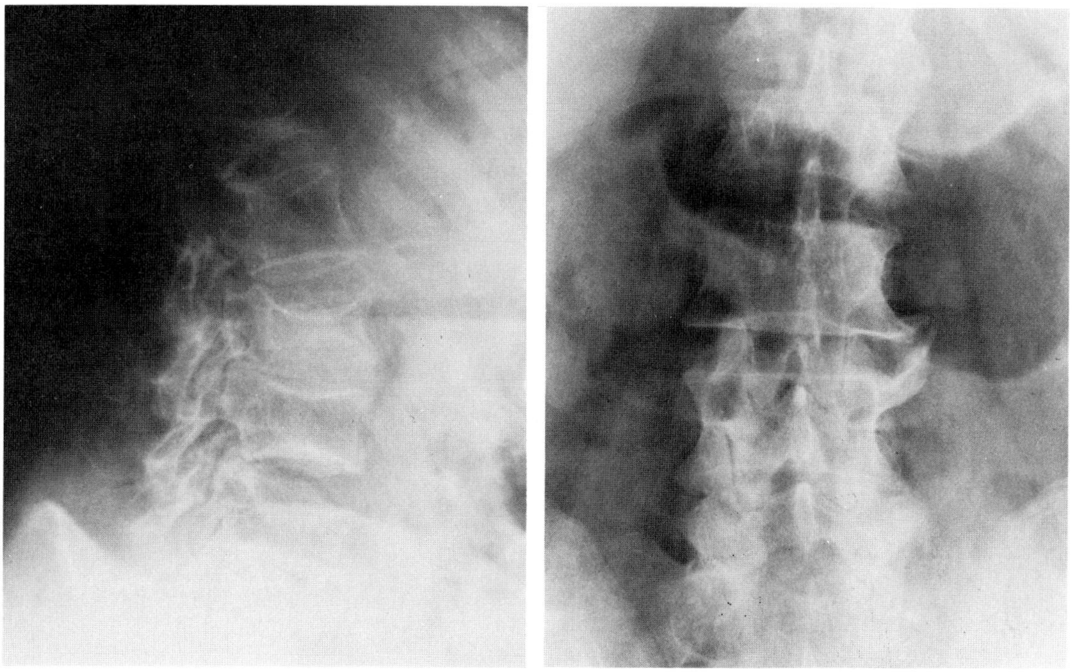

A *B*

Figure 2-6. Osteoporosis senilis of the vertebrae. *A*, A typical crush vertebral body fracture; *B*, coincidental hypertrophic vertebral body lipping.

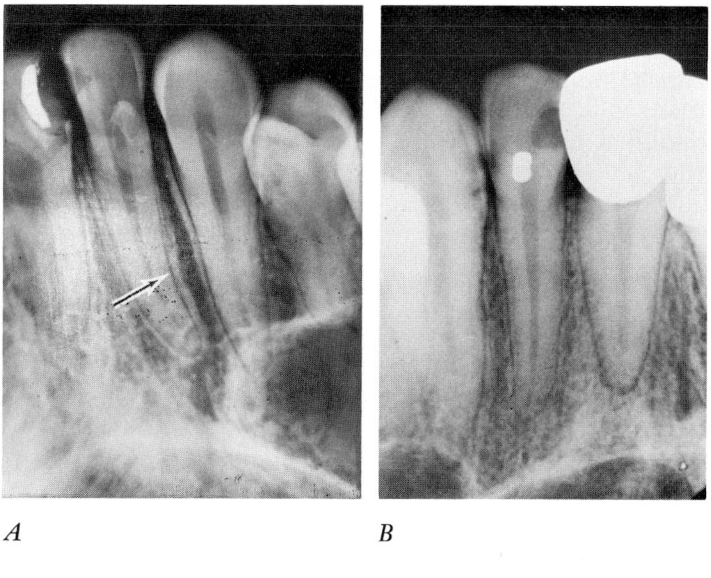

A *B*

Figure 2-7. *A*, Normal lamina dura; *B*, lamina dura are absent. (Courtesy of Eastman Kodak Co.)

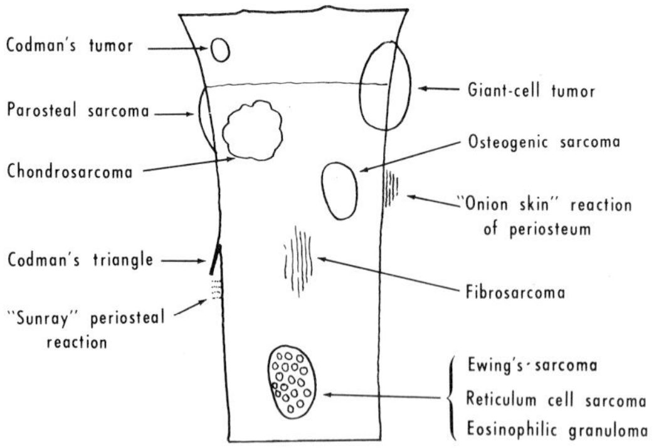

Figure 2-8. Certain characteristics of bone tumors. The majority are metaphyseal. Periosteal reactions to primary malignant tumors are indicated.

quainted with normal osseous anatomy. Congenital abnormalities of bone cause pain from malfunctioning of joints, from pathological changes in a pseudoarthrosis, or from the bone itself. For example, pain in the back from malfunction of joints may be the result of a unilateral lumbarization of the first sacral segment and arise from excessive movement of the facet joint on the unanchored side. Pain from a pseudoarthrosis arises when the anchored side of lumbarization is injured. Pain from a congenital abnormality of bone arises when a hemivertebra is present and there is mechanical dysfunction of neighboring joints.

In the extremities an example of pain from joints due to a congenital bone condition is that stemming from the presence of congenital bars in the tarsus (talar coalition). Pain arising from the bone itself may be due to congenital tibial fractures and the fractures associated with osteogenesis imperfecta.

OTHER PAINFUL BONE CONDITIONS

There are certain painful bone conditions which do not fit into our crossmatching exercise. These are discussed in Chapter 9, p. 196.

Hyaline Cartilage

Hyaline cartilage, having no nerve supply, is unlikely to be the primary source of pain. However, it is a space-occupying mass within a joint. If it thins, as in arthritic conditions, pain from stress on the other joint structures results because the cartilage occupies less space and the joint becomes unstable. This condition leads to sprain of ligaments and microtrauma to the capsule. It may also produce overstress on muscles acting on the joint and the development of fatigue in them. This may result in irritation of trigger points or an increased rate of musculotendinous degeneration, or fibrositis, any of which may cause pain.

TRAUMA

Chondral fractures occur particularly in athletes, but it is probably the associated effects of the trauma on other joint structures that causes the pain. However, the intraarticular derangement associated with this damage to the articular cartilage may be the source of the intrinsic joint dysfunction, impaired function, and pain.

Hyaline cartilage degenerates with repeated microtrauma from stressful occupations or activities. Rest from function encourages regeneration, which is a slow process. Nourishment is derived from epiphyseal blood supply and the synovial fluid. The synovial fluid itself is a dialysate of blood, but its contents are partially dependent on the normal destruction of the cartilage cells and on normal circulation or absorption, both of which are associated with normal activity, movement, and function.

Osteoarthritis, then, probably results from the excessive wear and tear of repetitive trauma. Acute traumatic arthritis may well be a misnomer since the immediate effects of the trauma involve any of the soft tissue structures of the joint rather than the hyaline cartilage. However, if the acute effects of trauma to a joint are not properly cared for and too early return to function after trauma is encouraged, changes within the joint characteristic of osteoarthritis quickly follow. Pain in an osteoarthritic joint is due to joint instability or to traumatizing or stretching the soft tissue structures. The pain in an osteoarthritic joint seldom arises from the changes seen radiologi-

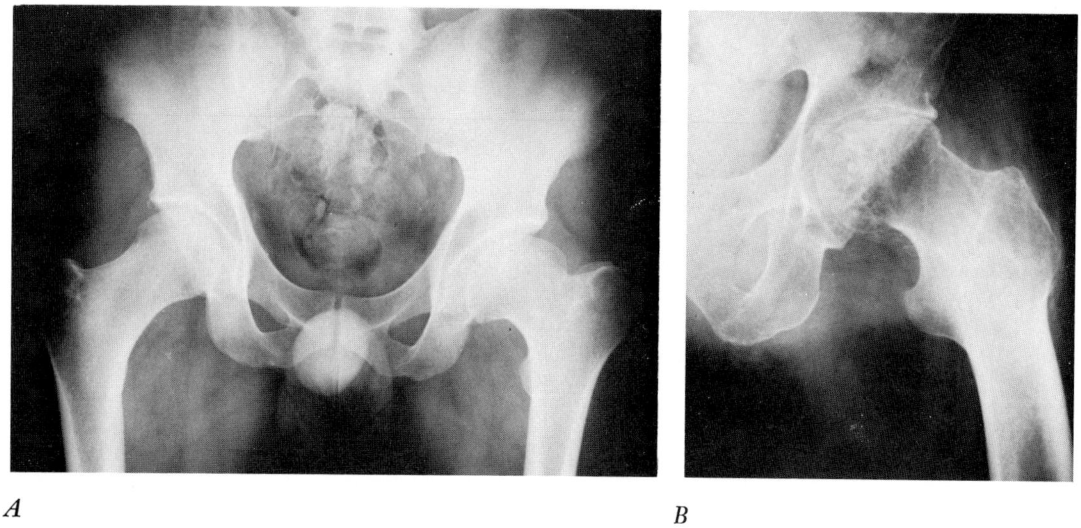

Figure 2-9. *A*, Early osteoarthritis of a left hip joint from trauma in early years (coxa plana). Note that both the head of the femur and the acetabulum are deformed. The joint space is remarkably preserved, and pain was arising not from the arthritis but from spasm of the supporting muscles of the joint, presumably largely because of the shortening of the femoral neck. *B*, Late osteoarthritis. Note irregular narrowed joint space, deformity of the head of femur, and subchondral cysts.

cally, unless the hyaline cartilage is so worn away that bone rubs upon bone. In that case it is the direct result of the condition. If the pain-producing components of an osteoarthritic joint are properly treated and the pain is relieved, the joint is still osteoarthritic. Figure 2-9 illustrates early (*A*) and late (*B*) osteoarthritis in hip joints.

INFLAMMATION

Hyaline cartilage is never the source of primary inflammatory change, but it is rapidly destroyed by lysins in blood and pus. Joint destruction is a frequent sequel to improperly treated pyarthrosis, and spontaneous fusion may occur.

Although hyaline cartilage splits away from the articular surface of bone in osteochondritis dissecans, the primary pathology is vascular—in the end arteries of the epiphysis of bone; the condition should not be considered a primary problem of hyaline cartilage. The cartilage plaque may, however, become a loose foreign body within the joint, thereby causing symptoms.

METABOLIC DISEASE, NEOPLASMS, AND CONGENITAL ANOMALIES

There are no metabolic diseases, neoplasms, or congenital anomalies of hyaline cartilage to be distinguished in our differential diagnostic consideration of causes of musculoskeletal pain.

Joint Capsule

Although there are visceral and parietal layers of the joint capsule, for our purposes the capsule may be regarded as an entity.

TRAUMA

A relatively minor traumatic insult to the capsule results in synovitis characterized by an excess of normal synovial fluid within the joint without palpable capsular changes. If the trauma is more severe, bleeding occurs into the joint, causing hemarthrosis. Severe trauma may tear the capsule.

It is important to differentiate clinically between excess synovial fluid and blood within the capsule. The reason is that in hemarthrosis it is mandatory to remove blood from a joint

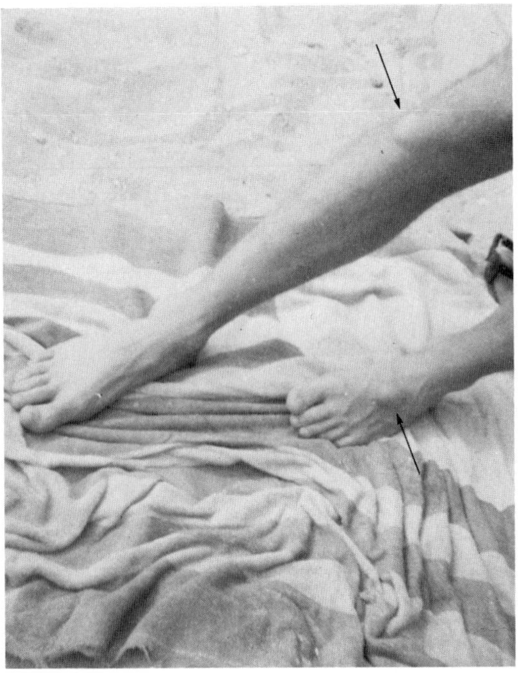

Figure 2-10. Surfer's ganglia, one on the dorsum of the left foot and another on the right knee (arrows).

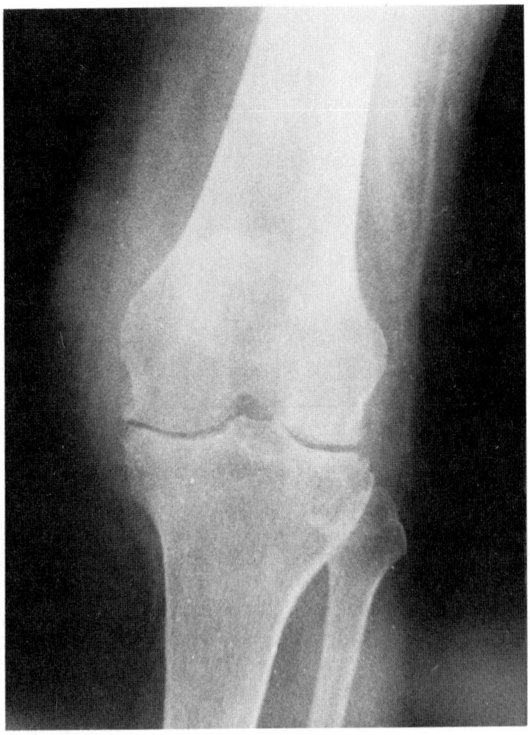

Figure 2-11. Changes typical of rheumatoid arthritis of a knee. This picture should be compared with Figure 2-12, showing changes characteristic of osteoarthritis. The soft tissue swelling, the generalized epiphyseal osteoporosis, and the relatively smooth articular surfaces of the adjacent bones with the symmetrical joint space diminution are characteristics not found in osteoarthritis.

as soon as possible to spare the hyaline cartilage from early destruction (see Chapter 8, p. 154). When there is an associated fracture through an articular surface into a joint, fat in blood aspirated from the joint is diagnostic of the fracture.

Ganglia arise from the capsule of joints following trauma; the Baker's cyst at the back of the knee is one example. Ganglia occur with rheumatoid arthritis, and a new cause of ganglia is found in surfers' feet and knees. Malignant changes in such ganglia have been reported but are rare. Figure 2-10 illustrates surfer's ganglia. Their mechanism of pain production is unknown but probably results from a stretching of the synovium.

INFLAMMATION

Pyarthrosis, gonorrheal and syphilitic infection (Clutton's joint), the granulomatous infections, and collagen vascular diseases primarily affect the synovial capsule. Figure 2-11 is an x-ray picture of rheumatoid arthritis of a knee joint showing the characteristic changes of soft tissue swelling around the joint, rather symmetrical and smooth subchondral bone on each side of a diminished joint space, and generalized osteoporosis of the femoral and tibial condyles. Figure 2-12 is an x-ray picture of an osteoarthritic knee to be compared with Figure 2-11. In osteoarthritis there is irregularity of the subchondral bone borders, and the joint space shows irregular diminution in its extent. Often the medial joint space is more diminished than the lateral joint space. There are subchondral cysts and marginal osteophytes. There is no osteoporosis in the femoral or tibial condyles and no generalized soft tissue swelling. Figure 2-13 is an x-ray picture of

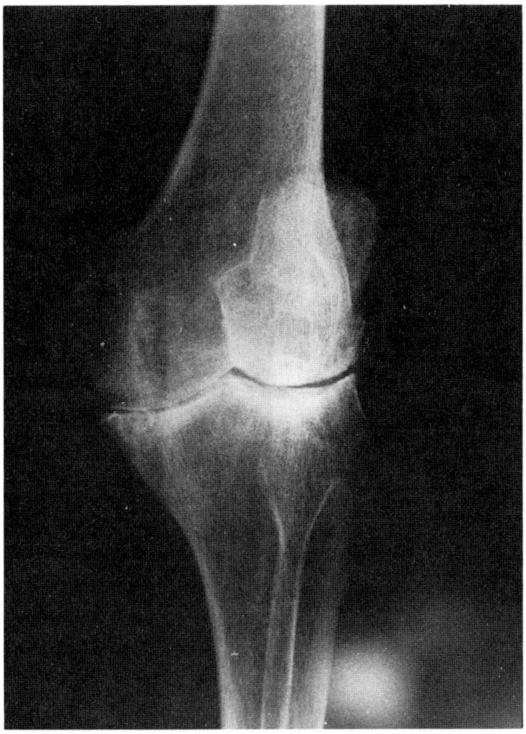

Figure 2-12. Knee, showing the classic changes of osteoarthritis.

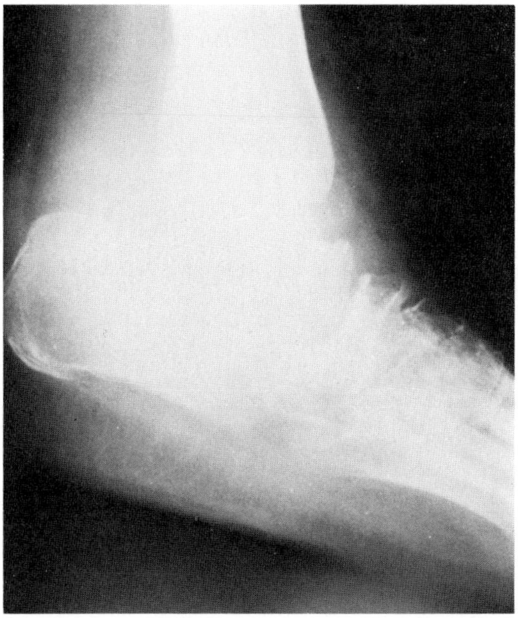

Figure 2-13. Ankle, with tuberculous arthritis of the subtalar joint. Characteristically there is not very severe pain but there is unexpectedly gross muscle atrophy. The osteoporotic changes in the visible bones are common to many inflammatory joint diseases.

a tuberculous ankle–subtalar joint. The joint is destroyed, but gross osteoporosis of the adjacent tarsal bones, intense but relatively localized soft tissue swelling, and an unusual degree of atrophy of muscle above the ankle are apparent.

Blood within the capsule of a joint sets up capsular inflammation and, if retained, destroys hyaline cartilage. Crystals in gout or pseudogout set up capsular inflammation. There are no characteristic joint changes in gout, but punched-out areas of bone are typical. Figure 2-14 is an x-ray picture of feet showing these classic changes.

However, in a large number of diseases "arthralgia" and/or "arthritis" are complicating manifestations, and it is not clear which anatomical structure is primarily involved. Typical of this situation are the neuropathic joints of Charcot, which are associated with syphilis, syringomyelia, and diabetes. Figure 2-15 is an x-ray picture of a diabetic neuropathic subtalar joint showing gross irregular destruction of the bones and joint and associated florid new bone formation without any intense soft tissue reaction. Contrary to some teaching, Charcot's joints may initially be painful joints. In the advanced stages of the involvement they are generally pain-free.

Joints may be involved in conditions such as ulcerative colitis, acromegaly, other endocrine disorders, parasitic infestations, serum sickness, salmonella and fungal infections, avitaminosis, and intermittent hydrarthrosis. In almost any systemic disease, such as bacterial endocarditis, hepatitis, influenza, rubella, and vaccinia, arthralgia may be a symptom. Figure 2-16 is an x-ray picture showing the changes in an adult's mortise joint resulting from epiphysitis in the patient's growth years. The ankle joint symptoms are, of course, those of structural stresses from the associated osteoarthritis.

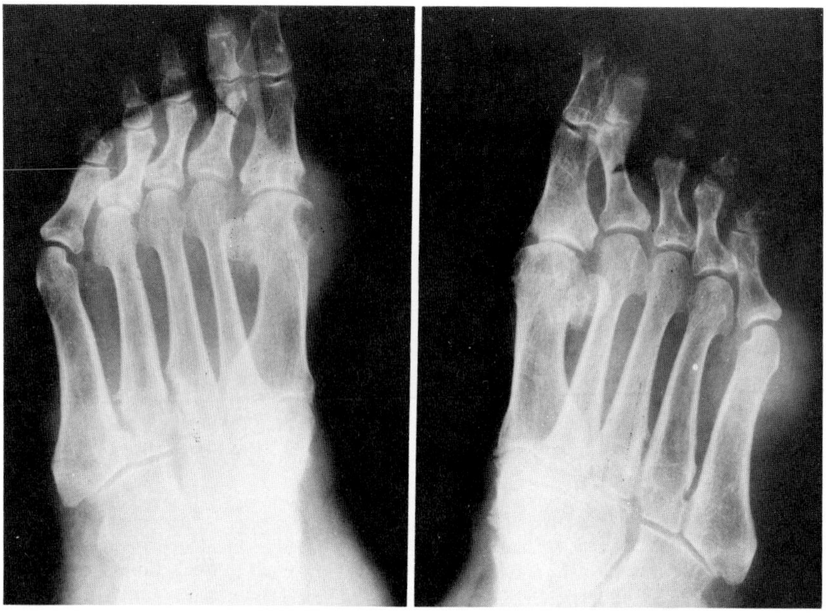

Figure 2-14. A pair of feet, showing classic changes of gout especially in the heads of the first metatarsal bones, but not confined to them.

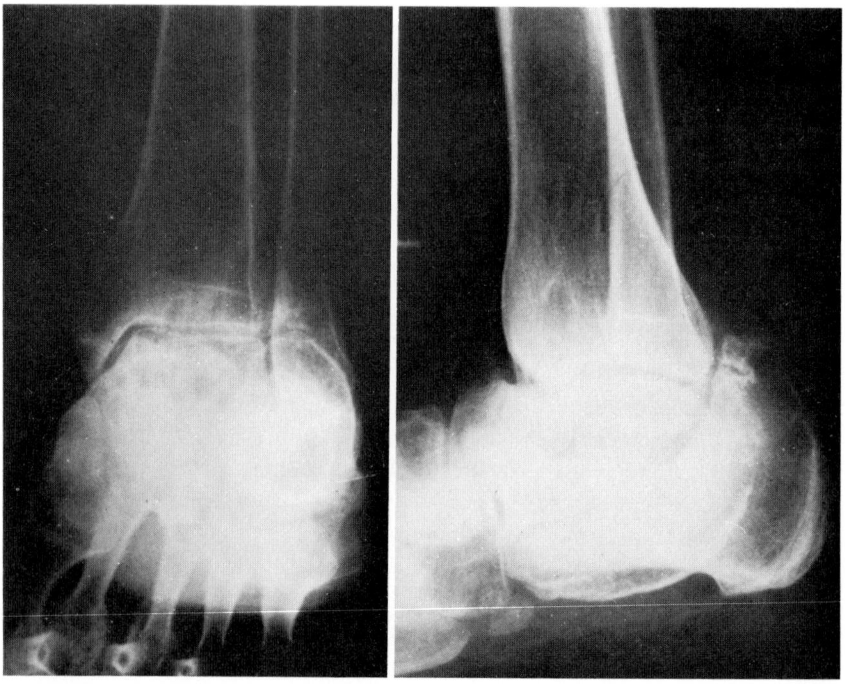

Figure 2-15. Ankle, showing changes typical of a Charcot's joint. The subtalar joint is primarily affected.

NEOPLASMS

The synovium is prone to neoplastic change. The neoplasms are mostly benign, but the malignant synovioma is particularly lethal. A capsular malignancy should always be suspected when pain arises from a joint whose capsule is palpable.

Osteochondromatosis is a common benign neoplastic capsular condition—a condition in which there are said to be "joint mice." If the calcified bulbous part of the tumor breaks off from its stalk, it becomes a foreign body within the joint and may block joint function. Figure 2-17 illustrates osteochondromatosis in a shoulder. Osteochondritis dissecans must be differentiated from this. Figure 2-18 illustrates osteochondritis dissecans in a knee. In a joint in which there is a meniscus, meniscal trauma will need to be differentiated. A chondral fracture in a joint after trauma may also have to be differentiated.

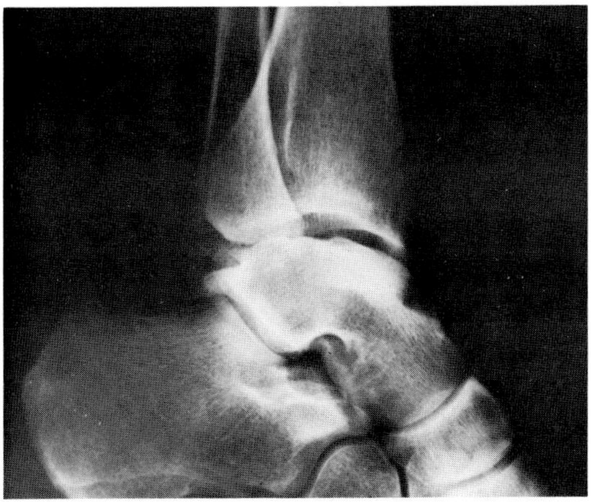

Figure 2-16. Changes of epiphysitis of the upper surface of the talus producing early symptoms of osteoarthritis.

PIGMENTED VILLONODULAR SYNOVITIS

Pigmented villonodular synovitis falls between inflammation and neoplasm. It is locally invasive but not metastatic. It is the only condition other than hemophilia that causes spontaneous bleeding into a joint.

METABOLIC DISEASE AND CONGENITAL ANOMALIES

There are no metabolic diseases or congenital anomalies of capsules to be differentiated in our differential diagnostic consideration of causes of musculoskeletal pain.

Ligaments

Ligaments are associated with one of the three major diagnostic "nevers" in musculoskeletal medicine (see Chapter 4, p. 51). A ligament is never tender on palpation unless the liga-

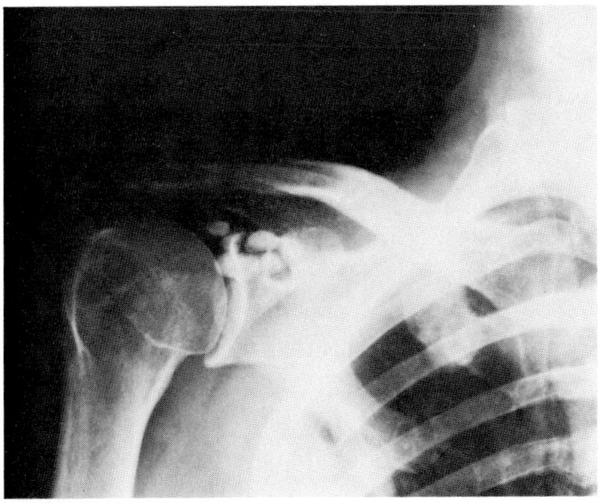

Figure 2-17. Osteochondromatosis of the shoulder, showing typical joint mice.

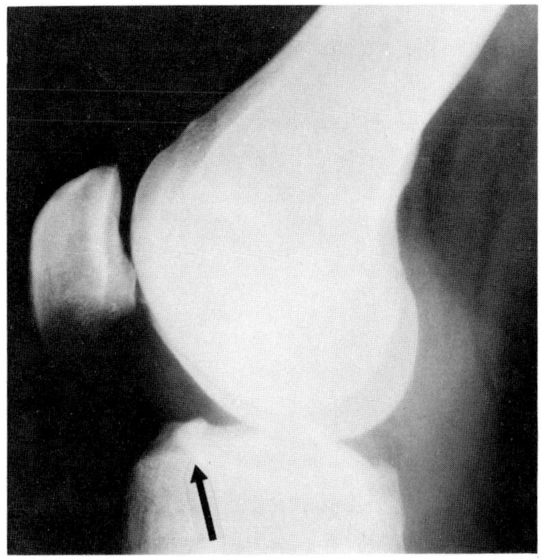

Figure 2-18. Osteochondritis of the tibial plateau in the knee, indicated by the arrow.

ment itself is injured or, and perhaps more importantly, there is something wrong with the joint (or in the back junction) that the tender ligament supports.

TRAUMA

Ligaments may be sprained by trauma; or, if the trauma is more severe, ligament fibers may be torn; or, where trauma is more severe still, a ligament may be ruptured. A ruptured ligament can be a more serious injury than a fractured bone for two reasons: First, the rupture may be overlooked; second, it may be difficult to bring the separated torn ends of the ligament together to allow healing. Healing of ligaments is notoriously slow and poorly achieved.

The diagnosis of a ruptured ligament depends on demonstrating abnormal and excessive joint movement, and a stress x-ray examination may be necessary to achieve this (Figure 2-19). The establishment of the diagnosis is so important that infiltration of the injured ligament with local anesthetic is often required to relieve the pain for this purpose. Even surgical repair of a ligament, if delayed, may not succeed in producing an effective ligament.

Avulsion fractures have to be considered under the heading of trauma of ligaments as well as muscles (see p. 29). The avulsion of the tibial malleolus, of the ulnar styloid, or of a spinous

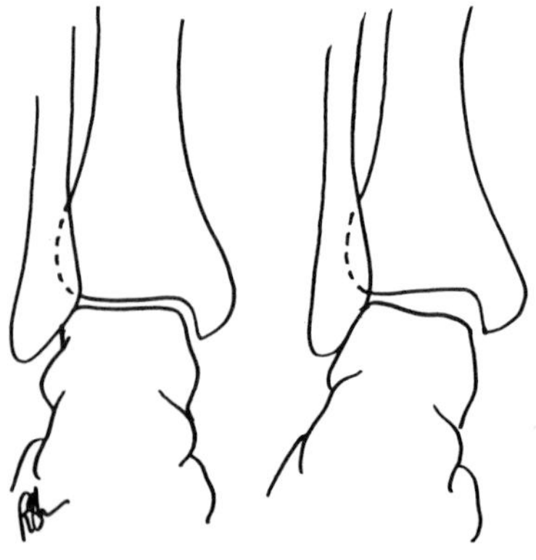

Figure 2-19. Tracings from x-ray pictures of an ankle. The left-hand picture is normal. The right-hand picture shows the effect of the stress, tilting the talus in the mortise away from the tibia. This is diagnostic of a ruptured deltoid ligament.

process of a cervical vertebra is indicative of ligamentous injury and should be treated as such. The fracture itself in these cases is insignificant from a clinical point of view, although the apposition of the bone fragments with a screw in the case of a malleolus avulsion may have to be undertaken.

NEOPLASMS

We have never seen or heard of neoplastic changes in ligaments. Fibrous tumors could appear in them, but we do not have to consider them in our differential diagnostic assessment of musculoskeletal pain.

INFLAMMATION, METABOLIC DISEASE, AND CONGENITAL ANOMALIES

There are no primary inflammatory conditions, metabolic diseases, or congenital anomalies of ligaments to be identified in our differential diagnostic consideration of causes of musculoskeletal pain.

Muscles

For convenience, we divide muscles into the fleshy muscle, the musculotendinous junction, the tendon, and, when present, the tendon sheath. In some areas there are tendon retinacula which may be torn or to which, following injury or inflammation, the tendon may become adherent, giving rise to impaired function and pain. In the fingers, vinculae may be an additional source of trouble.

TRAUMA

Fleshy muscle fibers may be contused by trauma, or they may be torn. Sometimes microscopic fiber tears are the result of repeated microtrauma.

Tendons may be strained, torn, or ruptured. When a tendon is ruptured, the function of the muscle is lost. Tendons that commonly rupture are those in the rotator cuff at the shoulder, the tendon of the long head of the biceps muscle, the tendon of the plantaris muscle, and the Achilles tendon. The injury causing tendons to rupture (in the absence of disease) is usually intrinsic (i.e., from functional stress) except with the rotator cuff, in which it may be due either to intrinsic or extrinsic trauma. In this case a blow on the point of the shoulder is the common extrinsic cause of rupture of the rotator cuff.* Tendons are likely to be severed in open wounds. Ruptured tendons, particularly the extensor tendons of the fingers, are often a complication of rheumatoid arthritis. Rupture tendons heal or reattach when treated conservatively by immobilization. This is true even for the Achilles tendon. Surgical repair of severed tendons may still be necessary for a good functional result.

Avulsion fractures have to be considered in connection with trauma and muscle tendons. Avulsion of the greater tuberosity of the humerus is caused by a blow or a fall onto the tip of the shoulder, the rotator cuff not being torn. In this case, though surgical apposition of the bone fragments by use of a screw is recommended, if the separation is less than ⅛ inch, the injury is treated as a muscle injury. In the back, avulsion of a transverse process is evidence of severe muscle strain and is treated as such. Figure 2-20*A* illustrates an avulsion fracture of the greater tuberosity of the humerus. Figure 2-20*B* illustrates an avulsion fracture of the greater trochanter of the femur. Excessive muscle pull in adolescents may avulse epiphyses; young athletes commonly avulse the epiphysis of the iliac crest (hip pointer) or the lesser trochanter of the femur (groin pull).

Tendon sheaths, when traumatized, react like synovial capsules and tenosynovitis results. Palpable crepitus on movement of the tendon within the sheath is diagnostic of this. The differentiation between traumatic tenosynovitis and inflammatory tenosynovitis is important, as tendon sheath inflammation may herald the onset of disease, particularly tuberculosis and rheumatoid arthritis. Figure 2-21 illustrates a tuberculous tenosynovitis.

* A study reported in the mid-sixties showed that on autopsy 40 percent of 200 subjects, all 50 years of age or more, who complained of shoulder pain in life had rotator cuff ruptures without a history of extrinsic trauma.

INFLAMMATION

Fleshy muscle fibers can be involved in acute infective processes, as in carbuncles, or following pyogenic invasion from wounds. They are excellent media for the growth of aerobic and anaerobic organisms, but in these gross conditions there should be no diagnostic problem.

It is the acute and chronic muscle pain without gross changes that presents problems. Here acute and chronic fibromyositis may have to be differentiated from the involvement of muscles due to systemic rheumatological diseases. At this point we mention the myopathies, muscle dystrophies, and muscle atrophies only to discard them as they do not usually produce primary musculoskeletal pain and therefore do not complicate the differential diagnostic assessment of musculoskeletal pain. Fibromyositis and trigger points, which we believe to be separate and definite entities, are discussed in Chapter 9, p. 189.

METABOLIC DISEASES

There are a number of painful metabolic diseases of muscle: McArdle's disease (see Chapter 9, p. 184), which causes aching muscles during exercise; phosphofructokinase deficiency, the symptoms of which are similar to McArdle's disease; acid maltase metabolic disease (Pompe's disease), which causes aching muscles after exercise; and long-chain fatty acid intolerance metabolic disease.

NEOPLASMS

Fleshy muscle neoplasms are not uncommon but are reasonably easily detected by palpation. Tendon neoplasms occur but are rare and, again, are detected on palpation. They are usually of the fibrous series of tumor.

Tendon sheaths are quite often the seat of neoplastic changes of synovium, and most commonly, the giant cell tumor or xanthoma is found in these structures. They are benign. Malignant tumors may occur.

CONGENITAL ANOMALIES

Muscles may be absent, but as substitution for an absent muscle usually takes place early, no

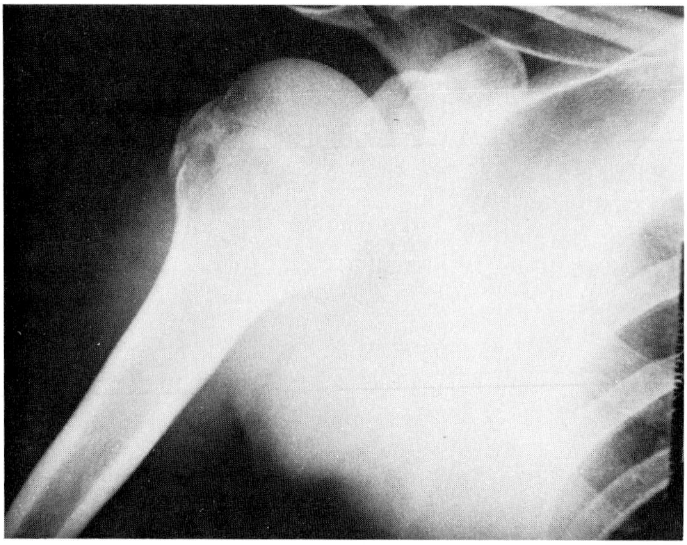

A

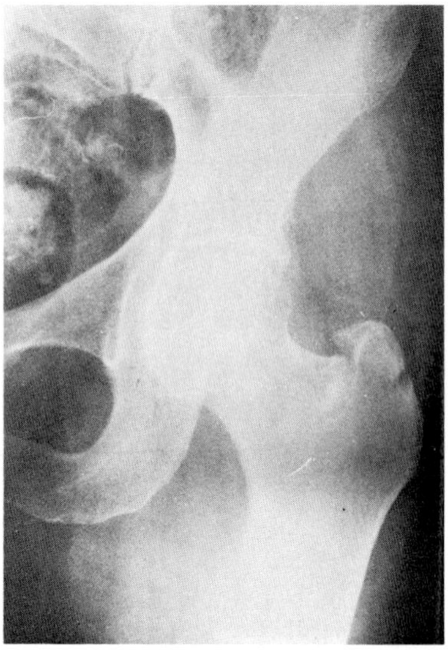

B

Figure 2-20. *A*, An avulsion fracture of the greater tuberosity with less that ⅛ inch separation, which would not be treated surgically. *B*, An avulsion fracture of the greater trochanter. (Courtesy of Andrew Dobranski, M.D.)

pain results. Thus the lack of a muscle does not have to be considered in our differential diagnosis assessment of musculoskeletal pain.

Menisci

As there are only five joints in the body in which intraarticular menisci are found (see p. 17), the menisci have to be considered as a potential source of pain only when those joints are involved in a pain syndrome.

The two characteristic symptoms of meniscal pathology are mechanical locking of the joint and a clicking within the joint. Locking, however, also occurs with osteochondromatosis in the knee, chondral fractures, and osteochondritis dissecans when the hyaline cartilage fragment has separated and become a loose body within the joint. These must all be differentiated. Radiographic examination except by arthrography does not reveal the presence of meniscal trauma (see Figure 5-25, p. 78). However, osteochondritis dissecans (Figure 2-18) may be revealed by the presence of a

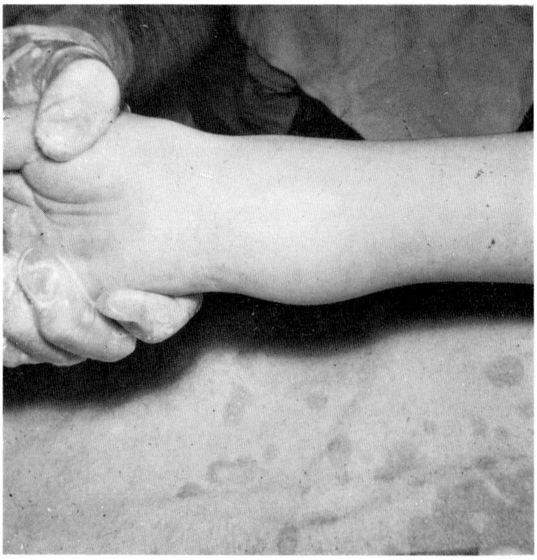

Figure 2-21. Swelling of tenosynovitis. In this case the etiology was tuberculosis. (Courtesy of Bruce Butler, M.D.)

rounded dark shadow in relation to the subchondral bone. X-ray photography reveals the "joint mice" of osteochondromatosis (Figure 2-17) when, as is usual, the caps of the tumors are calcified.

TRAUMA

Trauma may loosen the attachments of a meniscus allowing it to lock the joint mechanically, causing dysfunction and pain. In the knee the medial meniscus is subject to bucket handle tears and transverse ruptures presumably because it is anchored down by a third attachment to the medial collateral ligament. The lateral meniscus has only an anterior and a posterior attachment, and if crushed between the femoral and tibial condyles it may undergo degeneration and form a pseudomucinous cyst, but it rarely fractures. Baker's cysts are sometimes found in association with meniscal injuries.

A retained posterior pole of a medial meniscus in the knee after surgery may cause symptoms similar to those of a traumatized meniscus and should be recognized as a cause of pain. Also, menisci degenerate in rheumatoid arthritis. Menisci may regenerate after surgery and may be the underlying cause of pain. Here we wish to emphasize that if pain is thought to be associated with a meniscal injury, removal of the meniscus is indicated, whatever the age of the patient, unless it can be restored to its normal functioning position by manipulative or other conservative means.

INFLAMMATION

Inflammatory conditions do not primarily affect menisci, but after any inflammatory process has resolved, the meniscus may be loosened at its attachments and become involved in pain-producing dysfunction.

CONGENITAL ANOMALIES

The only known congenital anomaly is the discoid variant of the lateral meniscus in the knee joint. This must be considered in the differential diagnostic assessment of knee joint pain because it may produce an internal derangement of the knee.

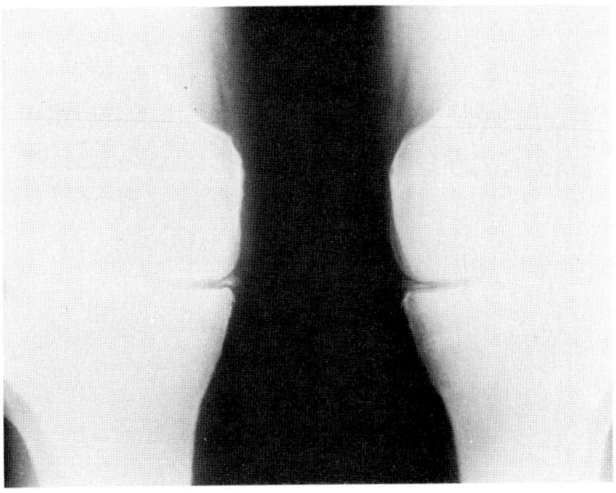

Figure 2-22. Chondrocalcinosis, showing calcification of the medial meniscus.

METABOLIC DISEASE

Pseudogout (chondrocalcinosis) and ochronosis affect the menisci. Figure 2-22 illustrates chondrocalcinosis of the medial meniscus of the knee.

NEOPLASMS

There are no neoplasms that primarily affect menisci, so they do not enter into our differential diagnostic evaluation of causes of musculoskeletal pain.

Bursae

The diagnosis of bursitis must be linked to pathological conditions which affect a bursa. Unless there is a bursa anatomically in a place from which pain appears to be emanating, the pain cannot be coming from bursitis. The prerequisites of the diagnosis of bursitis are the symptoms of pain and limitation of movement and, particularly, a swelling with or without skin redness and heat. The prerequisite signs on which the diagnosis may be based are tenderness, limited movement, and the presence of a swelling, which is fluctuant and warm or hot in a place where there is a bursa. Figures 2-23 and 2-24 show the locations of the major bursae in the extremities. There are no primary bursae in the back. Figure 2-25 illustrates classic subdeltoid bursitis in a patient's

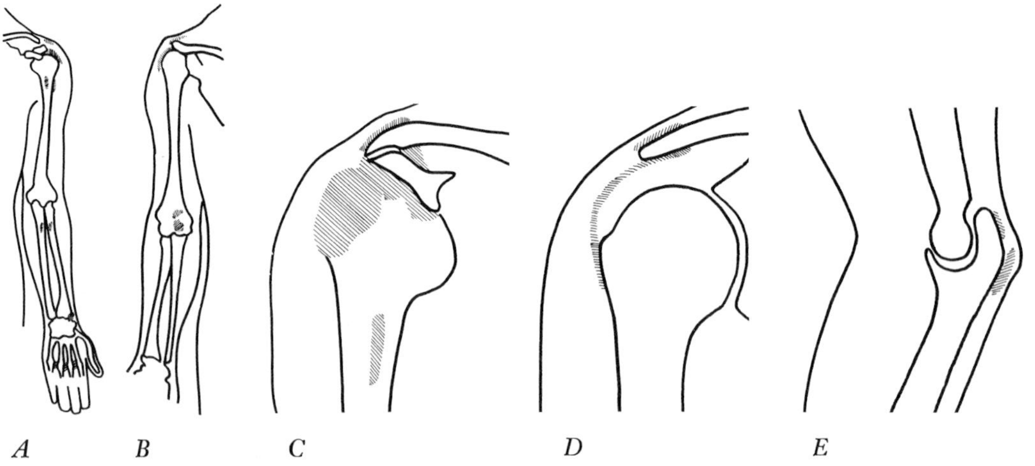

Figure 2-23. Location of the major bursae of the upper extremities. *A*, In the anterior aspect of the arm; *B*, in the posterior aspect of the arm; *C*, around the shoulder anteriorly; *D*, around the shoulder posteriorly; *E*, at the elbow.

right shoulder, with all the prerequisite signs described above.

TRAUMA

Traumatic bursitis is usually limited to the olecranon bursa, the ischial bursa, the trochanteric bursa, the calcaneal bursa, the bursae in relation to the metatarsophalangeal joints of the big and little toes, and the bursae between the metatarsal heads. It seldom, in fact, is found primarily in the subacromial bursae.

INFLAMMATION

A bursa can become infected, especially after a needle is introduced into it for some therapeutic reason. Inflammatory bursitis can be the heralding feature of rheumatoid arthritis and gout.

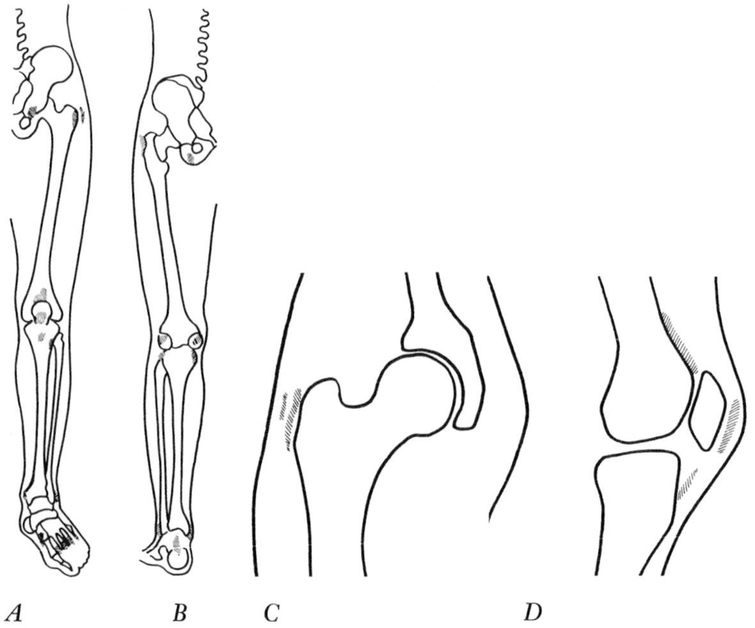

Figure 2-24. Location of the major bursae in the lower extremity. *A*, In the anterior aspect of the leg; *B*, in the posterior aspect of the leg; *C*, around the hip; *D*, in front of the knee.

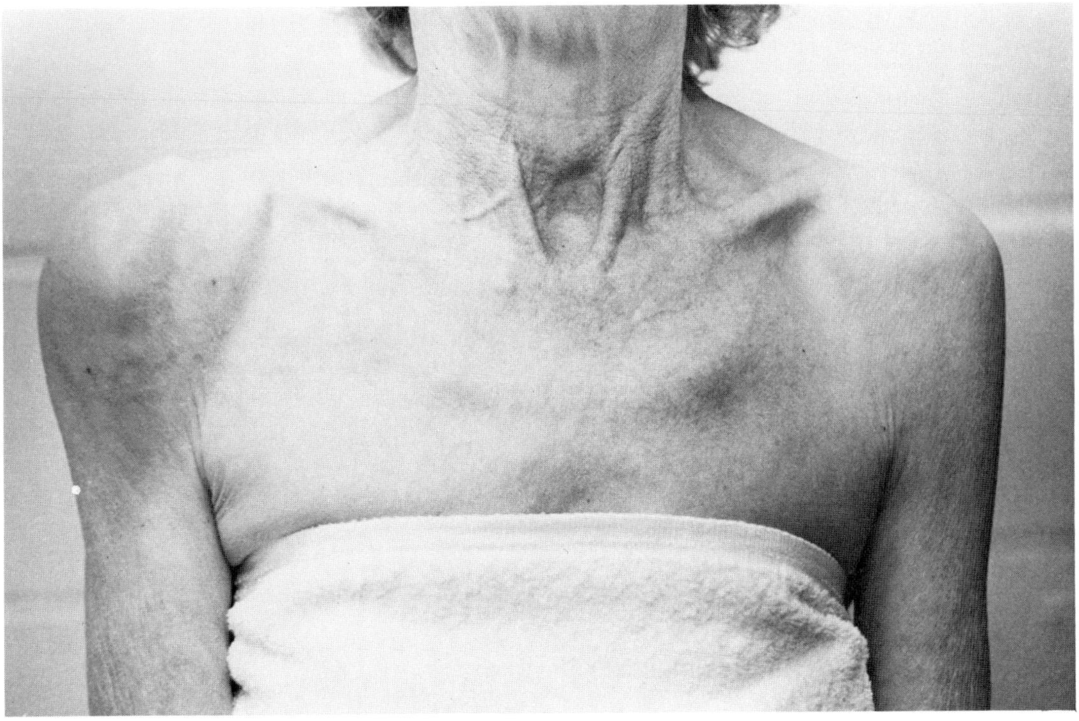

Figure 2-25. Classic bursitis of the right shoulder.

NEOPLASMS

As the bursae are lined with synovium, presumably they may be the seat of synovial tumors. We have never seen such a condition, but this should be kept in mind as a possible cause of pain in differential diagnostic assessment.

METABOLIC DISEASE AND CONGENITAL ANOMALIES

There are no metabolic diseases or congenital anomalies that primarily affect bursae. Therefore they do not enter into our differential diagnostic evaluation of causes of musculoskeletal pain.

BURSAE AND THE BACK

There are no anatomical bursae in the back. However, adventitious bursae may form, particularly, as mentioned earlier, between the vertebral scapula border and the rib angles and between impinging or "kissing" spinous processes, especially in the low lumbar spine.

Summary

If we now recapitulate our differential diagnosable causes of musculoskeletal pain, arrived at by crossmatching anatomical structures with possible pathological types of involvement, we arrive at a diagnostic possibility of 23.

There remains the twenty-third cause of musculoskeletal pain: mechanical. Joint dysfunction arises from the loss of joint play movements, a concept of pain new to most physicians. This has been defined earlier (p. 9) and is to be enlarged on later. It is a most common cause of pain. It must always be included in the differential diagnostic assessment of causes of acute musculoskeletal pain arising from intrinsic trauma or of chronic musculoskeletal pain remaining after resolution of more serious pathological conditions.

BIBLIOGRAPHY

Breimer, C. W., and Freiberger, R. H. Bone lesions associated with pigmented villonodular synovitis. *Am. J. Roentgenol. Radium Ther. Nucl. Med.* 79:618, 1958.

Bywaters, E. G. L., Dixon, A. St. J., and Scott, J. T. Joint lesions of hyperparathyroidism. *Ann. Rheum. Dis.* 22:171, 1963.

Dimich, A., Bedrossian, T. B., and Wallach, S. Hypoparathyroidism. *Arch. Intern. Med.* 120: 449, 1967.

Dodds, W. J., and Steinbach, H. L. Primary hyperparathyroidism and articular cartilage calcification. *Am. J. Med.* 104:884, 1968.

Dubin, A., Kushner, D. S., Bronsky, P., and Pascale, L. R. Hyperuricaemia and hypoparathyroidism. *Metabolism* 5:703, 1956.

Feldman, M. J., Becker, K. L., Reefe, W. E., and Longo, A. Multiple neuropathic joints, including the wrist, in a patient with diabetes mellitus. *J.A.M.A.* 209:1690, 1969.

Grahame, R., Sutor, D. J., and Mitchener, M. B. Crystal deposition in hyperparathyroidism. *Ann. Rheum. Dis.* 30:326, 1971.

Jaffe, H. L., Lichtenstein, L., and Sutro, C. J. Pigmented villonodular synovitis, synovial, bursae and tenosynovial lesions. *Arch. Pathol.* 31:731, 1941.

Kellgren, J. H., Ball, J., and Tutton, G. K. The articular and other limb changes in acromegaly. *Q. J. Med.* 21:405, 1952.

Lea, R. B. Non-surgical treatment of tendo-Achilles rupture. *J. Bone Joint Surg.* 54A:1398, 1972.

Martin, M. M. Charcot joints in diabetes mellitus. *Proc. R. Soc. Med.* 45:503, 1952.

Zucker, G., and Marder, M. J. Charcot spine due to diabetic neuropathy. *Am. J. Med.* 12:118, 1952.

3
Clinical Examination

Pursuing our thesis that there is no difference between the extremities and the back but that they are all components of the musculoskeletal system, we see no reason to have more than a single set of principles in undertaking the examination of any part of the system. Further, the principles of examination are exactly the same as those which are used with any diagnostic problem in any other specialty in medicine.

HISTORY

As in every other field of medicine, history taking is the foundation of diagnosis.

Onset of Pain

The onset of pain of musculoskeletal origin may be from an obvious cause such as trauma, or may arise insidiously without obvious cause.

ACUTE PAIN

When a patient complains of pain of sudden onset, it is obviously important to know whether or not the onset followed a traumatic episode. Here one must recognize that trauma does not solely imply some external force creating an injury. Trauma may be intrinsic, resulting from internal derangement of either muscle (examples are ruptures of the biceps tendon and the plantaris tendon) or joint (examples are meniscal injury or joint dysfunction). This is invariably associated with unguarded movement occurring during the performance of a normal functional activity.

Acute Pain and Trauma. If the traumatic episode producing the pain is the result of extrinsic trauma, the history of onset is quite clear. However, it should be remembered that extrinsic trauma to one part of the musculoskeletal system may be associated with intrinsic trauma in some other part. This may be unnoticed at the time of the more severe and obvious injury.

Acute Pain without Trauma. Rarely, acute pain in joints may have a sudden onset at rest. In this event systemic disease should be suspected, particularly with gonococcal arthritis, rheumatic fever, and rheumatoid arthritis. The same situation also arises with gout, which is certainly not confined to the big toe, and which may be precipitated by a dietary indiscretion, alcoholic excess, stress, including surgery, or the use of various medications, especially diuretics, low doses of salicylates, and penicillin. Hematogenous pyarthrosis in infants is another example in which the onset of acute pain is insidious.

SUBACUTE OR CHRONIC PAIN

Most pain-producing conditions other than trauma in the musculoskeletal system have a history of insidious onset while the patient is at rest. They start often as an aching and progress with time, which may be short or relatively long, to severe discomfort or even acute pain.

Nature of Pain

A deep throbbing, aching pain is often associated with serious bone or joint disease. A sharp pain is likely to be associated with simpler pathology; the great exception is the lightning pains of syphilis.

EFFECT OF REST ON PAIN

An important clue to the possible etiology of musculoskeletal pain is uncovered by the patient's answer to the question of how the pain is affected by rest. The pain of simple joint dysfunction is relieved by rest whereas virtually all other causes of pain in the musculoskeletal system except those related to a prolapsed disc produce stiffness after rest. Night pain is always suggestive of bone tumor but may be an early sign of ankylosing spondylitis or spinal cord tumor. The carpal tunnel syndrome is also worse at night, as are the thoracic outlet syndromes, but in these conditions the symptoms are more those of paresthesias than true pain. Night cramps are a perplexing problem discussed in Chapter 9, p. 185.

EFFECT OF ACTIVITY ON PAIN

Especially important in the patient's history is what happens to the pain with activity. Pain from simple joint dysfunction and disc prolapse is aggravated by activity but often is associated with only one functional movement. With other causes of pain initial activity eases the stiffening and pain, but continuing activity makes the pain worsen.

Joint disease, which in this context is anything more serious than joint dysfunction, has pain associated with all movements. In the joints of the lower extremity, weight-bearing may temporarily relieve pain just as pushing on a bad tooth may temporarily relieve toothache. In that case the pain is probably due to joint disease.

Muscle conditions that give rise to pain hurt more when the affected muscle is stretched than when it is not, and more when it is in use than when it is not. Although one expects a torn muscle to produce acute pain of sudden onset during movement, and indeed it does, microtrauma to muscle may manifest itself as a pain of insidious onset during rest. This can be a confusing development. Cold may precipitate muscle pain which can be sudden in onset and acute but removed in time from the precipitating cause. This too is a confusing situation. Barometric changes may produce varying symptoms in muscles as they do in joints.

Swelling

If there is swelling of the part, the history of how the swelling started is important. If it is in a joint and comes on quickly following injury, there is probably blood within the capsule. If it is in a joint and comes on slowly, it is likely to be due to an excess of synovial fluid. If it is in a joint and caused by pus, it may come on quickly or slowly. Swelling from edema tends to disappear with rest and elevation. Swelling from lymph is less likely to disappear with rest and elevation. Swelling of a bursa, whether it be a manifestation of trauma or disease, is usually sudden in onset, but in this instance the anatomical situation of the swelling should make the differentiation clear.

Involvement of One or More Joints

Monarticular pain is often traumatic, but any systemic disease involving joints may present as monarticular arthritis. In differential diagnosis the following conditions must be considered: Collagen vascular diseases may initially involve only one joint, as may gout, brucellosis, tuberculosis, gonorrhea, syphilis, tumors, pyarthrosis, and most of the other reactive or infective arthritides. Pain flitting from joint to joint, however, is characteristic of rheumatic fever, and multiple joint involvement is common with the other collagen vascular diseases, serum sickness, and reactive synovitis to systemic diseases.

Weakness

It is important to assess whether true weakness is present or the difficulty in using a certain portion of the extremity arises from pain. If weakness is present, is it of recent origin or long-standing? Is it progressive or static? If the difficulty is unilateral, one tends to think of a local problem or of a neuritic problem; if it is bilateral, a systemic disorder must be considered. Involvement of the proximal muscle groups bilaterally, including muscles of both the shoulder and pelvic girdles, in the presence of pain raises the suspicion of polymyositis or polymyalgia rheumatica. Bilateral involvement of distal muscle groups is not usually associated with painful musculoskeletal conditions. If pain is present, it is almost certainly neuritic in origin.

Atrophy

Atrophy may have been observed by the patient, although it is frequently overlooked. If it is observed, its duration should be noted. Atrophy of the interosseous muscles of the hand occurs rapidly following nerve injury or immobilization, as does vastus medialis atrophy following knee injuries.

Sensation

Inquiries about sensation should be included in history taking. Numbness is a common complaint but should be explored thoroughly, since the patient is frequently vague about this. Is the patient confusing numbness with weakness? Is it a feeling of numbness or objective loss of sensation? Is the patient describing loss of sensation or paresthesia?

If true sensory loss is present, its location and its extent, whether partial or complete, should be noted. If partial, it will be made worse by certain activities (as observed in the entrapment syndromes).

A more intense feeling than paresthesia is burning pain. This is typical of causalgia.

Limitation of Motion

Loss of joint motion may be due to pain, muscle weakness, or intrinsic joint pathology. Determination of which factor is primary should be undertaken. Motion may be lost in only one joint range or in multiple joint ranges. It is worthwhile to remember the principle that a painless joint has full motion or no motion at all; pain arises in a joint that has only partial movement in either the voluntary or the joint-play range.

System Review

In history taking, a general review of all the systems of the body may present clues to diagnosis. The following items have a bearing on possible causes of musculoskeletal pain:

General—weight loss or weight gain, appetite, fever, sweating, nervousness
Rheumatological—fatigue, skin problems, reaction to sunlight, Raynaud's phenomenon, urethral discharge, conjunctivitis
Gastrointestinal—indigestion, nausea, vomiting, pain or bowel difficulty, including bleeding and change in bowel habits
Genitourinary—bleeding, burning, frequency, hesitation, nocturia, venereal disease
Endocrine—hair and nail change, temperature intolerance, cramps, unexplained weakness, edema
Pulmonary—cough, sputum, night sweats
Ear, nose, and throat—sore throat, bad teeth, sinusitis, sore tongue, difficulty with swallowing
Gynecological—irregularity of menses, vaginal discharge, surgical procedures, miscarriages, stillbirths
Central nervous system—headaches, vision

changes, vertigo, paresthesias, radicular pains

Vascular—claudication, difficulty with erection, coldness of extremities, discoloration, response to cold (Raynaud's phenomenon)

Past History

As in all history taking, it is important to know the patient's age, marital status, occupation, and, if the patient is a woman, whether she has had any miscarriages or stillbirths.

The age is important because certain musculoskeletal problems are prone to affect people during different decades of life. The occupation is important because well-defined problems occur in people who engage in heavy laboring activities or manual activities and do not in people with more sedentary occupations. The work environment may be important. Information on marital status or extramarital activity may pave the way to information with regard to venereal disease, prostatic infection in males, or pelvic inflammatory disease in females. Past accidents may have been forgotten but must not be overlooked. A history of musculoskeletal pain in adolescence or childhood, or of chorea, allergies, asthma, and hay fever, may have a bearing on the cause of the present complaint. Inquiry should be made as to whether a patient ever drank raw milk, either cow or goat, or ate undercooked meat or fish, and whether he has ever had a pyrexia of unknown origin. Where the patient has lived may be of importance, as certain geographic areas have endemic diseases associated with arthralgia. All these things may provide clues to the causes of pain. In these days the history of foreign travel should also be noted because tropical diseases are frequently associated with arthralgia.

Family History

Certain rheumatic diseases that produce pain have a familial tendency. The family environment may be contributory in assessment of a patient with pain.

PHYSICAL EXAMINATION

General Features of Musculoskeletal Physical Examination

It is essential that the part to be examined be adequately free from covering. If there is a corresponding uninvolved part, it should always be examined too. In any complicated problem of the musculoskeletal system, a complete physical examination may have to be undertaken; for example, even distant skin changes may have a bearing on the diagnosis.

Temperature, Blood Pressure, and Pulse

If temperature, blood pressure, and pulse are not taken, useful clues to diagnosis may be overlooked. In feeling the pulse the examiner should not forget to compare one side with the other and also to take the pulse in varied locations and examining positions and postures. Differences in the blood pressure between limbs may be a diagnostic clue. It may be necessary to examine the blood pressure in all four extremities.

Examination at Rest

Examination of the involved part is first undertaken at rest before any study of movement is made. Skin color, consistency, and turgor are noted. Local skin temperature is felt. Swelling is noted and then palpated; the presence of fluctuance, pitting, or brawniness is observed, and whether the swelling is localized, generalized, unilateral, or bilateral. Is the swelling fluid or solid? If it is solid, is it firm or soft, free or attached to something? The lymph drainage areas of the part must be examined for palpable nodes.

Examination of Movements in the Voluntary Range

In the presence of signs of active inflammation within the joint, which can be detected by the examination procedures described up to this point, it may be unnecessary to examine joint movement at all. Examination of joint movements in infants under these conditions may indeed be disastrous. If undertaken, such an examination must be conducted with great gentleness and always within the limit of pain.

In the absence of signs of active inflammation within the joint, movement of the joint in its voluntary range should be noted as it is performed actively by the patient and then should be checked by passive examination. The degrees of movement in each normal range of movement should be observed not only as a baseline from which improvement or deterioration can be checked but also for comparison with the unaffected side. When the pain is in a joint of one of the lower extremities, any inequality of leg length is noted, since this may cause unnatural stresses of weight-bearing at otherwise normal joints and may result in joint dysfunction from relatively innocuous unguarded movements. It may also bring about early changes of traumatic osteoarthritis, or it may cause laxity of the supporting ligaments from constant repetitive strain in otherwise normal function. Any insufficiency, i.e., functional as opposed to pathological shortening, of the Achilles tendons should be noted.

Study of the gait may prove an important aid in diagnosis. For instance, a gluteal gait, which is a waddling gait due to weakness of the gluteus medius muscle (see Trendelenburg's test, p. 49), is commonly associated with pathological changes in the hip joint. It should be remembered that the examination of a painful knee joint is incomplete until the hip joint and ankle joint and foot on the same side have been examined also. Any child complaining of pain in the knee is presumed to have a pathological condition in the hip of the same side until it is proved that he does not. This is an application of the principle that a joint examination is incomplete unless the joints above and below it are also examined.

There are some special considerations on examining a knee joint. If a knee joint is by history limited in movement and is painful following the healing of a fracture of one of the bones of the leg, and if there is obvious fibrosis and binding down of the quadriceps expansion (invariably associated with atrophy of the quadriceps), it should be immediately apparent that the patella is fixed in its cephalad range of movement. In this position the patella blocks flexion of the knee, and restoration of movement to the joint will be impossible until the patella is mobilized. Nor, in these circumstances, can it be anticipated that normal physiology can be restored to the quadriceps muscle until the fibrotic changes in the quadriceps expansion are reversed. The prescription of mobilizing procedures to the joint or an exercise program for the muscle must fail unless the patella is mobile.

The ligaments of the joint being examined must all be palpated individually because it is a classic axiom that ligaments are never tender unless they are torn or ruptured or there is some pathological condition within the joint they support. It should also be remembered that the normal capsule of a joint cannot be palpated. If the capsule can be palpated, there is a pathological condition in the synovium itself.

Muscle Examination

The muscles of the involved joint must then be examined. Muscle atrophy may be perceived by inspection, but it can also be checked quite accurately in certain situations by measurements that are qualitative (range) or quantitative (strength). In considering knee joint pain, one must be particularly careful to assess atrophy of the vastus medialis accurately. In the average patient, circumferential measurement of the thigh at 2 inches above the patella will detect masked synovial swelling in the suprapatellar pouch. Circumferential measurement of the thigh at 4 inches above the patella will specifically detect atrophy of the vastus medialis, which is invariably present with pathological conditions of the knee joint. Measurement of the circumference of the thigh more proximal to this point detects group muscle atrophy, which is more commonly associated with pathological conditions of the spine or hip. As long as there is any atrophy or weakness of the vastus medialis, a knee will remain unstable, even after the successful treatment and eradication of any pathological cause of pain within the knee joint. Since the vastus medialis comes into full play only in the last 15 degrees of voluntary extension of the knee joint, reeducative exercises prescribed for the quadriceps muscles often fail to restore normal strength and function to the vastus medialis because of the inadequate way a patient is taught to do the exercises. Too often, the importance of the last 15 degrees of movement is not stressed, so the patient fails to concentrate on completing full extension.

Muscle weakness may occur without atrophy, because of pain or fatigue. A manual muscle test is often an important part of the clinical examination and provides another baseline from which improvement or deterioration of a pathological condition in the joint may be assessed. Also, alienation of a muscle may be detected. By alienation we mean the apparent loss of voluntary control of a muscle which, in fact, is a functioning muscle.

MANUAL MUSCLE TEST

A manual muscle test is a rather crude but valuable examination that serves as a baseline of weakness from which improvement or regression may be judged. There are six categories of strength: (1) normal, (2) good, (3) fair, (4) poor, (5) trace, and (6) zero. "Good," "fair," and "poor" may be reported plus or minus as well, but these are very individual assessments that are useless unless the same person reports on every muscle test on the same patient. There is no reason to perform manual muscle tests on patients whose weakness is due to upper motor neuron involvement, since they cannot be performed accurately. "Normal" means complete range of motion against gravity with full resistance. "Good" means complete range of motion against gravity with some resistance. "Fair" means complete range of motion against gravity with no resistance. "Poor" means complete range of motion with gravity eliminated. "Trace" means evidence of slight contractility with no joint motion. "Zero" means no evidence of contractility.*

At this point one must look for irritable trigger points and muscle spasm. If in history taking or early in the physical examination the patient has indicated a pattern of pain rather than a point of pain, a search has to be made for an irritable trigger point or points in muscle. This is not too difficult; our work has confirmed Travell [1952] that, given a pattern of referred pain, one can predict the location of the trigger point causing it. Having located it, one should be able on palpating it to reproduce the pattern of referred pain. If there is a predictable trigger point producing a predictable pattern of pain, the result of treatment by injection or the use of the vapocoolant spray and stretching is also predictable (see p. 190).

Considerable attention is paid to contractures and tightness of muscle, tendons, and fasciae in clinical examination. Tight hamstrings, tight heel cords, tight shoulder girdle muscles, tight lumbodorsal fasciae are frequently encountered. Myostatic contractures are implicated in the causes of pain of hip joint disease and in problems that arise in and around joints following immobilization treatment or after a patient has been at bed rest for any prolonged time.

We advocate the use of the term *myostatic spasm*. This brings the concepts of spasm and contracture more into focus of clinical facts and the realization that many "tight" muscle conditions are reversible by proper physical therapy—again particularly the use of injection therapy and the proper use of the vapocoolant spray.

True contractures do occur, but usually as a complication of surgery and severe trauma and as a complication of neuromuscular diseases. They are extremely difficult to overcome. Often they are not pain-producing but

* An assessment of factors such as shortening of bone following fracture healing or impairment of joint movement must be made in recording the results of a manual muscle test if it is to be meaningful.

require treatment because of severe limitation of function. Thus consideration of contractures per se does not further enter into our thesis of pain and its treatment.

Examination for causes of muscle tenderness must be undertaken. Palpation reveals fibrositic nodules which are unpredictable in location and produce unpredictable patterns of pain. (Fibrositis is discussed on p. 189.) Tenderness of muscles is also a sign of peripheral neuritis and phlebitis and a very distressing feature of poliomyelitis. It is a sign of parasitic infestation as in trichinosis. An aching discomfort in muscles is frequently seen in upper respiratory tract infections, gastrointestinal infections, and other infectious diseases. Acute muscle pain may be residual for some hours after cramping. Muscle tenderness is a feature of many rheumatoid diseases. Viral infection often causes acute pain.

Radiating pain must be differentiated from referred pain. Radiating pain invariably indicates some nerve involvement in the symptom complex. It is common with radiculitis and entrapment syndromes. In entrapment syndromes, Tinel's sign is additional confirmation of peripheral nerve involvement.*

Asymmetry of muscle groups as a residual from preexisting disease may also be a cause of current symptoms of pain, and their significance must be determined. With asymmetry of this kind, fascial tightness is often associated with the muscle tightness. Fascial tightness can be overcome by physical therapy, determination, and hard work. Tight iliotibial bands are a fairly common example of fascial involvement.

Examination of the Central Nervous and Vascular Systems

Examination of a patient with musculoskeletal pain is incomplete unless the related parts of the central nervous system are examined by assessment of all reflexes and sensation and, when necessary, the cranial nerves. The pulses are assessed on each side of the body, various postural adaptations being made as described in named tests. With neck injuries the eyes should be examined, since the dilatation of one pupil suggests irritation of the cervicothoracic sympathetic ganglion by, for instance, pressure from a hematoma. This is part of the reverse Horner's syndrome, which also includes a widening of the palpebral fissure and apparent proptosis of the eye (as distinct from a Horner's syndrome, which results from complete interruption or paralysis). It may be the only indication of a retropharyngeal abscess or tumor giving rise to neck pain. Fixed pinpoint pupils (dilated pupils also) may be the clue to drug addiction, of which intractable pain may be a pseudosymptom in an addict looking for a "fix." The Argyll Robertson pupil (the fixed irregular pupil which does not react to light) may be the clue to pain due to syphilis.

REFLEXES

Here we believe that a review of the available reflexes is useful. The tendon reflexes, though indicating fairly specific levels of pathology, may be inhibited or increased by voluntary effort. Unless very definite changes are noted, great reliance should not be placed on them in determining the level of a lesion. Further, a reflex may occur at more than one reflex arc level and may appear normal when in fact it is not.

Several additional points should be borne in mind with regard to reflexes. The root levels subserved by the reflexes vary greatly from individual to individual (and also from text to text) and can therefore be taken only as an approximation. Our presentation here is not complete but represents the most commonly used reflexes. Reflexes can be divided into superficial and deep (tendon and periosteal). Absent superficial reflexes accompanied by hyperactive deep reflexes or pathological reflexes indicate pyramidal tract involvement. Because of individual variation, comparison of the two sides is more important than a

* Contrary to some teaching, certain neurological conditions besides peripheral neuropathy may be associated with pain at least in their early manifestations. These include subacute combined degeneration of the cord, multiple sclerosis, syringomyelia, syphilitic diseases, and neuropathic joints.

numerical grade, and before a deep reflex is reported absent, it should be tested for in several different positions. Slow reaction of the reflex suggests the presence of hypothyroidism.

When dealing with problems in the cervical area, one should remember the following: The biceps tendon reflex arc passes through the fifth and sixth segments of the cervical cord levels. The brachioradialis reflex passes through the same levels. The pronator quadratus reflex passes through the sixth and seventh cervical cord levels. The triceps reflex passes through the sixth, seventh, and eighth cord levels. Hoffmann's reflex of the fingers indicates pyramidal tract involvement.

There is a gap in reflex testing between the first and sixth thoracic cord levels. The epigastric reflex passes through the sixth, seventh, and eighth thoracic cord levels and is elicited by stroking below and along the costal margin. This causes contraction of the upper fibers of the transversalis muscle with dimpling of the epigastrium on the side being tested. The usual upper abdominal quadrant reflex is controlled through the eighth to tenth thoracic cord levels. The lower abdominal quadrant reflex is controlled through the eleventh and twelfth thoracic cord levels.

To cross-check the abdominal reflexes one may use two less specific periosteal reflexes, namely, the costoabdominal and the pubic reflexes. The costoabdominal reflex is elicited by tapping the costal margin in the nipple line. This should cause the umbilicus to be pulled up on the same side. The arcs of this reflex are through the eighth and ninth thoracic cord levels. The pubic reflex, which is least specific, as it passes through the thoracic cord levels from the sixth to the twelfth levels, is elicited by tapping the pubis. This results in a general contraction of the abdominal muscles (but especially the recti) and contraction of the adductors of the thighs.

There are six reflexes to be examined in the legs. The cremasteric reflex arc passes through the first and second lumbar cord levels. Parenthetically, there may be a vestigial labial reflex in women. The patellar tendon reflex passes through the second, third, and fourth lumbar cord levels. The hamstring reflex passes through the fifth lumbar cord level. The Achilles tendon reflex passes through the first and second sacral cord levels. The anal reflex passes through the fourth and fifth sacral cord levels and is elicited by pricking the perianal skin on each side. Finally, there is the plantar reflex, which, when extensor in reaction, indicates pyramidal tract pathology.

Table 3-1 summarizes the cord levels through which the various reflexes pass. The gaps become self-evident, as do the overlaps.

Loss of muscle power and/or the presence

Table 3-1. Reflex Arcs

C1	
C2	
C3	
C4	
C5	Biceps and brachioradialis tendon reflexes
C6	
C7	Pronator quadratus reflex
C8	Triceps tendon reflex
T1	
T2	
T3	
T4	
T5	
T6	Epigastric reflex
T7	
T8	Upper abdominal quadrant reflex
T9	
T10	
T11	Lower abdominal quadrant reflex
T12	
L1	Cremasteric reflex
L2	
L3	Patellar tendon reflex
L4	
L5	Hamstring tendon reflex
S1	Achilles tendon reflex
S2	
S3	
S4	Anal reflex
S5	

of muscle atrophy may be more reliable signs than are equivocal reflex and sensory deficits. Analysis of the spinal fluid may be a necessary adjunct to the clinical examination. A raised protein in the cerebrospinal fluid is indicative of dural irritation from some epidural or intradural source.

Except for the two periosteal reflexes and the finger and plantar reflexes, all the reflexes mentioned are concerned with striated voluntary muscles having their centers within the cerebrospinal axis and may be inhibited, as we have said, by voluntary effort. Further, reflexes should not be deemed to be absent unless reinforcement has been added when one tries to elicit them.

Following spine surgery there may be a neurological deficit unconnected with the current problem. This makes postoperative signs the more difficult to interpret.

Finally, any area deriving its nerve supply from the part of the spine which is under consideration must be examined. For instance, an examination of the lumbar spine is incomplete unless the abdomen is examined and a rectal (and, if necessary, a pelvic) examination is undertaken. An examination of the thoracic spine is incomplete unless the chest, the chest wall, the heart and lungs, the upper abdomen, and the arms are examined. An examination of the cervical spine is incomplete unless the cranial nerves, the upper thorax with its outlets, the arms, and even the eyes, ears, nose, and throat are examined.

SENSORY EXAMINATION

The response to sensory tests tends to be subjective, and overlapping areas of sensory nerve supply are common. So again, anything less than definite patterns of sensory changes following known anatomical patterns should not be relied on too much. It is well to remember that sensory changes occur two segments lower than those at which pathological conditions in the thoracic part of the spinal cord are situated. The level of sweat changes coincides with the level of sensory changes and provides a useful check.

Examination for Joint Play

Joint dysfunction is one of the commonest causes of joint pain in clinical medicine. Examination for the absence of joint play must be undertaken.

RULES FOR JOINT-PLAY EXAMINATION

The patient must be recumbent; only then does the examiner have perfect control of the examining movements he is performing. The one exception to this rule is the examination of the joints of the fingers and at the wrist. The position of examination is vital to the accumulation of precise data relating to joint movement. The techniques of eliciting joint play must be adhered to. Since joint-play movements are, for the most part, small in range, their performance requires accuracy and precision.

In addition, there are certain rules of examining techniques which must be followed when manipulative maneuvers are used:

1. The patient must be relaxed, and each aspect of the joint being examined must be supported and protected from unguarded painful movement that may otherwise occur in the course of the examination. Pain produced by unguarded movements of painful joints puts the supporting muscles into spasm and prevents the performance of the examining movements for joint play.
2. The examiner must be relaxed. At no time should his examining "grip" be painful to the patient. His grasp must be firm and protective, but not restrictive.
3. Each joint of a topographical area must be examined separately. For instance, the wrist is not examined as such; rather, the radiocarpal joint, the midcarpal joint, the ulnomeniscocarpal joint, and, finally, the distal radioulnar joint are examined in turn.
4. One movement at a time is examined at each joint.
5. In the performance of any one movement, one facet of the joint being examined is moved upon the other facet of the joint,

which is stabilized. Thus, there should always be one mobilizing force and one stabilizing force exerted when a joint is being examined.

6. The extent of normal joint play can usually be ascertained by examining the same joint in the unaffected limb.

7. No forceful movement must ever be used, and no abnormal movement must ever be used.

8. An examining movement must be stopped at any point at which pain is elicited.

9. In the presence of obvious clinical signs of joint (or bone) inflammation or disease no examining movements need be or should be undertaken.

EXCEPTIONS TO RULES OF JOINT-PLAY EXAMINATION

Rules 3 and 4 above may be broken only if it is technically impossible to observe them. For example, the examining maneuvers for eliciting long axis extension of the midcarpal joint, the radiocarpal joint, and the ulnomeniscotriquetral joint, as well as the maneuver for pulling the head of the radius downward on the ulna, are the same. The movements of long axis extension at the midcarpal, radiocarpal, ulnocarpal, and proximal radioulnar joints are also performed when long axis extension is done at the glenohumeral joint. Similarly, the identical examining manipulative maneuver is used for long axis extension at the mortise joint and the subtalar joint and while performing the talar rock and the lateral tilt movements of the calcaneus on the talus.

With regard to the examination of the joints of the upper limbs, it is surprising how seldom both the midcarpal and the radiocarpal joints are affected at the same time by joint dysfunction from any cause. In searching for impairment of the long axis extension at either joint, one is simply taking up the slack of normal long axis extension of the uninvolved joint to determine its presence or absence in the other joint. If dysfunction is present in both joints, then one is simply examining both joints at the same time; no harm can come from this maneuver since the same movement is being used and is normal for each joint. In practice one seldom, if ever, finds evidence of loss of the joint-play movement of long axis extension at the ulnomeniscotriquetral joint, presumably because the articulating surfaces of the two bones are separated relatively widely by the intraarticular meniscus.

The downward movement of the head of the radius on the ulna, which is accomplished by the same technique, is legitimate, if there is no involvement of the midcarpal or radiocarpal joint. The performance of these movements is simply part of taking up the slack before exerting the manipulative pull to move the head of the radius. If all three joints happen to be involved, again, no harm can come from performing all three movements at one time since the pull for all is identical and normal.

For long axis extension at the mortise joint and the subtalar joint in the lower extremity the same rationalization must be allowed. Clinically, it is strange how often long axis extension is impaired at the subtalar joint with no impairment of the same movement at the mortise joint.

The only time the user of manipulative techniques may break these rules is when he knowingly does so for a specific reason. If the rules are broken unintentionally, the joint being examined may be damaged.

For details of movements of joint play the reader is referred to Mennell [1960, 1964].

The Back

While we stress that there are no basic differences in examining a patient with extremity pain and back pain and that topographical areas of the back are no different from each other, eliciting physical signs in the back presents technical difficulties that are not encountered in the diagnostic examination of an extremity. The functional movements in the back are an all-or-none phenomenon. No human being can move an isolated segment of the spine by using his voluntary muscles. Yet all pathological conditions in the spine (vertebral column) occur at intervertebral junctions, which are the functional units of the back consisting of two vertebrae, their joints, and the intervertebral disc. Thus little if any clinical information can be obtained from examining maneuvers that involve the patient in performing active or voluntary movements except forward bending and the recovery to the upright position.

A patient complaining of back pain complains also of loss of movement (function). These are the common dual symptoms of all painful musculoskeletal conditions. The less one asks a patient to do that causes more pain, the better. A patient with pain in the low back or thoracic spine, therefore, is asked only to bend forward and return to the upright position. The limitation of forward bending provides baseline information from which to judge the result of treatment. If a patient is complaining of cervical pain, limitation of voluntary movement in rotation, forward flexion, and extension is observed, again for baseline information only.

For the skeletal part of the examination, all other examining movements are performed passively by the examiner and are aimed at detecting loss of intervertebral movement to determine, by reproduction of pain, the junction(s) at which something is wrong. For the muscle part of the examination, palpation, strength or weakness, and atrophy are assessed with the patient at rest in either the supine or the prone position. Signs of muscle pathology may also be elicited by putting the muscles on stretch in the sitting position and rotating the trunk in the upright position to the right and left. The sacrococcygeal and coccygeal joints should be examined with the index finger in the rectum and the thumb posteriorly on the surface. The sacroiliac joints are examined by means of specific torsion techniques with the patient in the side-lying position.

Certain observations are helpful in the structural examination. First, the patient is asked to point with one finger to the spot from which he feels his pain is arising. One must then determine what structures lie under the pointing finger. If the finger points to a place under which there is only muscle, the pain is probably arising from muscle. He may be able to indicate only a pattern of pain which may be referred pain. In that case the examiner searches later, by palpation, for a predictably located irritable trigger point. If the finger points to a place well localized just to the medial side of one posterior superior iliac spine, the pain is probably arising from that sacroiliac joint. If he points to an intervertebral junction, something is probably wrong with either bone, joint, ligament, or disc at that junction.

On inspection, the examiner looks for a segmental deviation from a normal curve or a segmental flattening or increase in kyphosis. No human being can alter his spinal curves segmentally (i.e., two or three functional units) by voluntary muscle action because, as we have mentioned, any functional movement of the spine is all or none. If there is a segmental deviation from a normal curve and the patient has indicated by his pointing finger that he feels his pain in that area, it is likely that the pathological cause of his pain is located there. The only possibly innocuous cause of segmental mechanical deviation of a normal spine curve is preexisting disease or a congenital anomaly.

In normal forward bending and recovery to the upright position there is a synchronous smooth separating and coming together of the spinous process throughout the spine. If the examiner observes too much or too little movement at an intervertebral junction or in a segment of the spine, and if that is where the patient's pointing finger has localized the seat of his pain, the pathological condition causing pain is situated at or below that junction.

In the stooped position usually the individual spinous processes are easily identified in the thoracic and lumbar spines (Figure 3-1). The examiner percusses directly over each vertebra. Pain on percussion over a vertebra means there is something wrong with that vertebra or in one of the vertebral structures at the junction below that vertebra. A sharp, short well-localized pain on percussion indicates simple joint pathology, usually joint dysfunction. A deep aching and maybe throbbing pain on percussion indicates a more serious

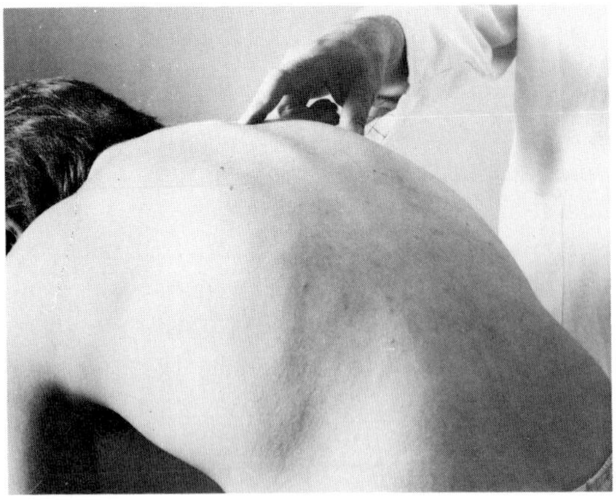

Figure 3-1. The spinous processes of the thoracic vertebrae, clearly identified in the forward bending (stooping) position.

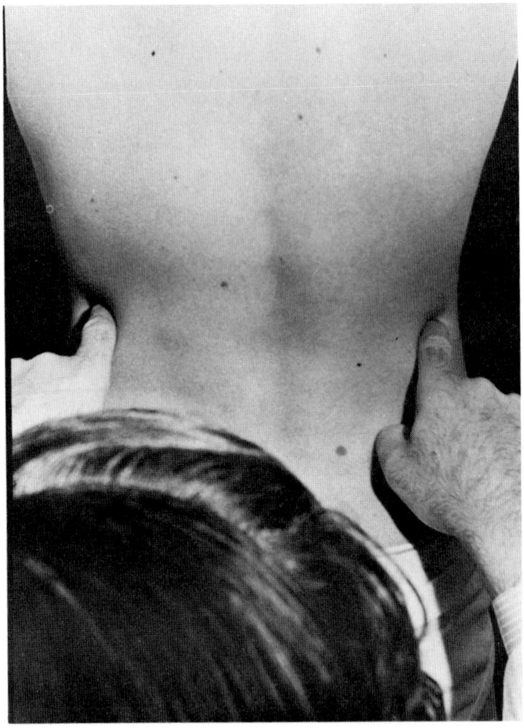

Figure 3-2. Method used to assess the relative heights of the iliac crests with the patient standing. Note that the left crest is significantly higher than the right.

pathological condition. A sharp short pain that reproduces a radiating pain is suggestive of a disc prolapse.

Clinical information may be gleaned by observing how the patient resumes the upright from the forward bending position, however limited the stooping may be. A patient with joint dysfunction resumes the upright position in a normal way. The patient with a more serious pathological condition resumes the upright position in a tortuous manner.

The heights of the iliac crests are assessed by placing the index fingers horizontally on them at eye level (Figure 3-2). No human being can make the iliac crests unequal in height by voluntary muscle action unless he stands with more weight on one leg than the other. The iliac crests may not be level if there has been preexisting disease in the lumbosacroiliac spine area or if there is a congenital anomaly in the spine or a gross shortening of one leg (more than ½ inch).

It is our clinical observation that if an iliac crest is raised, if the patient indicates that his pain is on that side, and if there is a segmental sciatic scoliosis convex on the painful side, the pathological condition causing the pain is joint dysfunction in the sacroiliac joint or the facet joint at the lumbosacral junction (or perhaps the fourth lumbar junction). If the pain is on the side of the lowered iliac crest and the concave side of the sciatic scoliosis, the cause is a disc prolapse or some other more serious pathological condition.

A gibbus deformity (segmental increased kyphos) is usually associated with bone disease or a vertebral fracture either from trauma or from some intrinsic pathological condition of bone.

Two measurements are useful in assessing the possible cause of low back pain. The distances between the posterior superior iliac spines with the patient sitting and prone should be compared (Figure 3-3). Normally there is a difference of measurement between these spines with change of posture. As a rule they are farther apart with the patient prone. They may be closer if the sacrum is horizontal instead of being in the more normal vertical position. Absence of movement between the

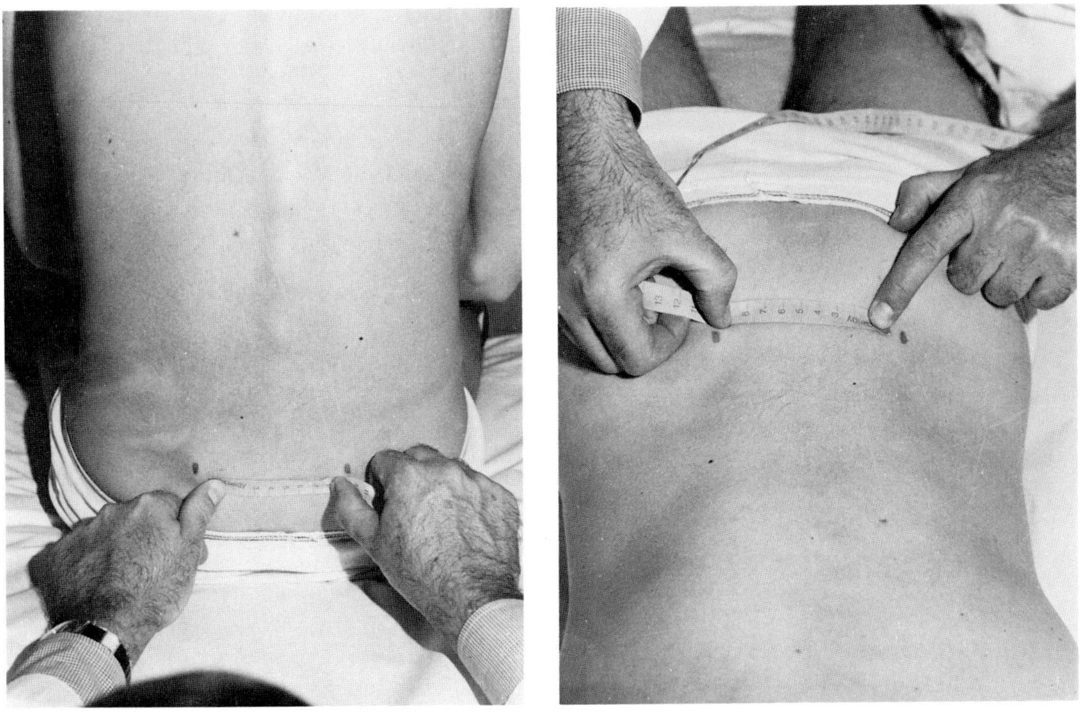

A *B*

Figure 3-3. Measuring the distance between the posterior superior iliac spines with the patient *A*, sitting and *B*, prone.

posterior superior iliac spines with change of posture means sacroiliac joint disease. If that disease is ankylosing spondylitis, there is likely to be a diminished respiratory excursion which can also be measured. The normal respiratory excursion in a young adult is between 2½ and 3 inches. In ankylosing spondylitis the excursion is often markedly limited early, usually to an inch or less. The other diseases which seem to show a predilection for the sacroiliac joint are tuberculosis, brucellosis, certain fungal diseases, Reiter's syndrome, and arthritis associated with ulcerative colitis.

Two other simple examining tests help to localize the intervertebral junction at which the pain-causing pathological condition is situated: the test of skin rolling and the test of vertebral springing. Both are performed with the patient in the prone position.

Normally the skin can be rolled over the spine freely and painlessly. In the presence of any spinal pathology there is tightness and pain (more than merely a pinching sensation) on skin rolling over the junction or over the vertebra where the pathological condition is situated (Figure 3-4). Joint dysfunction is unilateral in the back, and laterality can also be determined by skin rolling over the paravertebral troughs. If there is equal tightness and pain on skin rolling on each side of the spine, the cause of pain is likely to be more serious than simple joint dysfunction. Skin rolling also picks up areas of muscle pathology—an irritable trigger point, a muscle tear, or fibrositis. In such cases there is no tightness or pain on skin rolling over the spine. Parenthetically, if there is a constant area of tightness and pain on skin rolling and one is convinced that no musculoskeletal cause of a back pain exists, a pathological condition of a visceral organ under the area of tightness and pain should be suspected.

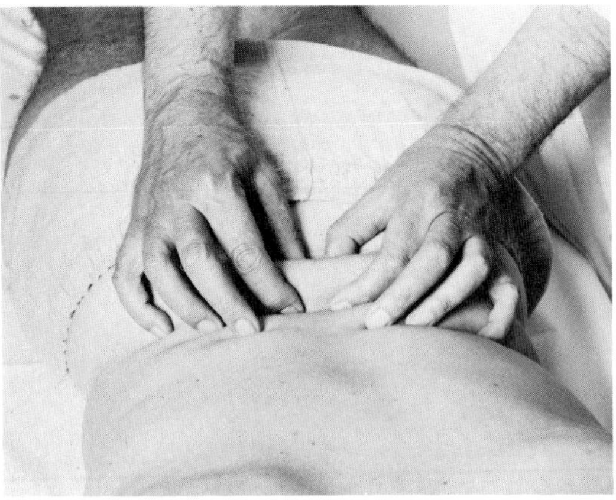

Figure 3-4. Manner in which skin is picked up between the examiner's thumbs and index fingers to perform the test of skin rolling.

Springing of a vertebra that produces pain indicates the vertebra in which the pathological pain-producing condition is situated, or it indicates that the pathological condition is at the junction below it. Figure 3-5 illustrates the method of springing a vertebra. Tightness and tenderness on skin rolling and pain on springing a vertebra at the same level clearly show where the cause of pain is situated.

There are other simple useful tests for assessing the cause of back pain. The bilateral jugular compression test increases the spinal fluid pressure. If the pain is intensified it is indicative of an extradural or intradural problem. A positive jugular compression test is commonly said to indicate a disc prolapse. However, one should remember that tumors and vascular anomalies also produce a positive test.

For an excellent and innocuous test, one can perform an epidural injection of 2 percent saline solution through the sacrococcygeal hiatus. Forty to 60 ml fluid can be introduced painlessly into the epidural space. If an epidural mass is causing symptoms, when the fluid, which is incompressible, hits the mass, the mass is compressed and symptoms from it are reproduced or exacerbated. The saline epidural injection has a second use. If there is radicular inflammation, the edema fluid in the nerve root is withdrawn into the hypertonic saline solution by osmosis, and relief of nerve pain is often immediate. For further details see Chapter 7, page 131.

Because the synovial joints of the spine and the vertebrae are so deep below the surface of the body, signs of inflammation, easily elicited in the bones and joints of the extremities, cannot be detected. Local heat and swelling cannot be palpated, nor can a thickened synovium be felt. That these signs are present though occult is not to be doubted, and they must be inferred by logical deduction. There are palpable ligaments at every intervertebral junction. We have indicated earlier that ligaments are never tender on palpation unless there is injury to the ligament or, more importantly, something is wrong with the joint (in the back junction) that the tender ligament supports. Synovitis in an extremity joint, whether from trauma or as an indication of joint disease, produces stiffness and aching. If it is traumatic synovitis, the swelling may take up to 24 hours to develop. Likewise in the back the presence of synovitis can be inferred by the history of the onset of the pain symptom.

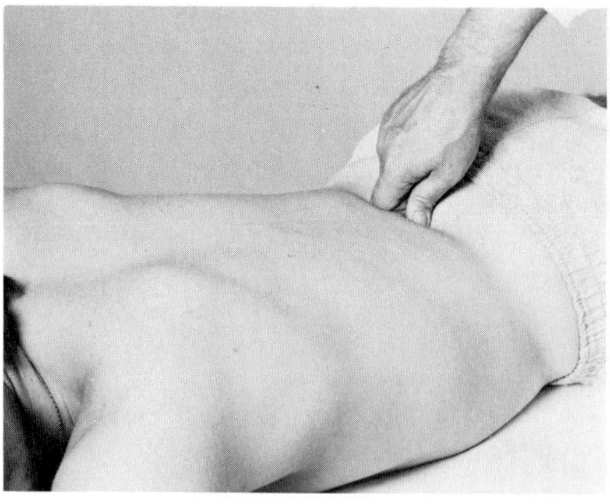

Figure 3-5. Springing a vertebra in the lumbar spine. The spinous process of the vertebra is cradled between the dorsiflexed terminal phalanx of the thumb and the flat middle phalanx of the index finger. The thumb and index finger lie over the lamina, and the thrust is equal on both sides avoiding any rotation.

If there is blood in a joint in an extremity and pain follows trauma, the onset is quick, acute, and increasing. On the same basis, hemarthrosis can be inferred in the back. If there is pus in a joint in an extremity, the symptoms are similar to those of hemarthrosis, but if the patient is obviously sick, pyarthrosis in the back may be inferred.

Stiffness with or without increased pain in the back following rest indicates more serious bone or joint pathology, especially if there is relief of pain on limbering up followed by more severe pain with continuing activity. Stiffness with rest and aggravation of pain on resumption of activities are suggestive of muscle pathology. Night pain is always suggestive of bone tumor, ankylosing spondylitis, or spinal cord tumor.

As examining techniques become more specific, an examining maneuver that causes pain on opening a facet joint suggests a diagnosis of joint dysfunction. If pain is produced on opening and closing a facet joint, i.e., if both facet joints at a junction produce pain on movement, a more serious pathological condition must be suspected.

Neck pain with limitation of movement superimposed on the chronic pain of Still's disease, rheumatoid spondylitis, or ankylosing spondylitis absolutely contraindicates the performance of any manipulative examining or therapeutic maneuver. The same danger should be recognized if a patient, especially a child, develops an acute torticollis associated with tonsillitis. Pain symptoms in the neck with or without paresthesias in the arms under these circumstances indicates ligament defect of the atlantoaxial junction with too much movement of the odontoid posing a potentially lethal situation.

The costovertebral joints in the thoracic spine must not be overlooked as a source of symptoms, and a method of examination to differentiate them from the interlaminar joints is available [Mennell, 1960]. Raney [1966] describes a syndrome of pain arising from traumatic changes in these joints and dysfunction or disease in them as a cause of acute and often bizarre pain.

Finally, the straight leg raising test merits close attention. It is one of the most important yet most often incorrectly performed and poorly interpreted tests in common use. Limitation of straight leg raising by pain in either the low back or the back of the leg or both is not specifically indicative of radiculitis.

Unilateral straight leg raising may be limited and painful because of any of four conditions: tight hamstring muscles, hip joint pathology, sacroiliac joint pathology, or radiculitis. Either of the first three conditions limits unilateral straight leg raising by pain that mimics sciatic pain. Tightness of the hamstrings can be determined by palpation, and pain from this is usually, but not always, felt behind the knees. This should have been assessed while the patient was examined standing and stooping, but it can be checked again. It should be borne in mind that the normal angle of straight leg raising in many individuals is less than 90 degrees. Hip joint pathology can be differentiated by use of the Trendelenburg test in the standing position and by impairment of passive range of motion, accompanied by pain, with the patient supine. Having assessed hamstring muscle tightness and the condition of the hip joint, one must then differentiate between sacroiliac joint pathology and radiculitis.

At a certain angle of straight leg raising the hamstrings pull on their origin at the ischial tuberosity. With the opposite leg stabilizing the other half of the pelvis, the innominate bone on the side being tested tends to be rotated backward on the sacrum through the sacroiliac joint as the muscle pull increases. At about the same angle, tension is being put on the sciatic nerve roots through pull on the sciatic trunk. Pathology in the sacroiliac joint or radiculitis produce pain in either the low back, the buttock, or the back of the leg. If the leg is then dropped ½ to 1 inch, both the stress on the joint and the pull on the nerve roots are relieved. With the leg stationary, the foot is dorsiflexed. No painful stress is added to the sacroiliac joint, but a pull is reinstituted on the nerve trunk. Exacerbation of pain on dorsiflexion of the foot is the true Lasègue's sign of radiculitis. If the pain is not reproduced by dorsiflexing the foot, sacroiliac joint pathology should be suspected.

A doubtful Lasègue's sign can be checked by using Naffziger's bilateral jugular vein compression test. This test increases the pressure of the spinal fluid, which in turn increases the pressure effect of a tumor in the epidural space. It may also be checked by the head-leg test, in which tension is exerted on the low spinal roots by pulling upon them from the head end during flexion of the head and neck.

When both legs are raised together with knees extended, the hamstring muscles pull equally at their origins on both sides, tilting the whole pelvis backward at the lumbosacral junction. With both legs raised, tilting tends to occur at an earlier angle of elevation than that at which sacroiliac joint movement takes place with only one leg raised; hence pain at an earlier angle with both legs raised and outstretched together, rather than with one leg raised by itself, almost certainly indicates lumbosacral junction pathology.

If some doubt should arise as to the validity of a patient's responses to the straight leg raising tests, the responses should be checked with the patient in the sitting position with the legs hanging. As the legs are straightened at the knee, the patient not only resists the straightening but also starts to lean backward.

Innumerable other clinical tests are described in the literature which supposedly give varied information as to the state of the back. We find most of them rather nonspecific and unhelpful. No test should ever be done unless the examiner knows exactly how each movement he makes affects some specific structure in the back. The procedures we have described are simple and clear. They do not inflict any undue distress on a patient in pain. Each supplies specific information.

A low back examination is incomplete unless abdominal, rectal, and vaginal examinations are performed and assessments made of the hips, knees, and feet and the peripheral pulses. It is also useful to remember that when a patient has back pain and abdominal pain, if the pains are at the same level, the cause is usually in the abdomen. If the abdominal pain is at a lower level than the back pain, the cause is usually in the back. After completing the examination, one at least knows that there is something wrong in the back, the location, and in which structure the primary pathology is located, even if the underlying pathological condition has not yet been discovered. One also knows whether physical treatment can be expected to help the patient and, by referring to the principles of physical therapy, one can choose the most helpful modalities and write a logical plan of treatment.

BIBLIOGRAPHY

Beetham, W. P., Polley, H. F., Slocumb, C. H., and Weaver, W. F. *Physical Examination of the Joints.* Philadelphia: Saunders, 1965.

Kellgren, J. H. On the distribution of pain arising from deep somatic structures with charts of segmental pain areas. *Clin. Sci.* 4:35, 1939.

Mennell, J. M. Manipulation and the treatment of low back pain. *Clin. Orthop.* 5:82, 1955.

Mennell, J. M. *Back Pain.* Boston: Little, Brown, 1960.

Mennell, J. M. *Joint Pain.* Boston: Little, Brown, 1964.

Raney, F. L. S. Radiculitis Arising from Abnormalities of the Costovertebral Joints. Paper and Exhibit for the American Academy of Orthopaedic Surgeons, 1966.

Travell, J., and Rinzler, S. H. The myofascial genesis of pain. *Postgrad. Med.* 11:425, 1952.

Williams, P. C. Examination and conservative treatment for disc lesions of the lower spine. *Clin. Orthop.* 5:28, 1955.

4
Clinical Observations Aiding Diagnosis

While deprecating the use of shortcuts to arrive at a diagnosis, we have found some clinical observations reliable in providing clues to correct diagnostic thinking. We record them for any help they offer the examiner.

FIVE "NEVERS"

Five "nevers" are helpful in assessing a problem of musculoskeletal pain. They are:

1. Fluid can never be palpated in a normal synovial joint. If fluid is palpated in a joint, something is wrong with the joint.
2. The normal capsule of a synovial joint is never palpable even when there is excess fluid within the joint. If a capsule of a joint is palpable, something is wrong with the capsule, usually of a serious nature or a joint disease is present.
3. Normal ligaments are never tender on palpation. If, therefore, palpating a ligament causes the patient pain, either the ligament itself is injured or (and we think this is much more significant) something is wrong with the joint supported by the ligament. In considering the back, the word *junction* is substituted for *joint* because at each junction there is more than one joint. Thus tenderness on palpation of an interspinous ligament, or of the posterior ligaments of a sacroiliac joint medial to a posterior superior iliac spine, clearly indicates some pathological condition at the level of the tender ligament or in that sacroiliac joint. There is one "catch" concerning ligament tenderness, and it is related to an often unrecognized ligament attachment at the knee. The superficial fibers of the medial collateral ligament of the knee attach to the tibia beneath the attachment of the pes anserinus (the common tendon of the sartorius, the gracilis, and the semitendinosus). Figure 4-1 illustrates the length of the superficial fibers of the medial collateral ligament of the knee joint. So, tenderness to palpation on the anteromedial aspect of the tibia about four fingerbreadths below its plateau may be a clue to the fact that leg pain is, in fact, arising from some pathological condition within the knee.
4. Osteoporosis senilis never affects the skull. Thus a patient with osteoporosis who has porotic lesions in the skull has a more serious metabolic disease, some metastatic involvement, or multiple myeloma.
5. Osteoporosis senilis never affects the lamina dura. Thus a patient with osteoporosis whose lamina dura has disappeared has one of the other metabolic diseases of bone. However, as alveolar infection and toothlessness are also associated with the absence of the lamina dura,

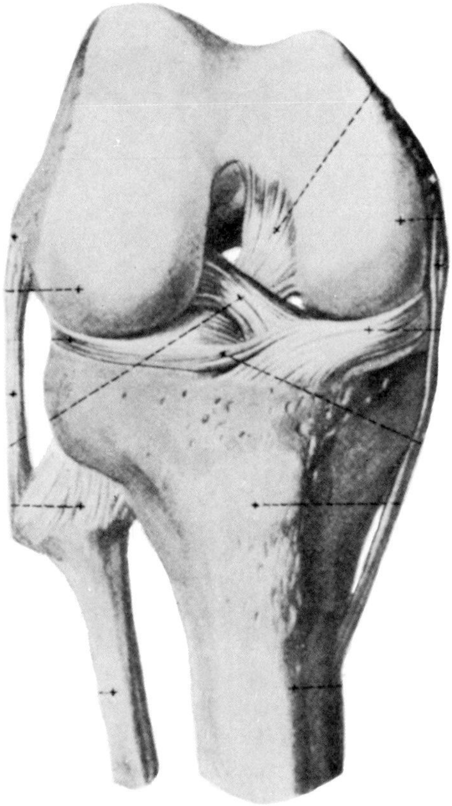

Figure 4-1. Length of the superficial fibers of the medial collateral ligament of the knee joint. Its insertion about four fingerbreadths down from the tibial plateau on the anteromedial aspect of the tibia should be noted. (The superimposed linear markings are of no significance to the purpose of the illustration.)

the clinical observation may have to be modified (see Figure 2-7, p. 21).

INFLAMMATION

There are two clinical signs to assist the direction of our thoughts: the presence of skin redness and increased local skin temperature. These signs are synonymous with inflammation and must be closely watched. In an elderly patient with brittle bones an undiagnosed fracture may be the underlying cause. Also, redness and warmth of the skin may be caused by acute gout and mistaken for an abscess or an infected joint.

INTERPRETATION OF ATROPHY

The strange relationship between muscle atrophy, joint swelling, and pain may be helpful in differentiating two special causes of joint pain. It appears to us that the diagnosis of joint tuberculosis is associated with unusually gross muscle atrophy, very little joint swelling, and a somewhat insignificant complaint of pain. The diagnosis of gonococcal infection of a joint is associated with an exaggerated pain response with very little joint swelling and muscle atrophy. The patient with a gonococcal joint withdraws the joint when the examiner approaches it to palpate it.

PSEUDOPARALYSIS IN CHILDREN

There are two painful conditions in children in which pseudoparesis is the predominant feature and the reaction of the accompanying parent, almost always the mother, is unusually hysterical. In the more serious of the two the infant or, less frequently, the child, develops hematogenous pyarthrosis. In the second the child has a pulled elbow, a condition which is readily correctable and discussed in greater detail in Chapter 9, p. 180.

MUSCULOSKELETAL NOISES

A frequent complaint by patients is that of noise of one type or another in or from their joints. The sounds may be heard by the patient, the examiner, or both and may simply be an annoyance to the patient or may be accompanied by pain.

Clicking is an interesting phenomenon in the diagnosis of joint problems. A subjective sensation of clicking in a joint, if it is one of the five joints in which there is an intraarticular meniscus, is often caused by meniscal damage. Objectively produced on examination of the knee, this is called McMurray's sign. Chondromalacia of the patella may also produce clicking in the knee and must be differentiated.

Snapping usually represents the passage of soft tissue over a bony prominence. An example is the iliotibial band snapping over the greater trochanter. The patient describes this as a "snapping hip." The tendon of the peroneus longus may snap over the lateral malleolus producing a "snapping ankle"; and the long head of the biceps may snap in the bicipital groove producing a "snapping shoulder." A different mechanism producing snapping is found in the trigger finger where there is either enlargement of a portion of the tendon or narrowing of the tendon sheath, often due to tenosynovitis or tenovaginitis. The inability of the tendon to glide smoothly with finger motion produces the snapping, frequently accompanied by pain.

Grating, sometimes described by the patient as grinding or crunching, may be severe and readily heard or felt by the examiner. This probably represents loss of articular cartilage in a joint and results from direct contact of bone on bone. Frequently patients will complain of a grating sound in the neck on motion which is obviously very loud to them but can seldom, if ever, be heard by the examiner. Its exact etiology is not clear.

Popping occurs spontaneously or is performed voluntarily as a nervous habit, as when a patient pulls on his finger joints. It is probably created by the same mechanism that occurs with removal of a rubber suction tip from the wall: sudden opening of an adherent crenation in the synovial lining of the capsule probably produces a vacuum effect and the accompanying popping noise. It has no clinical significance when performed by the patient or when unaccompanied by pain. In the back a popping sound followed by pain and locking is the usual history for facet joint dysfunction. Joint manipulation on the back or extremities performed as a therapeutic maneuver is frequently accompanied by a popping sound. This need not be present for the maneuver to be successful. Indeed, the "pop" may come from a normal joint when the joint that was supposed to be moved was not moved at all; i.e., a pop accompanying a therapeutic manipulation is no assurance of a successful manipulation.

Crackling and crepitus may both be heard and felt by the examiner. A fine crepitus is heard in a diseased joint, for instance, when it is afflicted by rheumatoid arthritis. A coarser crepitus may occur in a joint with osteoarthritis. Crepitus palpated over a tendon sheath indicates tenosynovitis, which may be traumatic or a manifestation of an infective tenosynovitis heralding a systemic disease, such as rheumatoid arthritis or tuberculosis.

LOCKING OF JOINTS

When a patient says a joint locks on him, there are five diagnoses for which evidence can then be sought. In joints having intraarticular menisci, a loose or damaged meniscus causes locking. But in any joint, osteochondromatosis, osteochondritis dissecans, or a chondral fracture may cause locking. Chondromalacia of the patella may also be a cause of locking of a knee joint. Each of the authors has seen an unusual cause of locking of the knee: lateral dislocation of the patellar tendon so that it functioned as a flexor of the knee preventing extension, and dislocation of the patella into the condylar notch of the femur.

In this context, it is our belief that the intervertebral disc, while not intraarticular, acts like an intraarticular meniscus in that, when it is injured, it acts like a damaged meniscus and locks the facet joints at the same junction. This would seem to be a logical cause of back pain in a patient with a herniated disc that is associated with segmental localized mechanical spinal derangement. Disc surgery commonly relieves the limb pain but leaves residuals of back pain. Residual back pain after surgery is commonly due to unrelieved joint dysfunction in the facet or sacroiliac joints and can be treated by manipulation.

CLINICAL DIAGNOSIS OF FRACTURES

Chapter 2, on crossmatching anatomical structures with pathological causes of pain, draws attention to the well-known fact that fractures are not always initially revealed by x-ray examination.

There are two clinical ways of diagnosing occult fractures without supporting x-ray evidence and without distressing the patient by unnecessary movement and resulting pain. Long bones have the property of conducting sound. A fracture in a long bone impairs sound conduction. When a fracture is suspected in a femur, the bell of a stethoscope is placed over the anterior superior iliac spine and the patella is sharply tapped. Normally a ringing sound is heard by the examiner. If a fracture is present, the sound is impaired or lost on the involved side when compared with the uninvolved side. To test a humerus, the bell of the stethoscope is placed over the acromion and the olecranon of the ulna is sharply tapped (Figure 4-2). To test the radius, the bell of the stethoscope is placed over the head of the radius and the radial styloid is sharply tapped.

When fracture of a small bone is suspected, diagnosis can be confirmed by testing for fat in blood aspirated from an involved joint. The fat can be seen in the blood with the naked eye when it is expressed from the syringe into a white dish. It floats like oil on water. If blood is aspirated from a hematoma in the area of an injury and there is fat in it, a fracture must be present.

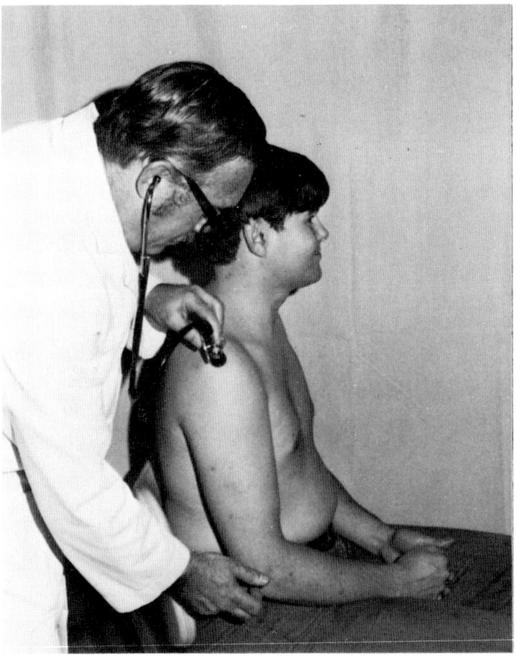

Figure 4-2. Testing for bone conduction of sound in the humerus. If a fracture is present, the sound is impaired.

LOWER EXTREMITY ABNORMALITIES AND BACK PROBLEMS

The legs and the feet may have a significant etiological bearing in any back problem, and they should always be examined. The type of shoes normally worn by the patient should be noted, as should evidence of abnormal or uneven wear. Of special importance is the sufficiency (relative shortness without contracture) of the Achilles tendons, or their relative insufficiency. The presence of a marked forefoot drop when the foot is at rest suggests tendo Achillis insufficiency. Unless the tendon is resilient, the heel cannot touch the ground at the same time as the ball of the foot without abnormally stretching the muscles, nerves, and blood vessels in the back of the leg. This, in turn, produces abnormal tension on the back muscles and abnormal weight-bearing stresses on all the joints of the foot, leg, and back.

Thus, an uncompensated tendo Achillis insufficiency is likely to produce knee and hip extension and a flattening of the lumbar lordosis. Overcompensation (the wearing of high heels) does the reverse and causes the wearer to stand with knees and hips flexed, lumbar lordosis accentuated, thoracic kyphosis exaggerated, cervical lordosis increased, and head thrust forward, which materially alters the center of gravity. The balance is thus disturbed and maintenance of normal tonus in supporting muscles prevented. Such abnormal posture produces excessive wear and tear in every joint, from the occiput to the joints of the toes. Muscle pain is the usual complaint because of constant muscle spasm and fatigue. Joint dysfunction quickly ensues. Such posture predisposes to acute symptoms whenever

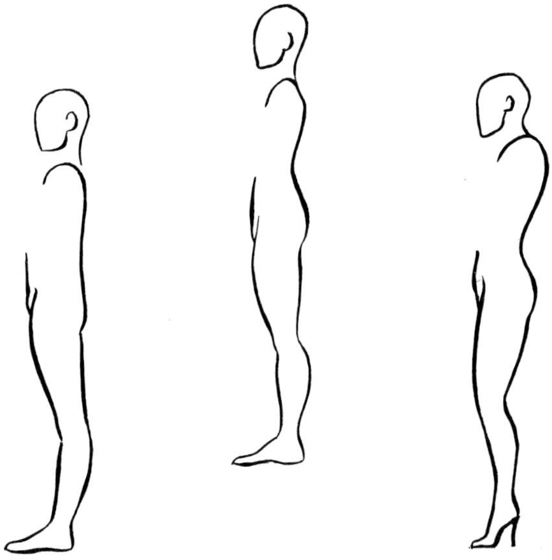

Figure 4-3. *A,* Normal posture and normal spinal curves with normally resilient Achilles tendons. *B,* The effect of Achilles tendon insufficiency: extension of knees and hips and flattening of normal spinal curves, especially the lumbar lordosis. *C,* The effect of wearing too high heels: flexion of the knees and hips and accentuation of all the spinal curves.

minor trauma is inflicted on the joints. Figure 4-3 illustrates the stress effects that arise from deviations from normal in the Achilles tendons. Two things must be remembered in adapting shoes to compensate for tendo Achillis insufficiency:

1. Correction of the full heel height must never be attempted or whatever resilience is left in the tendon will be lost.

2. Store shoes can have only a maximum of half an inch added to their heels without producing foot symptoms from the alteration of the slope of the soles of the shoes.

ANKLE PAIN

Pain felt in the ankle on walking uphill usually arises from the mortise joint; pain felt in the ankle when walking downhill usually arises from the subtalar joint. Pain in the ankle following immobilization of the leg for any cause is usually due to subtalar joint dysfunction. Pain in the lateral side of the ankle in the absence of trauma specifically to the ankle may be the result of joint dysfunction of the proximal tibiofibular joint.

MUSCLE SPASM

Muscle spasm is probably the commonest manifestation of musculoskeletal pathology, yet it remains the least understood phenomenon and the most abused diagnosis. When muscle spasm is (rarely) a primary phenomenon, it is a characteristic of poliomyelitis and occurs in the nonparetic muscles, usually those antagonistic to the paralyzed muscles. It is a feature of viral infection. It may be a primary reaction to cold. It is commonly associated with muscle contusion and muscle tears. For the most part, muscle spasm is a secondary guarding or splinting phenomenon to prevent painful movement in some other structure in which a primary painful pathological condition is present.

Muscle spasm is universally acknowledged as a protective phenomenon when it is associated with acute abdominal pathology, with an untreated fracture, and with an unreduced dislocation. All of these are acutely painful conditions. It is only logical to recognize that somewhat less painful conditions will also be associated with muscle spasm to a greater or lesser degree, and indeed this is our experience. To us, this is the explanation for the somatic component of visceral pain, such as is seen in coronary artery occlusion. That a lesser degree of muscle spasm cannot be detected by palpation does not mean it isn't there, although at times this must be an inference from the clinical symptoms. Particularly in examining for irritable trigger points, one must keep in mind the nonpalpable muscle spasm; a myofascial pain syndrome presumes muscle spasm.

There is no universal description of muscle spasm, but the term should be recognized as meaning a prolonged continuous contraction of muscle. The terms *stitch, cramp, charley horse, contracture,* and *spasticity* are loosely used sometimes to denote spasm and sometimes to denote a quite different pathological condition. Myostatic contracture is often myostatic spasm, and this is commonly found in association with painful hips and the frozen shoulder.

Muscle spasm need not involve the entire muscle but may involve only a segment of it, particularly in the back. For instance, springing or percussing a vertebra often elicits a transient localized muscle reaction at the same level which can be seen and felt. Also, it is quite common to find the shawl segment of the trapezius muscle in spasm in conjunction with neck pain, while the rest of the muscle appears to be normally resilient. The same thing is obvious in a sternocleidomastoid muscle in which there is an irritable trigger point. The predictable patterns of pain to be discussed later are all associated with muscle spasm, and the pain is relieved when the spasm is relieved (see Chapter 9, p. 190).

Medicolegal Implications

A common question asked physicians by attorneys in personal injury suits is: "What evidence did you have for muscle spasm?" The objective signs which we think are pertinent are derived from inspection, palpation, and examination of active and passive movement.

On inspection, unilateral and/or segmental contraction of a muscle is apparent. This cannot be simulated by voluntary muscle action. In the back, there is possibly room for mistaking a normal muscle for one in spasm when atrophy on the opposite side is present but this is a rare occurrence. On palpation, muscle in spasm has lost its resilience and is locally tender. On examination for active movement, spasm seldom limits all movement to the same degree. On examination for range of motion passively, the limitation of movement is usually less than that demonstrated actively.

Patterns of referred pain from trigger points, producing a localized muscle spasm not always detected clinically, are predictable and when the patient's complaints conform to the prediction, muscle spasm is present. This pattern is relieved when the spasm is relieved. Subjective signs such as pain on movement are of no value as evidence.

X-Ray Examination and Spasm

The appearance of flattening of normal curve in the spine is not an objective sign of muscle spasm. Rather, it denotes a process of dysfunction in the joints, giving rise to architectural changes which are painful and which cause secondary muscle spasm. We cannot believe that muscle spasm alone can alter the architecture of the spine; indeed, if it did, one would expect increased lordosis in the cervical and lumbar spines rather than flattening of the curves, since the major muscle groups act to increase the curves. Figure 4-4 is an x-ray picture of a cervical spine of a patient who had no symptoms or signs in the neck. Flattening of the cervical lordosis in the patient refutes the idea that muscle spasm causes the architectural change from the nor-

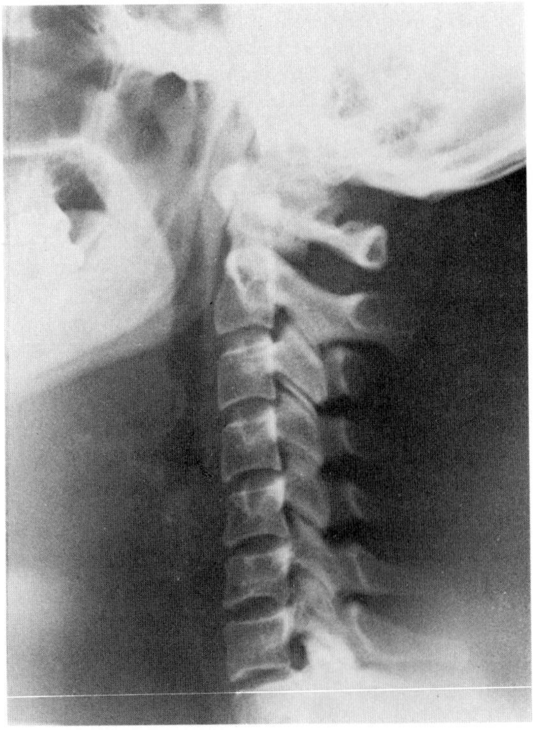

Figure 4-4. Flattening of the cervical lordosis in an asymptomatic patient.

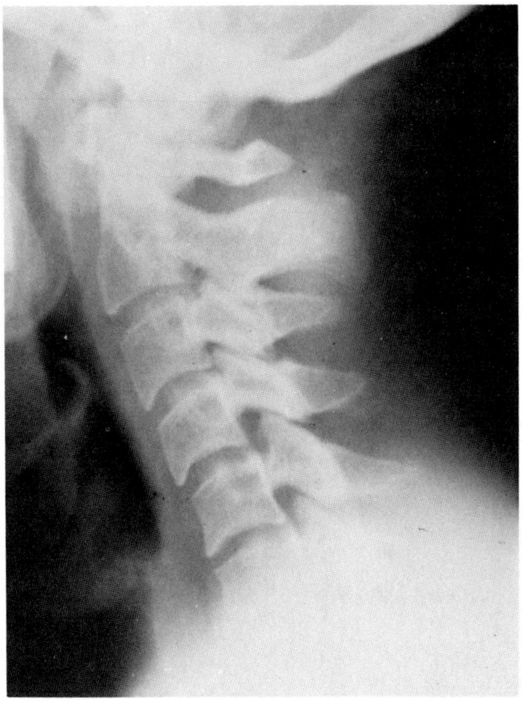

Figure 4-5. Segmental flattening in the cervical lordosis, a finding common with joint dysfunction.

mal. This is not the segmental reversal of the lordosis (illustrated in Figure 4-5) that is a common finding with mechanical joint dysfunction.

Muscle spasm is one of the most useful diagnostic signs that something is wrong in a muscle or an underlying structure. No human being can fake it, and it is the generally accepted diagnosis when a patient complains of musculoskeletal pain in an extremity, i.e., reflex muscle guarding. This is true of the back too, with a few exceptions.

AIDS TO DIFFERENTIATING CAUSES OF BACK PAIN

Having long used the same routine examination of patients who complain of back pain, a routine based on the principles we have espoused earlier, we find that certain clinical patterns of symptoms and signs evolve that reliably point to different pathological diagnoses.

We feel that being able to arrive at an early decision as to whether back symptoms arise from a musculoskeletal problem or are referred to from visceral pathology or systemic disease is a vital part of treatment. It is equally important to be able to differentiate serious disease states local to the spine from the simpler and less serious problems arising from trauma (intrinsic as a rule) that are daily presented in a doctor's office.

The recognition and acceptance of the fact that the back, from occiput to coccyx, is just another part of the musculoskeletal system and not some obscure system of its own is essential for the correct diagnosis of the cause of any back pain.

One of the authors (Mennell) has developed from clinical experience a method of tabulating significant symptoms from a patient's history, signs from the routine examination, clues from x-ray and laboratory findings, and general observations which differentiate between syndromes (Table 4-1). It should be borne in mind that no such table can be 100 percent accurate, that reliance on it as a shortcut to diagnosis can easily become a dangerous habit, and that obviously a table of this kind cannot be comprehensive. However, in spite of these limitations, the table can help the examiner to identify what is wrong with the patient, although he may not be able to determine immediately the cause. The table attempts to link each common diagnosis with the features common to the syndrome, drawn from the foregoing observations.

In the column under the heading "Joint" we refer to joint dysfunction. It should be noted that if the patient is first seen when the joint dysfunction has become chronic, the guarding muscle spasm may be a cause of atypical symptoms and may give rise to a mixed clinical picture ("Joint" and "Muscle" columns). Particularly, the joint may stiffen with rest rather than be relieved.

In the column under the heading "Muscle" the onset of symptoms may be insidious rather than sudden, if the muscle condition results from repeated microtrauma or prolonged poor

Table 4-1. Clinical Features for Differential Diagnosis of Back Pain

Symptoms and Signs	Trauma			Disease	
	Joint	Muscle	Disc Prolapse	Local	Systemic
Onset	Sudden During movement With snap or lock	Sudden With lifting With tearing feeling	Sudden With stressful move With lock	Insidious At rest —	Insidious At rest —
Effect of rest	Relieving	Stiffening	Relieving, especially with partial flexion	Stiffening and pain	Stiffening
Effect of activity	Aggravating —	Aggravating —	Aggravating —	Relieving, then aggravating Night pain suggests tumor	No special effect —
Location	Over junction or sacroiliac joint	Over muscle	Over junction	Over junction(s)	No localization
Architectural changes	Segmental alteration of normal curve Curve convex on side of pain Iliac crest raised on side of pain	No change in spinal curve — —	Segmental alteration of normal curve Curve concave on side of pain Iliac crest lower on side of pain	Segmental alteration of normal curve Gibbous contour —	No change in spinal curve — —
Mobility	Localized loss Recovery straight	Less well localized Recovery tortuous or straight	Localized loss Recovery straight	Localized loss Recovery tortuous	Possible generalized loss Recovery tortuous
Percussion	Short sharp pain	—	Short sharp pain, often with radiation	Dull throbbing ache	—

Obvious physical signs	One joint One level Unilateral Pain may be referred No neurological signs	No joint No area Unilateral Pain in muscle No neurological signs	One junction Same level joint Bilateral Radiating pain Mixed neurological signs	One or more junctions One or more levels Bilateral Pain may radiate Neurological signs, possibly changing at onset	None or many joints Many levels Bilateral — No neurological signs
General observations	Patient lies still No signs of systemic disease No fever	Patient varies position No signs of systemic disease No fever	Patient lies still No signs of systemic disease No fever	Patient lies still Possible signs of systemic disease Often fever	Patient restless, may sit up or roll up Signs of systemic disease Fever common
X-ray findings	Negative	Negative	Negative except with special techniques	Positive but maybe not at onset	Negative in back or osteoporosis phenomena
Laboratory findings	Negative	Negative	Negative except CSF protein often increased	Blood count changes, increased ESR; possible changes in blood chemistries	Blood count changes, increased ESR, altered chemistries

CSF = cerebrospinal fluid; ESR = erythrocyte sedimentation rate.
Source: J. McM. Mennell, in *Journal of Occupational Medicine* Sept. 1966.

work habits and posture. Following back surgery, persistent back pain may present a confusing clinical picture because of deep scarring, deficient vascularity, and even impaired nerve function.

Though this section deals with the differential diagnosis of back pain, it can be adapted to problems of extremity pain. Signs of disease in the extremities, local heat, joint swelling, and discoloration are, of course, obvious in the extremities whereas they must usually be inferred in the back.

History

CLUES FROM ONSET OF PAIN

Patients give three fairly consistent stories when relating how a back pain starts: (1) It starts suddenly during the performance of some normal movement or activity. (2) It starts insidiously while the patient is at rest. (3) It starts immediately after extrinsic trauma and is followed by a latent period of painlessness and then onset of somewhat different symptoms that persist and are the cause of the patient's seeking advice.

CLUES FROM EFFECT OF REST ON THE SYMPTOM

Commonly, rest does one of three things to the pain symptom: (1) It relieves the pain. (2) It causes stiffness. (3) It causes stiffness as well as an increase in the pain. There is also the pain that awakens the patient at night, *night pain.*

CLUES FROM EFFECT OF ACTIVITY ON THE SYMPTOM

Activity may have three effects on pain: (1) It reproduces pain if rest has relieved it. (2) It aggravates pain (all movements may be painful or only one movement may be painful). (3) It initially relieves pain, pain is aggravated when activity continues, is relieved again when the activity ceases, and is replaced by stiffness with rest.

CLUES FROM NOISES ASSOCIATED WITH ONSET OF SYMPTOMS

A snap or a pop is often associated with the onset of pain, in which case the patient states, "Something seemed to lock in my back (or neck or wherever)." The "locking" may be described without an associated noise, or the onset of pain may be associated with a tearing sensation in the back.

Clinical Signs

CLUES FROM LOCALIZATION

Localizing the seat of pain by having the patient point with one finger is frequently very accurate; where the patient points is indeed usually where the pain is. But such localization may be misleading, as the patient may point to the area of referred pain rather than to the source of his symptoms. Then the examiner has to decide what structure is under the pointing finger and what pathological changes can occur in that structure according to the principles outlined in Chapter 2 on crossmatching anatomical structures with pathology. On the other hand, the patient may not be able to localize the site of his pain. Instead, he may indefinitely place his hand over an area, which may indicate a visceral problem.

Further, the patient often clearly indicates a path or radiation of pain or maps out a pattern of referred pain. It is important not to confuse radiating pain, which usually follows a distribution of a nerve, and referred pain, which is a pattern rather than a pathway.

CLUES FROM ARCHITECTURAL CHANGES

Architectural changes are segmental deviation in a normal curve of the spine—*segmental* indicating the involvement of two, three, or four vertebrae—without a compensatory change in curve above or below the deviation. Figure 4-5 shows segmental flattening of the cervical lordosis between the fourth and sixth cervical vertebrae, which is commonly found with joint dysfunction. Preexisting disease and congenital anomalies of a vertebra are the only causes of segmental deviation from the normal spine curve unrelated to the patient's presenting symptoms. In low back pain, it is important to note the relative heights of the iliac crests, which are normally level except when there is gross difference in leg length or there is preexisting trauma, disease, or congenital abnormality in the back. The so-called short leg syndrome, in which the shortness of ½ inch or less is usual, does not alter the height of an iliac crest.

CLUES FROM MOBILITY

Observation of the patient stooping forward from the hips and recovering the upright from the stooped position is a revealing and very simple examination procedure. All musculoskeletal conditions that produce pain are associated with limitation of function. In the back, forward flexion is the common function that is impaired or lost. If on forward flexion or recovery from this position there is a localized decrease or an apparent increase of intervertebral movement, especially at one level, some pathological condition exists, either at the junction where the abnormality of movement is noted or below it.

Patients with simple joint dysfunction recover the upright from the stooped position normally, however limited the movement may be. Those with more serious pathology recover the upright position in a tortuous manner and often have to be helped.

CLUE FROM PERCUSSION

With the patient in the forward flexed position, it is a simple matter to percuss accurately over each individual vertebra. In simple joint dysfunction the patient experiences a short sharp pain on percussion; in disease conditions he experiences a deep, aching, throbbing pain from which it takes a little time to recover. With a disc prolapse in the low back the short sharp pain is experienced on vertebral percussion, which often reproduces the pain in the leg.

CLUES FROM LATERALITY OF SIGNS

Intrinsic joint trauma results in unilateral physical signs involving one joint at one level; or, in the case of the sacroiliac joints, one joint shows signs of loss of only one torsion movement. Pain may be referred to a locale distant from the involved joint, but there is no true radiation nor are there positive neurological signs associated with simple joint dysfunction. With a disc prolapse the signs are usually at one level but may be bilateral if the neurological signs are on one side and the associated joint signs on the other.

CLUES FROM PALPATION

There are 14 local areas in which tenderness on palpation exists in certain conditions. Five areas on each side of the back, one area in each buttock, and one in the back of each thigh must be examined. The back areas are illustrated in Figure 4-6.

1. One fingerbreadth medial to the posterior superior iliac spine is the most superficial posterior ligament of the sacroiliac joint. Tenderness here suggests sacroiliac joint pathology.
2. One fingerbreadth lateral to the posterior superior iliac spine is the puny part of the gluteus medius muscle origin, which may be torn by minor twisting trauma. Tenderness here suggests the pathology of a muscle tear; however it may be a local area of pain referred from the fourth lumbar junction.

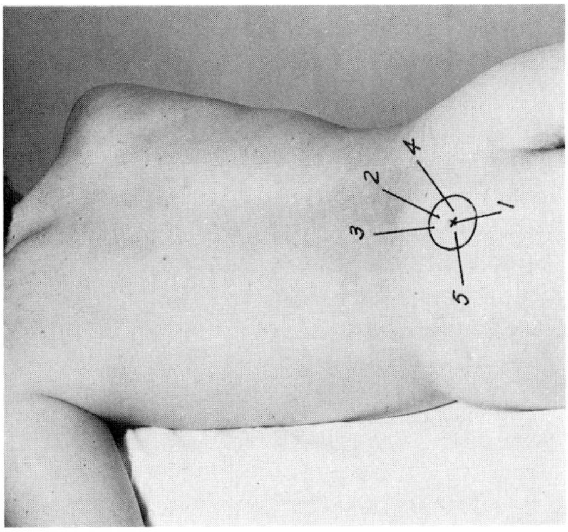

Figure 4-6. Points of palpation that are significant if they are tender. *1*, The palpation point for the posterior-superior iliac spine. *2*, The palpation point for the lumbosacral apophyseal joint pathology. *3*, The palpation point for a sacrospinalis tear. *4*, The palpation point over the posterior ligaments of the sacroiliac joint. *5*, The palpation point over a common area of gluteal muscle tear. This point of tenderness may be a referred pain from a pathological condition at the fourth lumbar junction.

3. One fingerbreadth above the posterior superior iliac spine is the point where the sacrospinalis muscle joins its tendon. Muscle fiber tears occur frequently at this junction with minor lifting trauma. Tenderness here suggests this pathology.

4. One fingerbreadth above and medial to the posterior superior iliac spine is the area over the lumbosacral interlaminar facet joint, where tenderness may be felt if dysfunction is present.

5. One fingerbreadth medial and inferior to the interlaminar facet joint is the area where local tenderness may be felt from lumbosacral disc pathology.

6. Tenderness in the buttock halfway between ischial tuberosity and the greater trochanter of the femur where the sciatic trunk emerges from beneath the piriformis muscle, suggests either spasm of the muscle or radicular pathology.

7. Tenderness on rolling with the fingers deeply over the sciatic trunk in the back of the thigh indicates neuritis.

CLUES FROM SIGNS OF SYSTEMIC DISEASE

In trauma, as opposed to disease problems there are no signs of systemic disease. In this respect, if patients with back pain always had their temperature taken, many errors in diagnosis would be avoided. Also, patients with back pain should always have an appropriate visceral examination, as pain is often referred to the back from visceral disease, enlarged lymph nodes (retroperitoneal, cervical, or mediastinal), and tumors.

CLUES FROM LABORATORY AND X-RAY DATA

Negative radiographic findings certainly do not rule out bone or joint diseases in their early manifestations. Special radiographic techniques such as myelography, epidurography, discography, tomography, and scanning techniques are by no means necessary for arriving at a probable diagnosis. These procedures may be very satisfying when they reveal positive evidence of pathology, but a negative finding does not rule out pathology, nor do positive findings always indicate the true cause of the patient's current symptoms. For instance, myelography may demonstrate a disc prolapse in a patient whose current pain is a symptom of Leriche's syndrome or aneurysm.

Laboratory data are not particularly helpful with the diagnosis of back pain arising from musculoskeletal pain, at least initially. Negative laboratory results should not dissuade one from making a diagnosis of a serious disease condition. Symptoms following trauma may draw attention to or trigger off pain from some systemic disease. An example would be trauma initiating symptoms in an asymptomatic patient with preexisting or masked metabolic disease of bone or metastases.

CLUES FROM GENERAL OBSERVATIONS

Certain additional things that the clinician observes about a patient may afford clues to direct his diagnostic thinking. A patient with joint dysfunction, a disc prolapse, or local disease in the spine tends to lie still. A patient with a muscle problem tends to vary his position. A patient whose back pain is caused by visceral or systemic disease tends to be restless and may sit up or even roll up into a ball.

Just looking at a patient may reveal that he appears sick and perhaps overanxious. He may obviously have lost weight. The examiner's suspicion should then be aroused that the symptoms are arising from local or systemic disease. Similarly, the presence or absence of fever should alert the physician to a local or generalized disease process.

Failure of the symptoms of back pain to respond favorably to appropriate physical therapy within two weeks at the most is very suggestive that the cause of the pain has been misdiagnosed. The problem merits reexamination. The diagnosis, the treatment, or both may be incorrect.

Two incidental signs are sometimes helpful clues in diagnosis of back pain. Atrophy—i.e., flatness of one extensor digitorum brevis muscle—may be the only motor sign of a low lumbar disc prolapse. The absence of one cremasteric reflex may be the only physical sign of a cauda equina tumor.

SUMMARY

We repeat that the observations in this section can never and should never replace a complete examination of a patient. They are offered simply as clues to whether or not there is something wrong, where, and the probable nature of it.

BIBLIOGRAPHY

Leach, R. E., and Zohn, D. A. A simple method for reducing a dislocated patella. *Am. J. Orthop.* 5:261, 1963.

Mennell, J. M. Differential diagnosis of visceral from somatic back pain. *J. Occup. Med.* 18:477, 1966.

5

Ancillary Aids to Diagnosis

The usual ancillary aids to diagnosis may be helpful and necessary but often are unhelpful and even misleading. We therefore take pains to stress the latter situations as well as comment on the former.

X-RAY EXAMINATION

Too great reliance should not be placed on radiographic appearances in the diagnosis of pain in the musculoskeletal system.

X-RAY PICTURES, DISEASE, AND TUMORS OF BONE

Serious bone conditions are revealed late radiographically. As a rule of thumb, 50 percent of cancellous bone or 25 percent of cortical bone must be destroyed before its deficit becomes grossly visible. Radiographic signs of acute osteomyelitis may take three weeks or more to appear. In contrast, the Codman's triangle associated with malignant tumors of bone, if carefully sought, may be a relatively early sign. Codman's triangle is simply the elevation of the periosteum over cortical new bone formation developed too quickly to be calcified. It is an attempt by nature to wall off the tumor mass which has not destroyed enough cancellous bone to be visible radiographically. It is illustrated diagrammatically in Figure 2-8, p. 22. In children especially, early differentiation of serious conditions may be vital.

Most bone tumors occur in the metaphyseal area, but three types are characteristically diaphyseal: Ewing's tumor, reticulum cell sarcoma, and eosinophilic granuloma. The first two are highly malignant, while the eosinophilic granuloma is usually benign. Since the typical cell in all three tumors is a round cell, the histological differentiation is difficult and becomes all the more important (see Figure 2-8, p. 22).

The giant cell tumor never crosses the epiphyseal line of bone until it has closed, a fact that may be of help in diagnosing tumors adjacent to the epiphyseal line. The appreciation of periosteal reaction and of bone expansion radiographically may give the clue to whether or not a tumor is malignant. Early periosteal reaction, as opposed to the later periosteal reaction of osteomyelitis and hypertrophic pulmonary osteoarthropathy, is characteristic of malignancy.

X-RAY PICTURES AND FRACTURES

Even fractures are not always revealed by x-ray examination at the time of injury. They may take as long as two weeks to show up. There are five classic areas in which fractures may be revealed late radiographically: (1) the

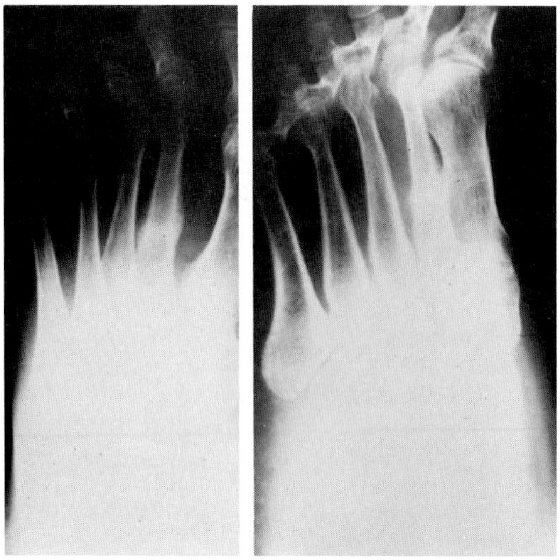

Figure 5-1. *A*, Anteroposterior and *B*, oblique views of stress fracture at the base of the second metatarsal bone. Initial film three weeks earlier revealed no evidence of bony injury.

neck of the femur, (2) the neck of the humerus, (3) the waist of the carpal navicular, (4) the waist of the talus, and (5) the head of the radius. In addition, there is a group of fractures—stress fractures—which also may escape early detection. The more common sites of stress fractures are: (1) the neck of a metatarsal bone (march fracture), (2) the calcaneus, (3) the lower third of the tibia, (4) a rib, (5) a vertebra, (6) the fibula, (7) the pubic ramus, and (8) the femur. Figure 5-1*A* is an x-ray photograph of a typical stress fracture at the base of the second metatarsal bone. An initial x-ray picture taken three weeks earlier showed no evidence of fracture.

With regard to vertebral body stress fractures, if only the upper vertebral plate is involved, the etiology of the fracture is probably trauma. If both upper and lower plates are involved, a pathological fracture should be suspected.

In viewing x-ray films of vertebrae one may find the clue to the presence of neurogenic tumor in erosion of the medial aspect of a pedicle. In children the pedicles separate; the increased distance between them is measurable at the level of a tumor.

It is helpful to remember that some conditions involve the vertebral body only, some the vertebral body and the disc, and some the disc only. For instance, a primary malignancy involves the body only. Tuberculosis and pyogenic organisms produce destruction of the body and the disc. The *Salmonella* genus of organisms shows a predilection for the disc. Heroin addicts may show disc infections.

Oblique views of the spine reveal lack of fusion in the pars interarticularis, especially in the lumbar spine, where spondylolysis shows as a radiolucent collar around the neck of the "scottie dog." Figure 5-2 illustrates the normal "scottie dog" in an oblique view of a normal lumbar spine. The nose is a projection of the transverse process, the eye is the pedicle, the neck is the pars interarticularis, the forefoot is the inferior laminar facet, the ear is the superior laminar facet, and the body of the dog is the lamina.

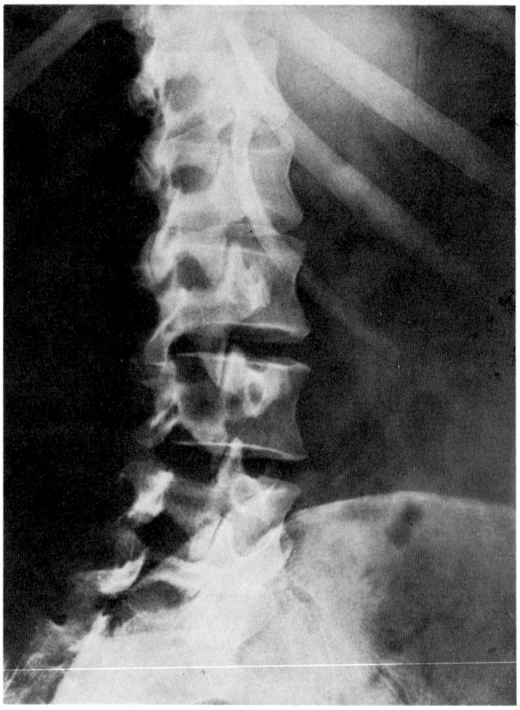

Figure 5-2. Normal "scottie dog" in an oblique view of the normal lumbar spine.

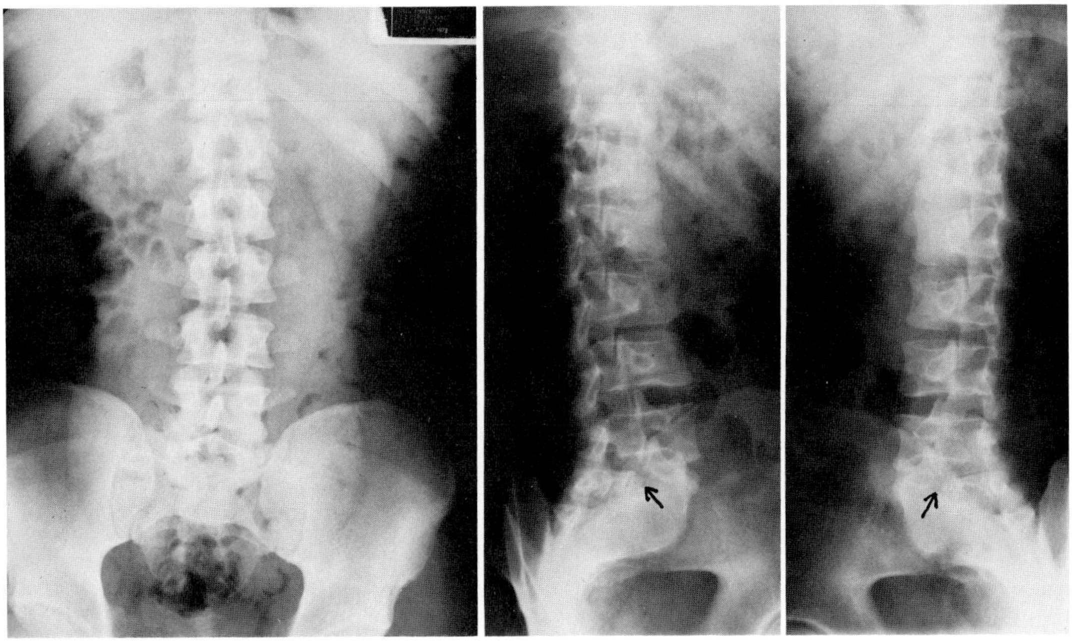

A *B*

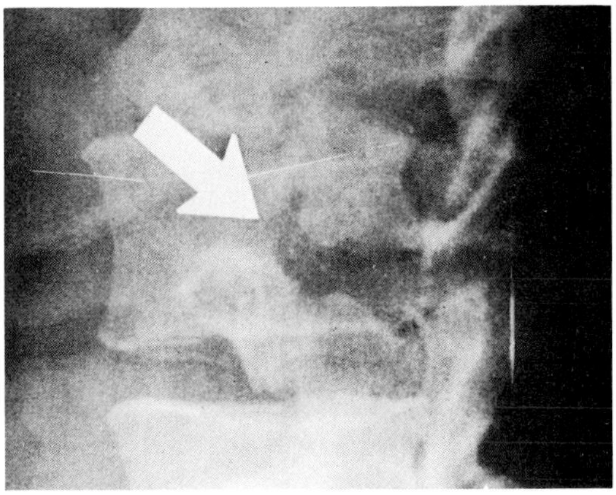

C

Figure 5-3. *A*, Long swayback with six vertebrae having characteristics of lumbar vertebrae. No other anomaly is apparent unless oblique views are taken. *B*, Right and left oblique of the spine shown in anteroposterior view in *A*, revealing bilateral spondylolysis in the pars interarticularis of the anomalous lumbarized first sacral segment. *C*, Gross spondylolysis in a higher lumbar vertebra. This is the radiolucent collar around the neck of the "scottie dog."

Figure 5-3*A* illustrates an apparently normal lumbosacral junction in a spine having six vertebrae with lumbar characteristics (i.e., the long swayback). Bilateral spondylolysis is revealed only when oblique views are taken (Figure 5-3*B*). Figure 5-3*C* illustrates a very gross defect in the pars interarticularis of a lumbar vertebra (spondylolysis). This could be mistaken for a fracture. Sometimes it is difficult to determine whether a vertebral body fracture is fresh or not unless there is a clear history of recent symptoms and on clinical examination the level of pain coincides with the level of change on the x-ray film. A step-down defect of the superior anterior pole of a vertebral body suggests recent injury, as does a zone of compaction. Both features are illustrated in Figure 5-4. Parenthetically, changes in the adult resulting from adolescent Scheuermann's disease, which involves the upper and

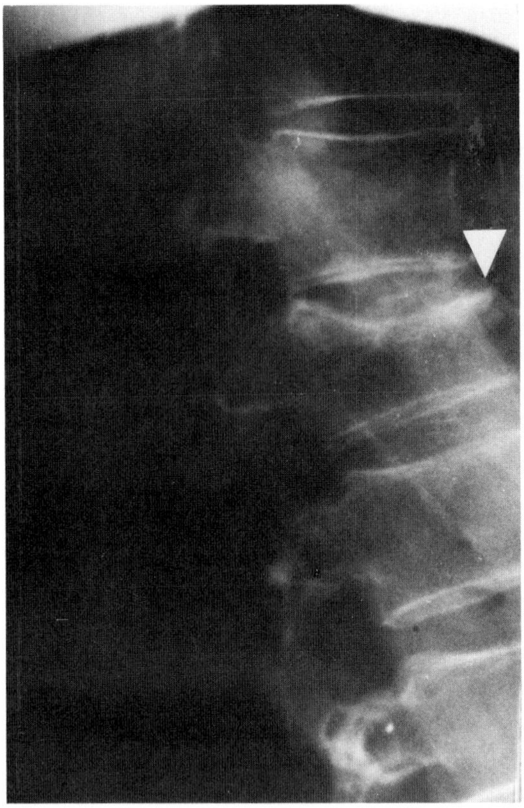

Figure 5-4. Step-down defect (arrow) and zone of compaction in a fresh vertebral body fracture differentiating it from an old fracture.

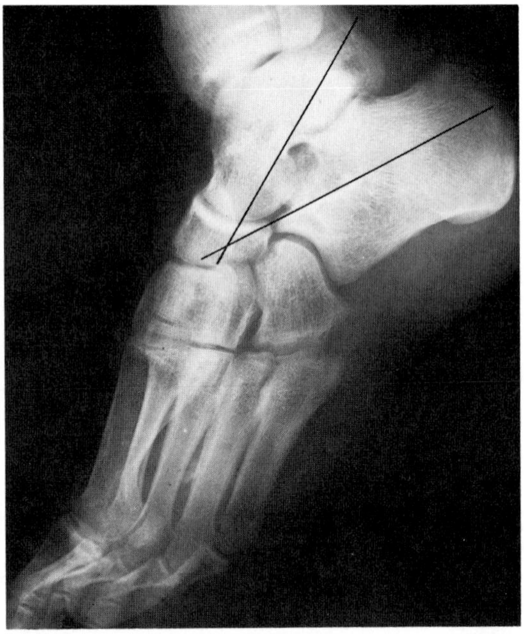

A

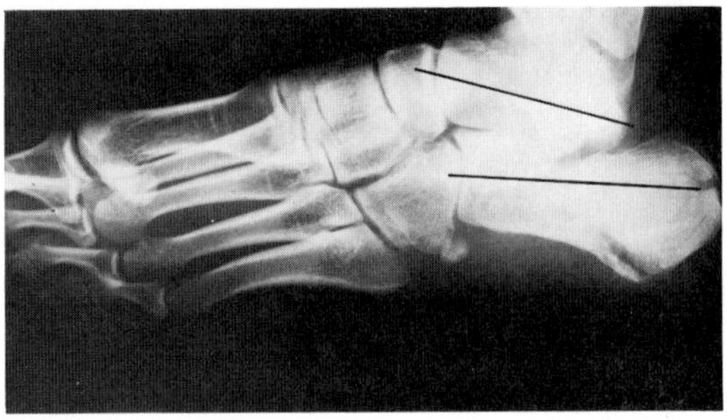

B

Figure 5-5. *A*, Lines drawn through the axis of the talus and calcaneus subtending an angle of 35 degrees. *B*, A grossly disorted angle subtended by the same lines indicating a calcaneal fracture (i.e., the angle is obviously less than 35 degrees).

lower vertebral plates, may be mistaken for fracture lines. However, widening and flattening of the whole vertebral body should provide a differential clue.

A calcaneal fracture may be difficult to detect. There is a special way of arriving at the diagnosis. In a lateral radiograph of the foot, lines drawn through the long axis of the calcaneus and the talus subtend an angle of 35 degrees where they cross each other (Figure 5-5*A*). If this angle is less than 35 degrees, a calcaneal fracture should be suspected. Figure 5-5*B* illustrates this in an exaggerated manner.

Other fractures that are difficult to demonstrate radiographically may be the cause of quite severe pain of long standing. These are fractures of the scapula, the sternum, the sacrum, and the lamina of a vertebra.

X-RAY PICTURES AND DISLOCATIONS

A dislocation is usually detectable in routine radiographs, but there are five areas in which special signs have to be sought or special views must be examined.

1. The Wrist. The appearance of the lunate should be noted following wrist injuries. If the shadow of the lunate is triangular instead of

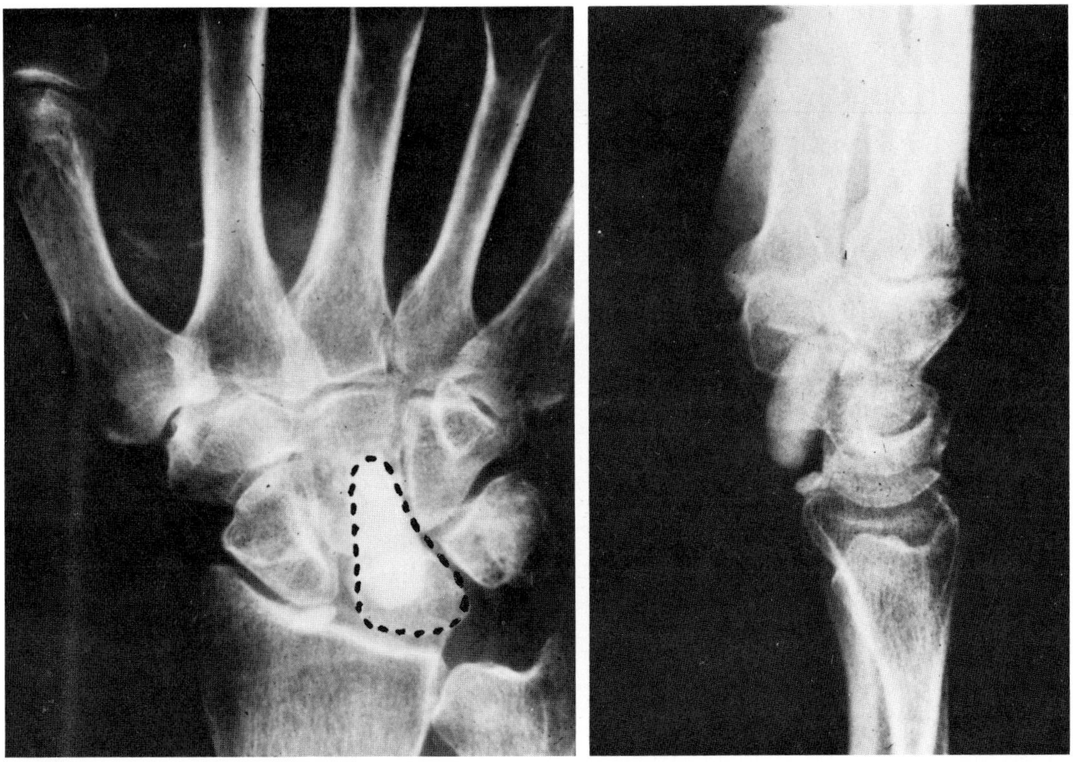

A B

Figure 5-6. *A,* Triangular shadow of the lunate instead of the semilunar shadow a normal x-ray film would show (see Figure 1-1, p. 5); *B,* dislocated lunate.

semilunar there is a perilunate dislocation (Figure 5-6).

2. The Shoulder. When there is a posterior dislocation of the shoulder, the usual films show no change in the appearance of the glenohumeral joint. Axillary views reveal the condition but entail extreme and unnecessary pain in positioning of the tube in relation to the joint. Instead, a true lateral transthoracic view of the glenohumeral joint reveals the diagnosis. In this radiograph the shadow of the axillary border of the scapula is clearly seen, as are the head and shaft of the humerus. If the arc of the border of the scapula is projected, the projected line should pass smoothly into the outline of the adjacent cortical shadow of the humeral shaft just below the head. This line is variously called Mallory's line or Murphy's line (Figure 5-7). If the projected line of the arc of the border of the scapula comes to a T junction with the cortical shadow of the neck of the humerus, there is a posterior dislocation of the glenohumeral joint. The importance of the early diagnosis of this condition lies in the fact that first, the dislocation is invariably associated with a fracture in the head of the humerus, and second, if reduction is delayed, it can be achieved only by surgical means. A posterior dislocation of the shoulder should always be suspected when the etiological factor predisposing to symptoms of pain and loss of function begins with *e* (for *etiology*). There are three such etiological factors: epilepsy, electric shock (accidental or therapeutic), and ethanol (the drunk).

3. The Hip. There is a line described to show dislocations of the hip, especially in children. Shenton's line is illustrated in Figure 5-8. If

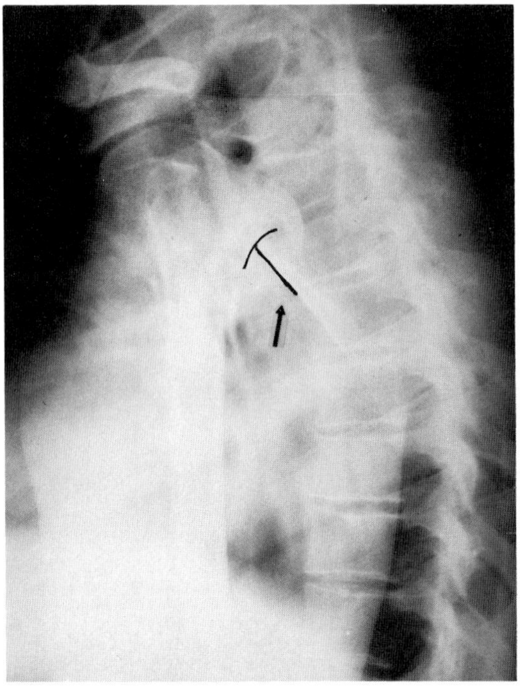

Figure 5-7. Mallory's (Murphy's) line. The axillary border of the scapula (marked with an arrow), if projected in its arc, comes to a T junction with the head of the humerus, indicating a posterior shoulder dislocation.

the line is broken, it indicates a frank dislocation.

4. Femoral Capital Epiphysis Slip. There is an unnamed line associated with the so-called preslip condition which, if recognized, may save the patient lasting disability. As 40 percent of patients suffering from a slipped femoral capital epiphysis are said to have the condition bilaterally, it is especially important to study an allegedly normal hip in a patient with the condition on one side (Figure 5-9).

5. Spondylolisthesis. Brailsford's bow line, when present, in an anteroposterior view of the lumbosacral spine, is diagnostic of spondylolisthesis at the lumbosacral junction. Figure 5-10 illustrates the appearance of a normal lumbosacral junction with two lines—one outlining the body of the fifth lumbar vertebra and the

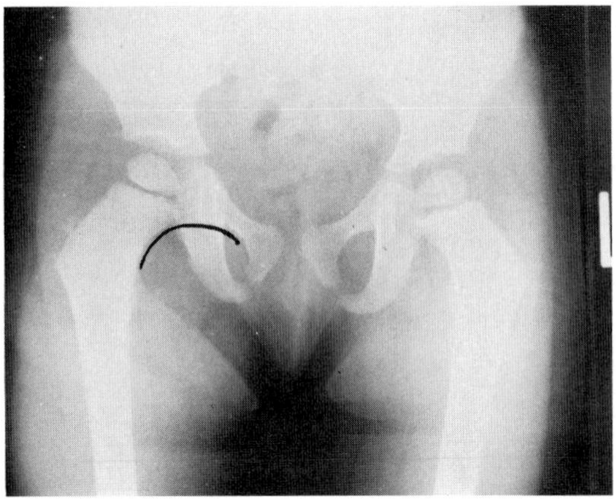

Figure 5-8. Shenton's line. If the line is not smooth when a projection is made joining the arc of the neck of the femur to the arc of the inferior border of the pubis, there is a dislocation of the hip.

other the body of the first sacral segment. The superimposition of these lines is Brailsford's bow line.

STRESS X-RAY PICTURES

Stress x-ray films may be essential to arriving at a diagnosis of ligament rupture at a joint, especially after certain ankle and knee injuries. As ligaments are notoriously poor healing structures and lack of healing may produce severe disability, we advocate resorting even to anesthesia if the pain is too great to allow good stress x-ray photographs to be taken (see Figure 2-19, p. 28).

X-RAY PICTURES AND THE CERVICAL SPINE

The soft tissue paravertebral retropharyngeal shadow in the lateral view should never exceed in width the thickness of the body of the fifth cervical vertebra. If it does, bleeding (hematoma) or a retropharyngeal tumor should be suspected (Figure 5-11).

Further, the anterior aspect of the odontoid should never be separated from the body of the atlas by more than 3 mm, although in children 5 mm may be accepted. If it is, there is either a fracture at the base of the odontoid or ligament injury posterior to it.

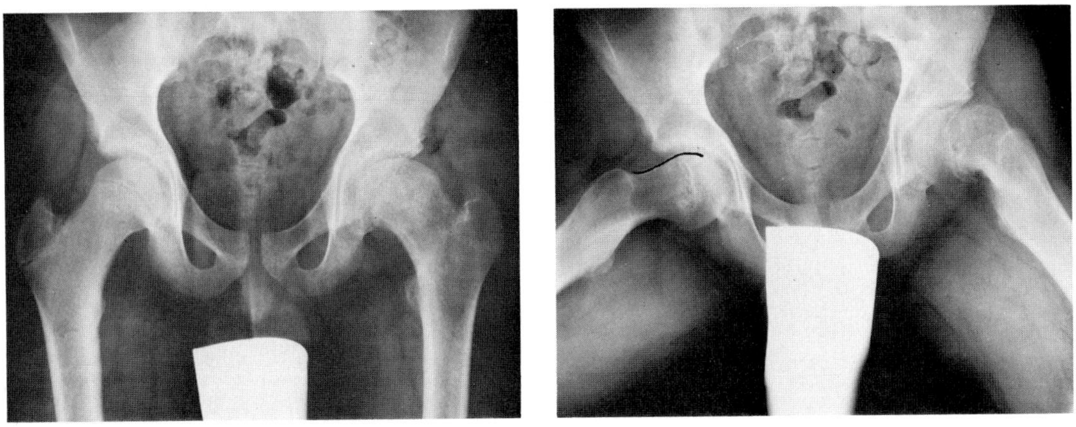

A *B*

Figure 5-9. *A*, An early slipped femoral capital epiphysis on the left with an apparently normal hip on the right. *B*, The same two hips in the frogleg position showing the preslip condition on the right. The line of the superior aspect of the neck of the right femur projects smoothly over the head of the femur superiorly instead of there being a step-up onto the femoral head.

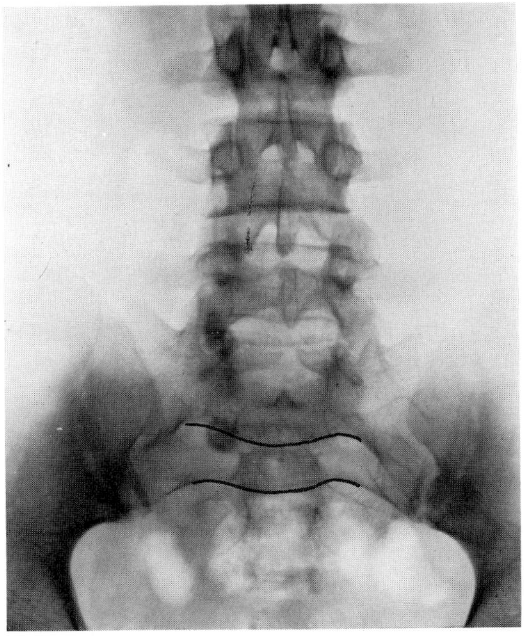

Figure 5-10. Normal appearance of the anteroposterior view of the lumbosacral junction showing the double line–one the body of the fifth lumbar vertebra and, below it, the other the body of the first sacral segment. These lines overlap to form Brailsford's bow line of spondylolisthesis.

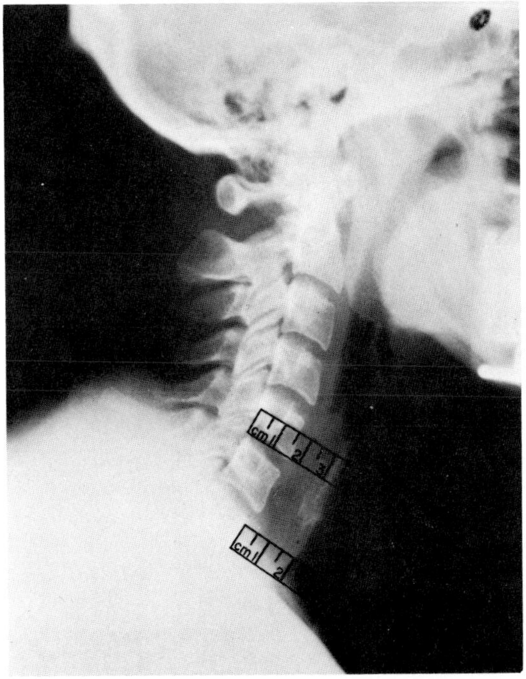

Figure 5-11. Lateral view of a cervical spine. The width of the body of the fifth cervical vertebra is 1.5 cm; the width of the retropharyngeal shadow is 2 cm. The increased width of the retropharyngeal shadow indicates soft tissue pathology–in this case probably bleeding from trauma.

In the anteroposterior view of the cervical spine, bossing (an enlargement which tends to be triangular in shape) of the transverse processes on the seventh cervical vertebra may be associated with an invisible scalene band. This may be the cause of a thoracic outlet syndrome.

X-RAY PICTURES AND THE HANDS

X-ray photographs of the hands provide useful diagnostic information even when they do not appear to be involved clinically. They are particularly helpful in the differential diagnosis of the rheumatic diseases. Small juxtaarticular erosions suggest rheumatoid arthritis while larger punched-out lesions suggest gout or, more rarely, sarcoidosis. (Figure 5-12 is an x-ray picture of the hands of a rheumatoid arthritic patient. Note the punched-out areas. Were the hands not so obviously those of a rheumatoid patient, the punched-out areas might lead to a mistaken diagnosis of gout.) Soft tissue calcifications are commonly associated with scleroderma but may also be seen with Raynaud's phenomena and hyperparathyroidism. Subperiosteal bone resorption is an early characteristic of hyperparathyroidism. Resorption of the terminal tuft occurs with psoriatic arthritis, scleroderma, hyperparathyroidism, and thromboangiitis obliterans. Narrowing of the first carpometacarpal joint is typical of osteoarthritis, while diffuse osteoporosis, particularly of the carpal bones, is typical of rheumatoid arthritis.

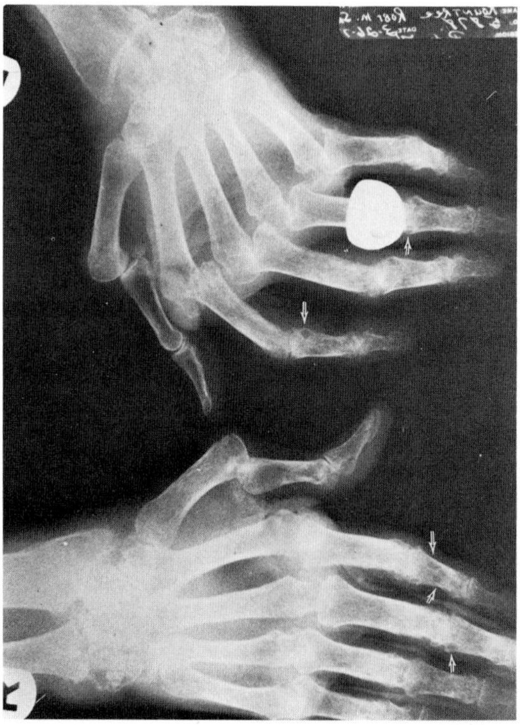

Figure 5-12. A pair of hands showing classic changes of rheumatoid arthritis with osteoporosis, subluxation of the joints, and bony erosions. The bony erosions should not be confused with gout. (Courtesy of Andrew Dobranski, M.D.)

X-RAY PICTURES AND EXTRA OSSIFICATION CENTERS

Extra ossification centers and extra sesamoid bones should be recognized as they may be the cause of diagnostic confusion. These are quite common (Figure 5-13). The former occur

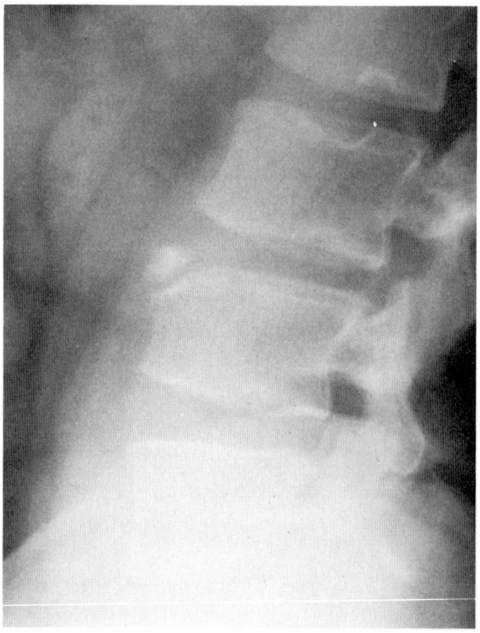

Figure 5-13. An accessory ossification center of a lumbar vertebra which might be confused with a vertebral body fracture.

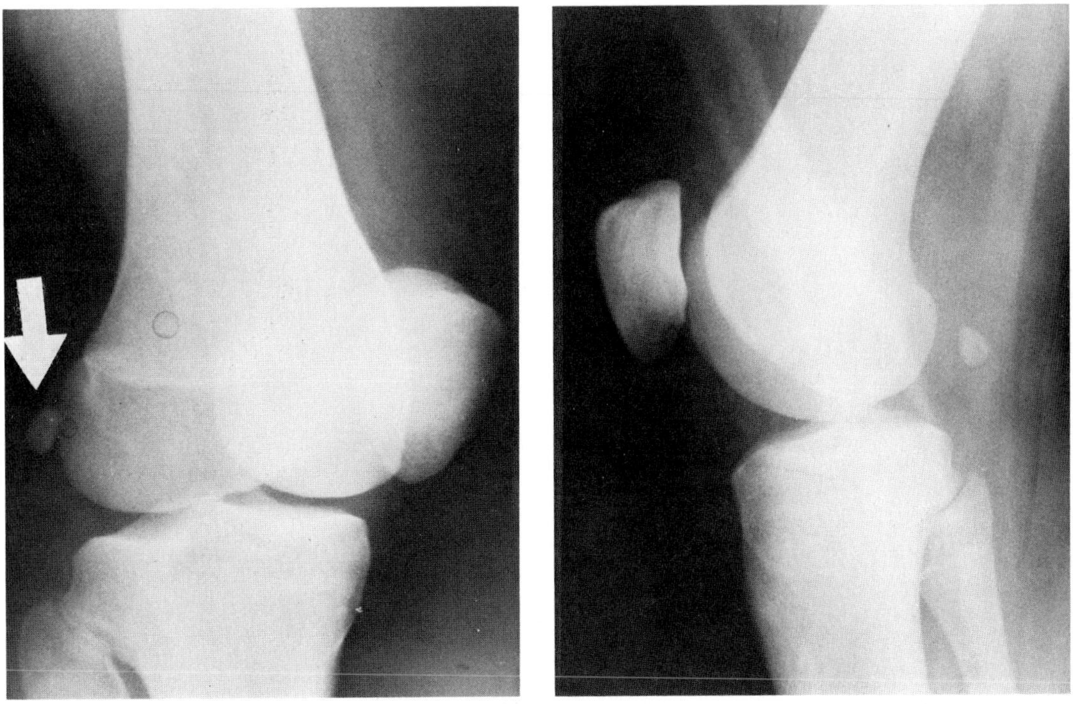

A *B*

Figure 5-14. *A*, Oblique and *B*, lateral views of a fabella located behind the lateral condyle of the femur.

rather frequently in a vertebral body, and an example of the latter is the fabella located behind the lateral condyle of the femur (Figure 5-14).

MISLEADING CHANGES SEEN IN X-RAY PICTURES

Hypertrophic lipping of a vertebral body anteriorly or laterally is usually of little clinical significance though it suggests reparative processes in weakened ligaments or some other case of junction instability. Figure 5-15 shows gross degenerative changes throughout an asymptomatic cervical spine. Another common finding is the Schmorl's node, which we also believe is of little clinical significance. These changes are not osteoarthritic, since osteoarthritic changes are singular to synovial joints.

Other misleading changes observed on radiographs are those of ectopic calcifications. Calcifications around the shoulder joints may or may not have any bearing on a patient's current symptoms. Calcification of the vertebral discs may be associated with the rare disease of ochronosis but more often is merely a finding in asymptomatic individuals, particularly in the elderly population. Similarly, calcification of the cartilages of the joints (chondrocalcinosis) may be related to the syndrome of pseudogout, or even to hyperparathyroidism, but more often is merely a radiographic finding in asymptomatic individuals who have had x-ray photographs taken for other purposes.

X-RAY PICTURES AND MISCELLANEOUS OBSERVATIONS

In adolescents and young adults with musculoskeletal symptoms that do not fit any pattern ankylosing spondylitis may be suspected. The earliest changes of this disease seen on x-ray

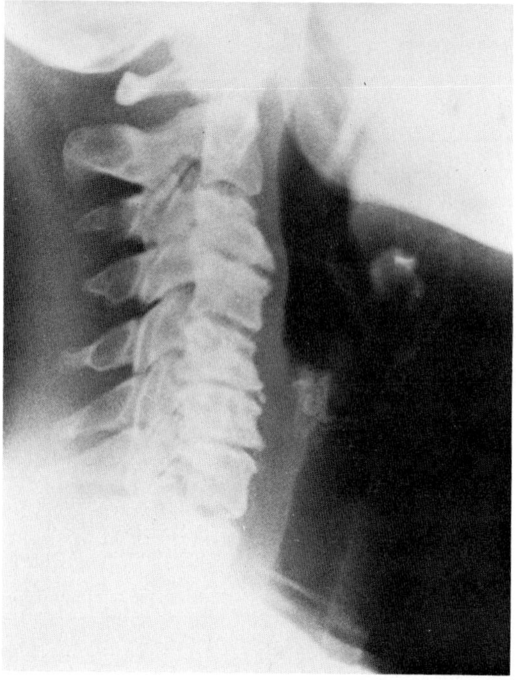

Figure 5-15. Severe degenerative changes in the cervical spine in an asymptomatic patient. X-ray pictures were taken in a routine survey of the spine after the patient was involved in an accident and had nothing to do with the patient's symptoms acquired less than two hours earlier.

films are found in the sacroiliac joints. The initial change is widening of the joint spaces. Later, irregularity and sclerosis of the subchondral bone are diagnostic. The appearance of sacroiliitis (see Figure 5-16) and osteitis condensans ilii (Figure 5-17) may be confusing, but in the latter case the sacral aspect of the joints is never involved. It should be remembered that Reiter's syndrome presents changes similar to sacroiliitis in the sacroiliac joints. A still later radiological sign of ankylosing spondylitis is squaring off of the vertebral bodies while the disc spaces are unchanged (Figure 5-18).

X-RAY PICTURES AND SOFT TISSUE CHANGES

Although we have drawn attention to only one abnormal soft tissue change shown in cervical spine radiographs, we cannot sufficiently stress the importance of noting abnormal soft

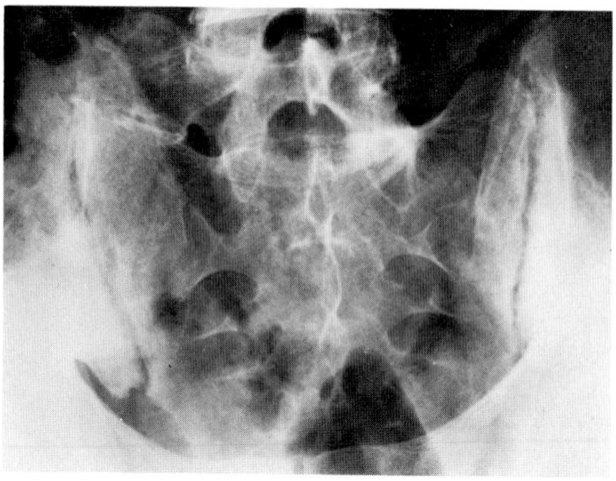

Figure 5-16. Sacroiliitis in ankylosing spondylitis. (Courtesy of David L. Berens, M.D.)

tissue shadows from which diagnoses can be made or inferred. Capsular swelling in joints of the extremity without bone change can be detected. A paravertebral tuberculous abscess without bone change can be seen. Abnormal organ outlines may suggest associated musculoskeletal diagnosis, which account for pain symptoms; for example, an enlarged liver might suggest metastatic disease. Disc spaces, of course, are soft tissue shadows. The fourth

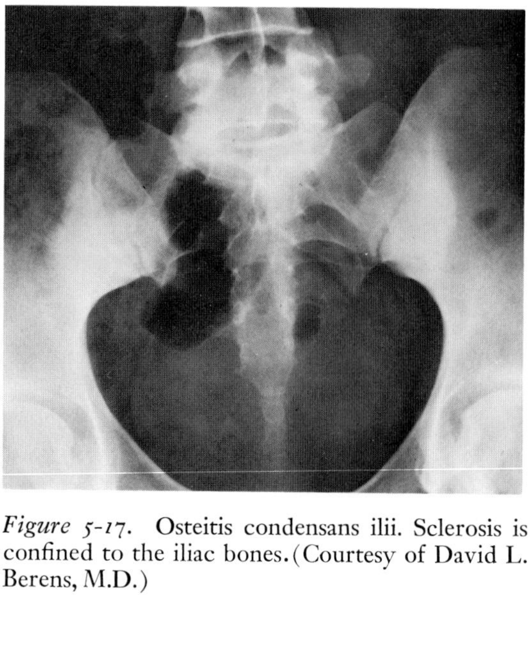

Figure 5-17. Osteitis condensans ilii. Sclerosis is confined to the iliac bones. (Courtesy of David L. Berens, M.D.)

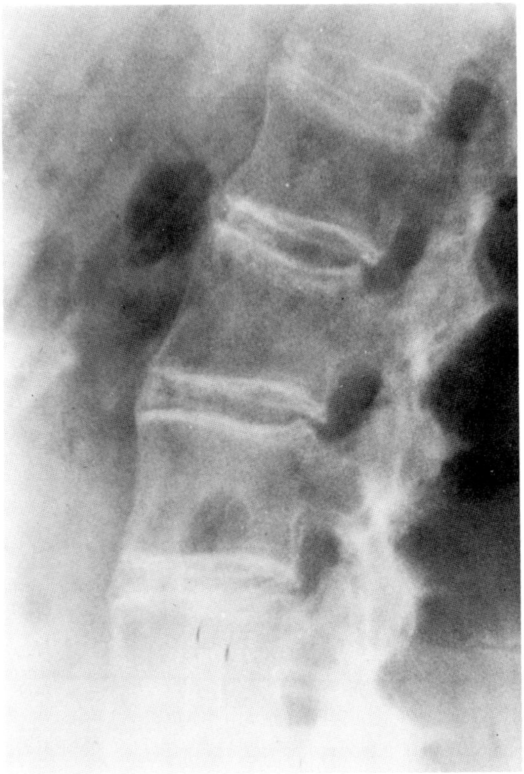

Figure 5-18. Typical bamboo spine of ankylosing spondylitis with squaring off of the vertebral bodies and sparing of the discs.

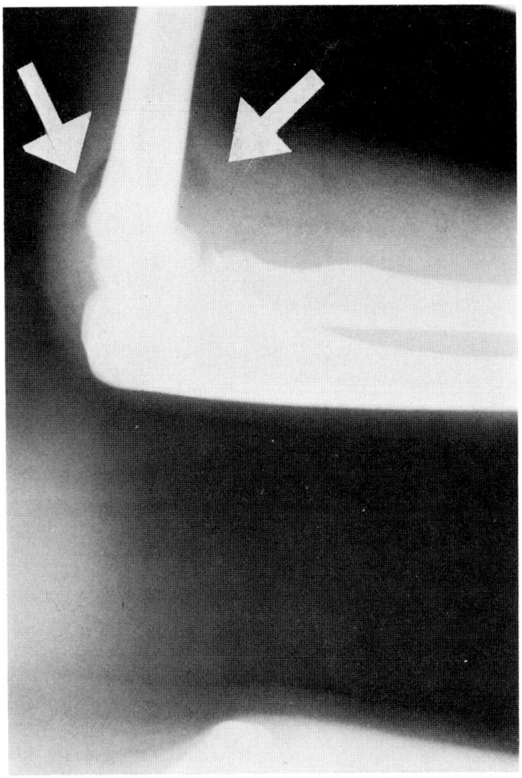

Figure 5-19. Elevation of the fat pad at the elbow of a child, indicative of a supracondylar fracture which is not revealed radiographically.

lumbar disc space is normally larger than the third and fifth disc spaces; unless this fact is recognized, a diminished fourth lumbar disc space might be viewed as being normal. Observation of elevation of the shadow of the fat pad over the lateral condyle of the elbow in children provides an early clue to the presence of a fracture (Figure 5-19).

SPECIAL RADIOGRAPHIC AND OTHER DIAGNOSTIC TECHNIQUES

There are innumerable special radiographic techniques in the differential diagnosis of musculoskeletal pain. That each has its place is undoubted, but they can never take the place of a proper clinical diagnostic evaluation. Positive results are very reassuring; negative results may be very unreliable.

Myelography

A contrast medium is introduced into the subarachnoid space and is run up and down by proper positioning of the patient while defects are sought in the column of dye. Although defects are commonly interpreted as disc prolapses, they may equally though less commonly be due to vascular anomalies, tumors, or occasionally an epidural abscess. Figure 5-20 illustrates a defect due to a neurofibroma. Figure 5-21 illustrates three defects in the lumbar spine that cannot be diagnostic of the disc prolapses causing the patient's symptoms. Figure 5-22 illustrates a defect in the cervical myelogram due to a cord tumor but initially diagnosed as a disc prolapse. In a patient whose dural sac narrows at a higher level than usual, a significant defect at the lumbosacral junction may not be demonstrated.

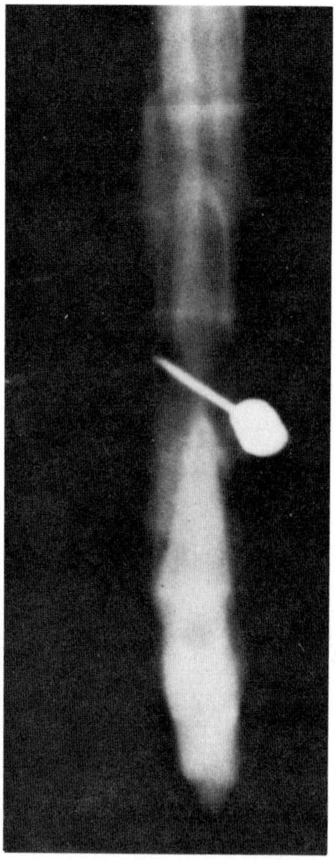

Figure 5-20. A well-localized circular defect in the myelographic column below the needle, typical of a neurofibroma.

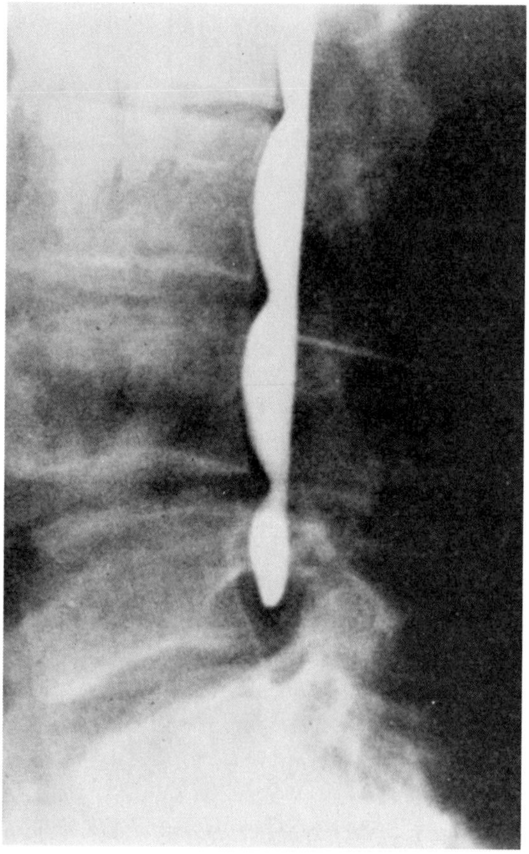

Figure 5-21. A myelogram demonstrating three indentations, any or all of which could be due to a disc prolapse. The clinical level of pathology was between the fourth and fifth lumbar vertebrae. The correctness of the clinical observation was proved at surgery.

Spinal fluid removed when a myelographic study is performed should always be examined for cellular and protein content. And, because there is now a generation of physicians who have never seen central nervous system syphilis, serological studies should be done routinely.

Epidurography

Epidurography gives the same information as myelography, but the contrast medium is introduced into the epidural space through the sacrococcygeal hiatus instead of intrathecally. A water-soluble iodine preparation is used. The advantage of this procedure over myelography is that narrowing of the dural sac does not give false negative results. Epidural adhesions may be visualized, and the procedure can be done on an outpatient basis. The disadvantages are that it is convenient only for lumbar studies and no specimen of cerebrospinal fluid is obtained.

Discography

In discography a contrast medium is introduced into the disc itself with or without hydrocortisone admixed. Certainly a spill-out of the contrast medium into the epidural space clearly indicates a rupture in the annulus fibrosus. The procedure also clearly demonstrates a degenerated disc or a normal disc. We do not advocate discography for two reasons: (1) The needle may inflict an initial trauma to the annulus which may weaken it and later

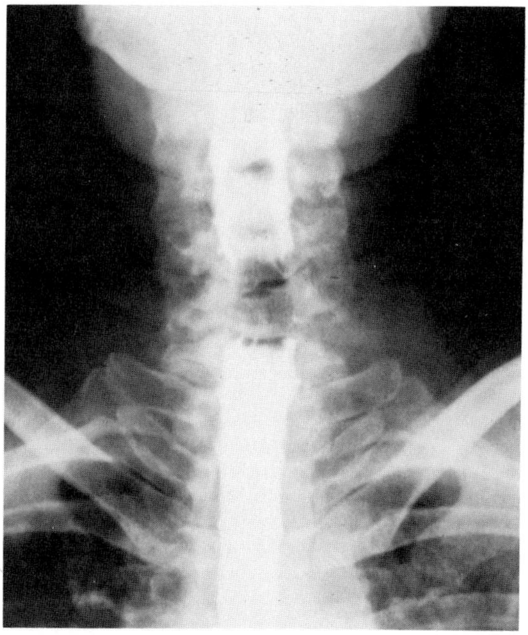

Figure 5-22. A large defect from an intradural tumor mistakenly diagnosed initially as a disc prolapse.

cause a disc prolapse; (2) the intradisc pressure resulting from injection of fluid into a disc may promote degeneration or even precipitate a prolapse of the nucleus pulposus. Proponents of discography initially claimed relief of symptoms following the procedure. This could well be if the hypertonic solution spilled into the epidural space where it would act like an epidural injection (see p. 131) when used therapeutically. Figure 5-23 illustrates a positive discogram with dye flowing out of the fifth lumbar disc into the epidural space. The fourth lumbar disc space is diminished and the disc is degenerated, but the dye is confined. In Figure 5-23*B* the dye is shown confined within the fourth lumbar disc space.

Arthrography

Dye or air (or both) is injected directly into the capsule of a joint. This technique is used to demonstrate capsular leaks, meniscal damage, or the presence of "loose bodies." Figure

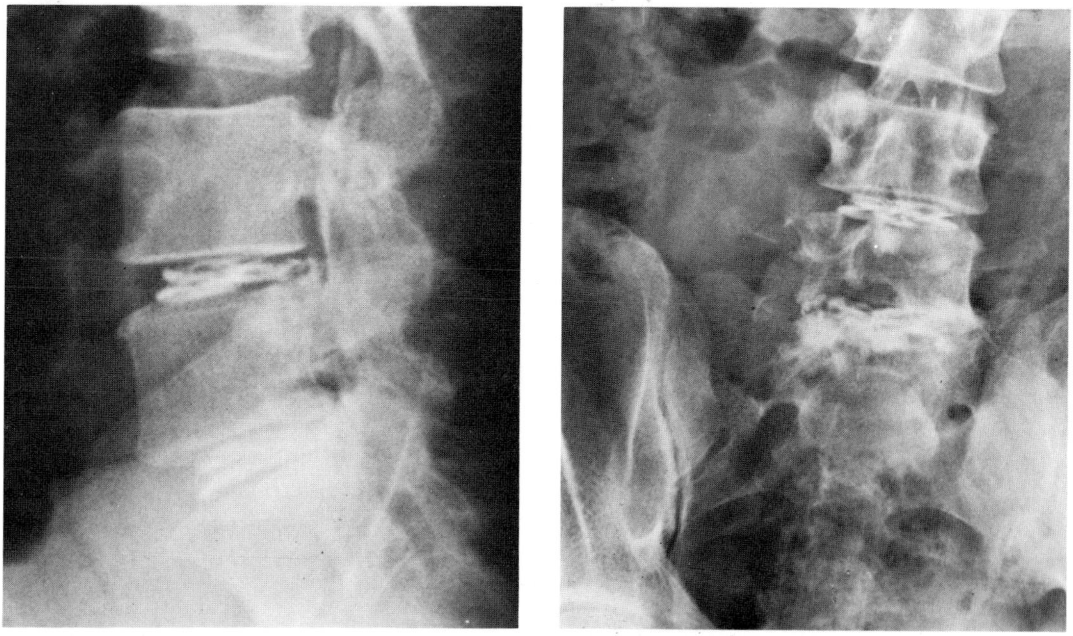

A *B*

Figure 5-23. *A*, Discogram showing a rupture of the fifth lumbar disc and free flow of dye into the epidural space. The fourth lumbar disc space is diminished, the disc is degenerated, but the dye is confined. *B*, An anteroposterior view of the same patient showing the more normal fourth lumbar disc with the dye contained. (Courtesy of David L. Berens, M.D.)

5-24 is an arthrogram of a shoulder joint demonstrating a tear in the rotator cuff by leakage of the dye into the subacromial bursa. Figure 5-25 shows arthrograms of a knee demonstrating a torn medial meniscus.

Tomography

Tomography is a radiographic technique by which it is possible to bring into focus bone and soft tissue at predetermined depths; that is, each exposure reveals tissue detail in slices. By this procedure otherwise hidden fractures, early bone defects from tumors and disease, and soft tissue masses may be detected. Figure 2-4*B*, p. 19, shows typical tomograms demonstrating Brodie's abscess.

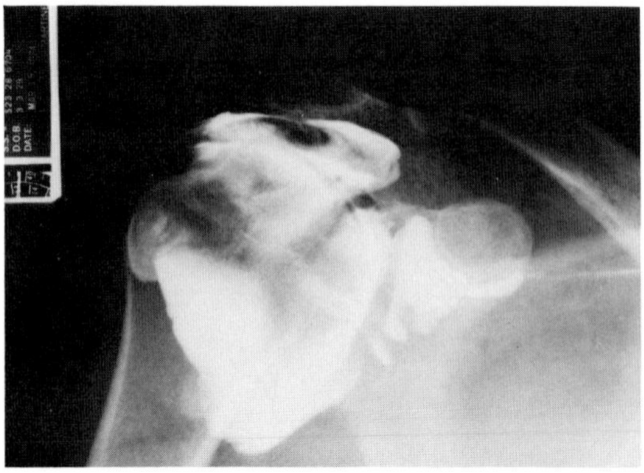

Figure 5-24. Arthrogram of a shoulder joint demonstrating a tear in the rotator cuff by leakage of dye into the subacromial bursa.

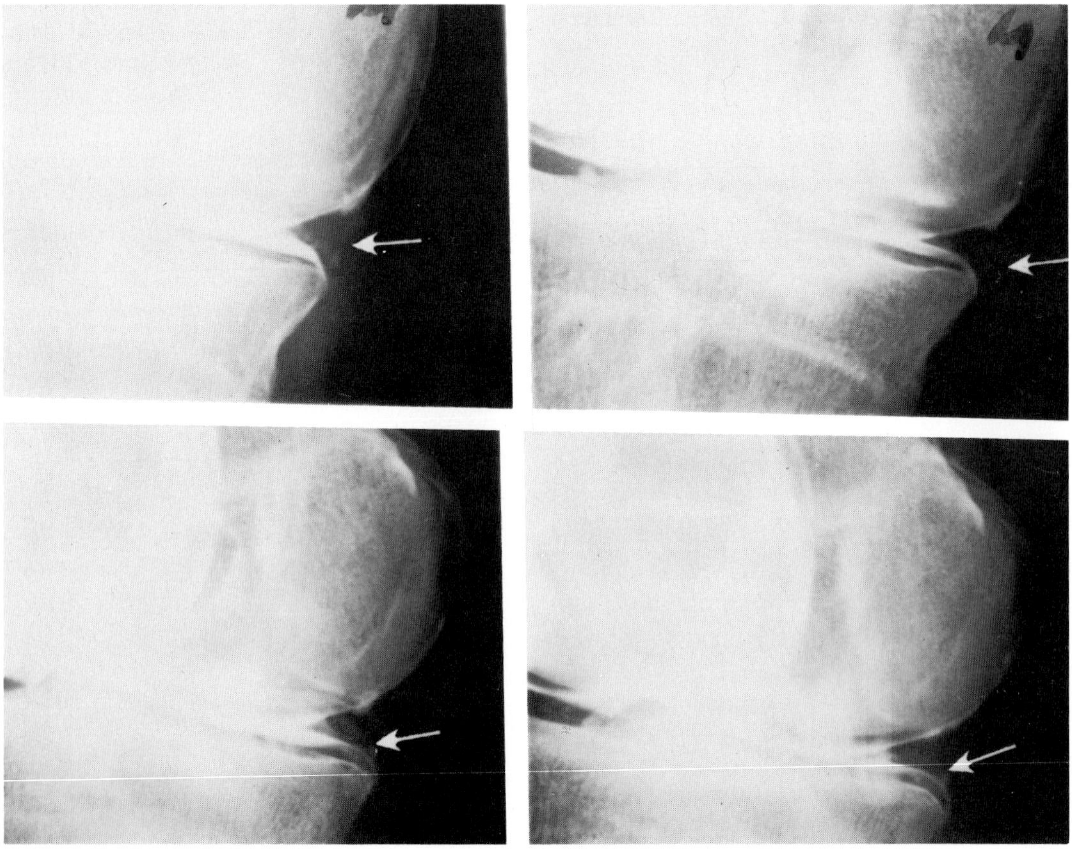

Figure 5-25. Posteromedial arthrogram of the knee. The dye seeping into the medial meniscus shadow (arrow) indicates a tear of the meniscus.

Bone Scans Using Radioisotopes

Different radioisotopes are taken up selectively by different tissues, and bone has a predilection for technetium 99m pyrophosphate, which is used in the following illustrations. A scan reveals increased or decreased tissue activity, but the cause of the change cannot be determined; that is, bone scans can determine that there is something wrong but not what is wrong. Thus, an increased radioisotope uptake may occur with fractures, tumors, or a disease process in which there is increased formation of bone. A decreased radioisotope uptake indicates decreased bone activity, and this may be found in conditions in which osteoporosis is a feature, as in the metabolic diseases of bone.

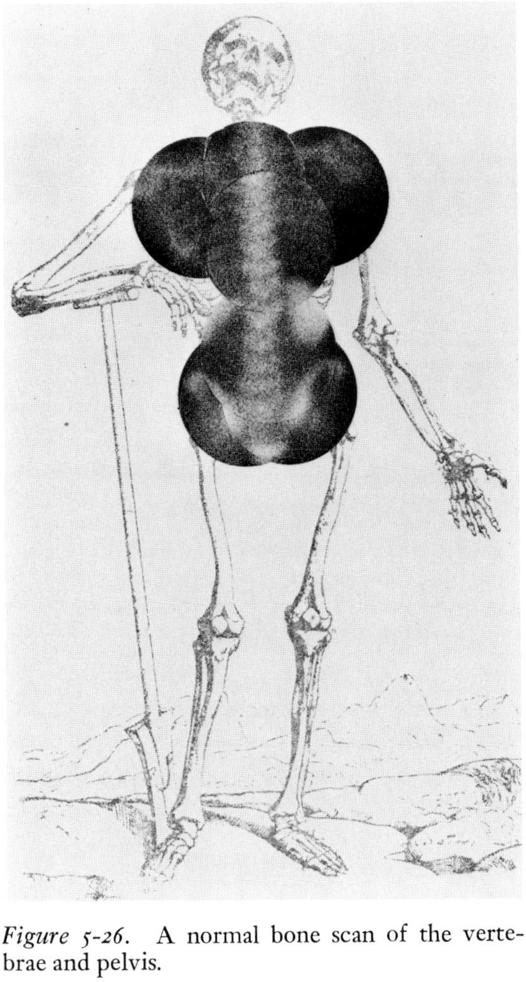

Figure 5-26. A normal bone scan of the vertebrae and pelvis.

Figure 5-26 illustrates a normal bone scan. Figure 5-27*A*, *B* shows a lumbar spine with a possible abnormality indicated by an arrow in the lateral view. Figure 5-27*C* is a bone scan of the same patient on the same day revealing diffuse metastatic skeletal infiltration. Figure 5-27*D*, x-ray photographs taken two months later, shows changes well established. This series of pictures demonstrates the importance of not relying on negative x-ray reports if disease is suspected clinically.

Thermography

By the use of infrared photography, areas of increased vascularity in tissue are detected. The procedure allows one to determine the degree of involvement of, for example, a limb in peripheral vascular disease (Figure 5-28), to note the activity of diseases such as rheumatoid arthritis, and to detect soft tissue tumors such as breast cancer. Figure 5-29 shows a black (cold) tumor in the supraclavicular area in a patient's chest wall. It was a benign cystic growth. It is our feeling that thermography might be used to show localized temperature changes around joints in which joint dysfunction is present, but to our knowledge this application has not been made. To be able to determine such temperature changes would be particularly useful in assessing the cause of back pain arising from joints, when dysfunction is the cause of pain. We would expect immediate normalization of a change seen thermographically following treatment by joint manipulation. Figure 5-30 shows thermograms of a patient with a history suggesting muscle strain in the right thoracolumbar junction area. The patient presented some diagnostic difficulty. The thermograms demonstrate a well-localized hot (white) spot on the right, exactly where he indicated pain, which anatomically had to be in muscle. Figure 5-31 is a thermogram showing vapocoolant spray patterns (wide paraspinous black streak on the right with oblique black lines from it) 15 minutes after treatment of the patient in Figure 5-30. The deep black streak in the middle of the spine shows the immediate effect of one sweep of the vapocoolant spray.

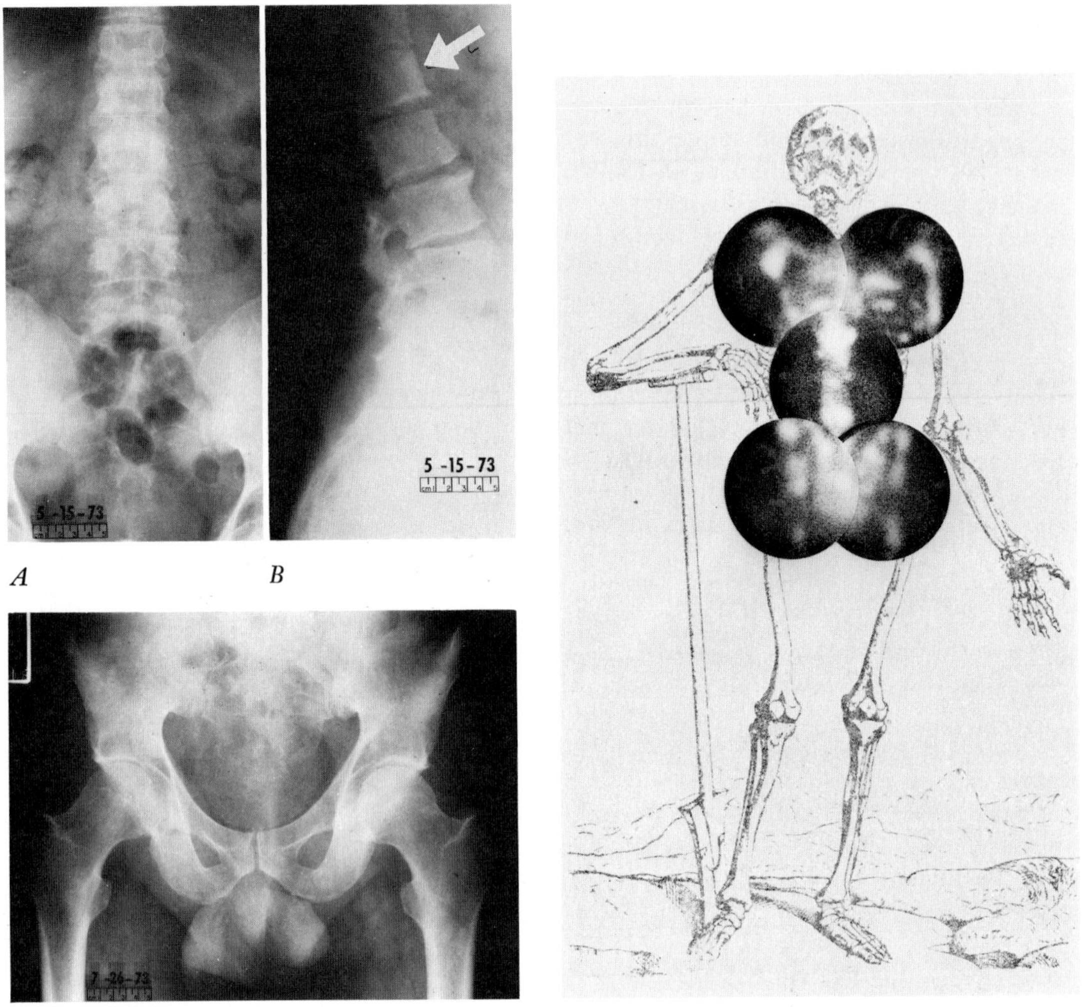

A B C D

Figure 5-27. *A*, Anteroposterior x-ray picture of lumbar spine and sacroiliac joints showing no gross changes in the bones and joints. *B*, Lateral view of the same spine. Arrow at the first lumbar vertebra indicates suspicions of the radiologist that something is wrong. *C*, Radioisotope scan of the same patient on the same day revealing extensive and unequivocal generalized metastatic involvement. (Courtesy of Jack Mangum, M.D.) *D*, Anteroposterior x-ray film of the same patient two months after the one illustrated in *A*, showing extensive metastatic changes in the pelvis and lumbosacral area. This demonstrates the importance of not relying on negative x-ray photographs if disease is suspected clinically.

Arthroscopy

It is possible now to introduce an arthroscope into a joint, thereby enabling one to visualize disease processes and allowing the performance of closed tissue biopsies from specific areas or of specific lesions. Pathological lesions can be photographed through the arthroscope. The procedure is not in common use and provides little, if any, more information than an arthrotomy. Arthrotomy would probably be required anyway in cases in which arthroscopy might be used.

LABORATORY STUDIES

Data derived from studies of blood and other material in the laboratory are peculiarly disap-

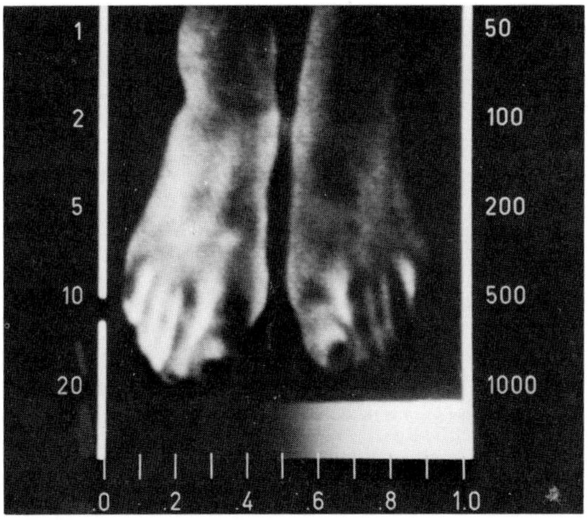

Figure 5-28. Thermogram of feet showing vascular deficiency. There were no clinical signs to suggest the rather widespread involvement. The history was one of impaired ambulation without noticeable pain other than occasional night cramps. (Courtesy of AGA Thermovision.)

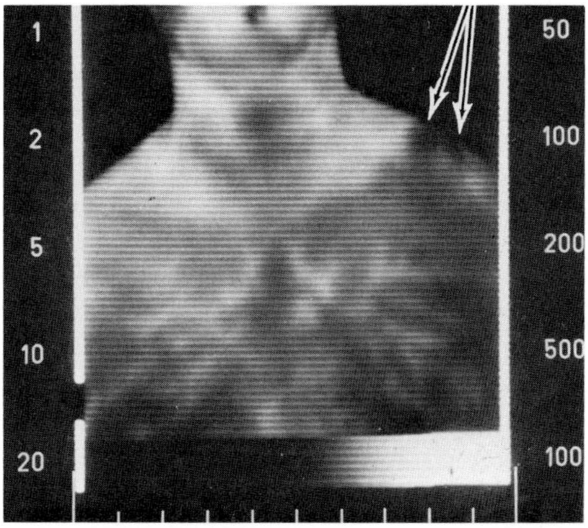

Figure 5-29. Thermogram of a patient with recurrent cystic tumor over the left clavicle. There is a clearly cold (black) area in the tumor location suggesting that it is benign. Histologically, it proved to be a hygroma. (Courtesy of Jack Mangum, M.D., and AGA Thermovision.)

pointing as an aid to diagnosis of the common causes of musculoskeletal pain, but these studies may be invaluable in the differential diagnosis of more serious pathological conditions. This section is in no way meant to be a laboratory manual but merely a jog to the memory and a reminder of the limitations of many laboratory studies.

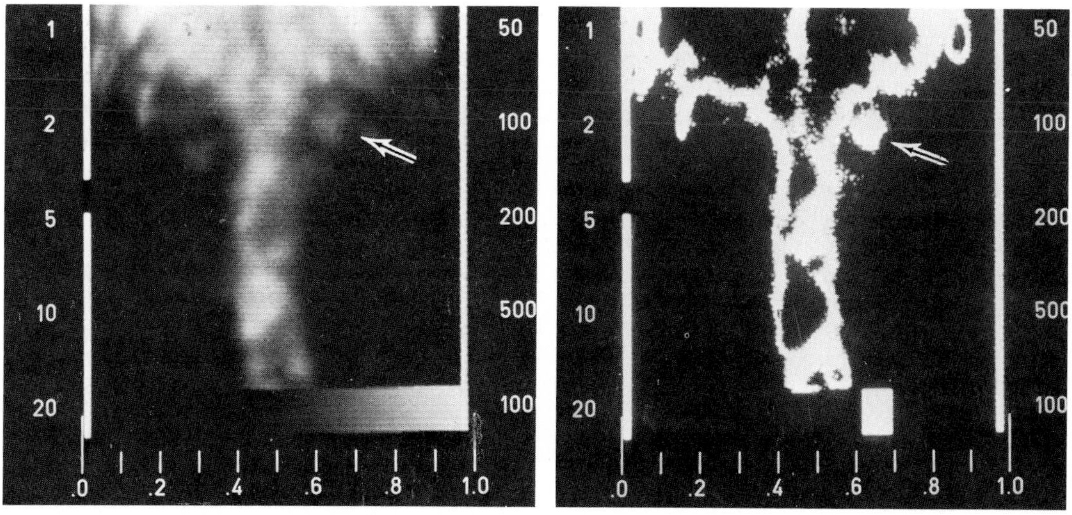

A *B*

Figure 5-30. Thermograms showing the hot spot in the right paraspinous area in the patient described in the text. (Courtesy of AGA Thermovision.)

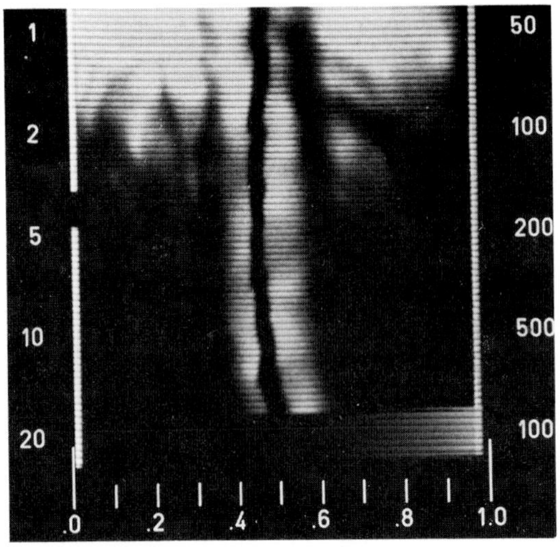

Figure 5-31. Thermogram showing the patterns of cooling using the Fluori-methane spray in treatment of the patient as illustrated in Figure 5-30. In addition the deep black streak in the middle of the spine shows the immediate effect of one sweep of the vapocoolant spray. (Courtesy of AGA Thermovision.)

Hematology

ERYTHROCYTES

In musculoskeletal pain problems, anemia tends to be associated with chronic disorders. It may occur in rheumatoid arthritis or other collagen vascular diseases with chronic infections such as tuberculosis and chronic osteomyelitis, with neoplasms, and with osteoporesis senilis.

LEUKOCYTES

A change in the leukocyte count is nonspecific in itself but may be helpful. Leukopenia occurs with chronic collagen vascular disease states, and with viral infections. Leukocytosis may be seen in active rheumatoid arthritis or gout. Eosinophilia occurs in polyarteritis, parasitic infestation, and allergic drug reactions. Monocytosis sometimes occurs in collagen vascular diseases, tuberculosis, mononucleosis, and other chronic inflammatory conditions.

Immunohematology of Leukocytes. Detection of the HL-A (human leukocyte locus A) 27 antigen offers an additional diagnostic tool in ankylosing spondylitis and Reiter's disease. The antigen is found in approximately 90 percent of patients with ankylosing spondylitis and 75 percent of patients with Reiter's disease, whereas in the normal population it is found in less than 10 percent.

PLATELETS

Thrombocytosis occurs in inflammatory and neoplastic disorders. Thrombocytopenia has many causes and can be associated with a painful hematoma or hemarthrosis.

ACUTE PHASE REACTANTS

Acute phase reactants are factors in nonspecific blood tests that become abnormal with any acute or chronic inflammatory reaction. The most commonly used tests are the erythrocyte sedimentation rate (ESR) and the carbohydrate-reactive (C-reactive) protein (CRP).

Sedimentation Rate. Many acute and chronic inflammatory conditions elevate the serum globulin and/or fibrinogen level and hence the sedimentation rate. In the elderly, an elevated sedimentation rate is often unrelated to any specific disease process, but if this is the only abnormal study then cancer should be suspected and ruled out. In a few select diseases–ankylosing spondylitis, polymyalgia rheumatica, and intervertebral disc space infections–the sedimentation rate may be the only abnormal laboratory finding. Otherwise, it is nonspecific but may provide evidence of active disease when the patient's complaints outweigh the objective findings. The Westergren method is the more reliable test. The upper limits for men is 10 mm per hour and for women 20 mm per hour. The sedimentation rate may be used to follow the response to drug therapy in ankylosing spondylitis and the course of other chronic inflammatory diseases.

Carbohydrate-reactive Protein. Carbohydrate-reactive protein is a protein in the serum of patients with inflammatory conditions that precipitates upon the addition of an antiserum to the C-polysaccharide of the pneumococcus.

Its principal use in joint pain problems is as an indicator of the presence of rheumatic fever.

ANTISTREPTOCOCCAL ANTIBODIES

Antistreptococcal antibodies are those that rise in response to a streptococcal infection. Usually one can detect a titer of over 200 cc per milliliter in children within three weeks of untreated infection. The antistreptolysin O titer is usually measured. The addition of the antihyaluronidase and antistreptokinase titers increases the likelihood of a positive result from 80 percent to 95 percent. In adults one must demonstrate a rising or falling titer consistent with the clinical condition because a single elevated titer may have been due to an unrelated previous infection. A new test, the streptozyme, measures several antistreptococcal antibodies and is a useful indication of unresponsive or spreading streptococcal infection.

Serology and Immunology

Serological testing for syphilis should be part of every diagnostic workup, particularly with the recent upsurge in the incidence of venereal disease. A nonreactive serum may give the clinician a false sense of security, and it should be remembered that the spinal fluid can be reactive when the blood is not.

The rheumatoid factors are anti–gamma globulin antibodies found in the serum of a patient with rheumatoid arthritis. Positive tests may also occur in patients with other collagen vascular diseases, chronic infections, and noninfectious diseases with hypergammaglobulinemia, and in a small percentage of apparently normal individuals. A simple agglutination slide test is available as a screening device. Because of false positive results, it should be followed with the more quantitative tube dilution latex or bentonite flocculation test. Patients with rheumatoid arthritis usually exhibit a positive test within a year of the onset of disease. Those with higher titers often have more severe articular destruction and extraarticular manifestations such as nonspecific pulmonary lesions, or, rarely, amyloidosis. The tests are likely to be negative in juvenile rheumatoid arthritis and the rheumatoid arthritis variants.

The demonstration of antibodies to nuclear material of cells is a hallmark of systemic lupus erythematosus. Laboratory testing reveals the lupus erythematosus cell phenomenon, in which leukocytes phagocytize nuclear material, and serum antinuclear antibody determinations show, by fluorescent staining, the reaction of the serum globulins with nuclear material. Demonstration of the lupus erythematosus cell is a good screening device; it may be performed on blood or joint fluid aspirates. However, repeatedly negative results do not exclude the diagnosis, and the test may be positive in other conditions such as rheumatoid arthritis and reactions to various drugs such as hydralazine and procainamide. Serum fluorescent antinuclear antibody (FANA) determinations are more sensitive. They may be positive despite a negative lupus erythematosus cell preparation. Since both the lupus erythematosus cell phenomenon and the antinuclear levels may fluctuate in the course of disease, repeated testing may be necessary to confirm the diagnosis.

SERUM PROTEINS

Measurement of the serum proteins gives only nonspecific diagnostic information in patients with musculoskeletal pain. Electrophoresis of the serum proteins, however, may show increases in various albumin or globulin fractions in specific diseases. These changes are nonspecific in the rheumatic diseases in which elevations of the alpha-2 or gamma globulin fractions occur. In multiple myeloma one may find a peak in the electrophoretic pattern caused by an overproduction of light chains of antibody proteins. In some cases of myeloma and other hyperglobulin diseases, only a diffuse increase of globulins is found by electrophoresis. Here the relatively new technique of immunoelectrophoresis may be helpful in making the diagnosis by showing the increased synthesis of a specific immunoprotein.

Chemistry

Of the many chemical determinations available in assessing undiagnosed musculoskeletal pain, the following have proved most useful in common clinical situations:

- Postprandial glucose, 2-hour
- Glucose tolerance
- Uric acid
- Calcium
- Phosphorus
- Alkaline phosphatase
- Acid phosphatase
- Muscle enzymes

GLUCOSE

The fasting glucose may be normal in latent diabetes mellitus, so a 2-hour postprandial specimen is required as a screening test. Before making the diagnosis definitely, one should perform a glucose tolerance test, since neuritic pain from diabetes may be the presenting symptom of the disease and often must be differentiated from musculoskeletal pain.

URIC ACID

Hyperuricemia occurs frequently, and the higher the level, the more likely the chance of gout developing. Hyperuricemia associated with an acute monarticular arthritis usually confirms the diagnosis of gout, but it should be recognized that the serum uric acid level may be normal during an acute attack of gout. Further, a person may have asymptomatic hyperuricemia and a painful joint from another cause. For this reason, examination of the joint fluid (and even a biopsy) may have to be undertaken before a diagnosis is made.

CALCIUM, PHOSPHORUS, AND ALKALINE PHOSPHATASE

These determinations are routinely obtained in the evaluation of patients suspected of having metabolic or neoplastic bone disease. Calcium values tend to be variable, so repeat examinations are sometimes necessary to detect an elevation. Since calcium is partially bound to the serum albumin, it is important to know the albumin value as well. In the more serious conditions the changes of the levels of these elements are characteristic. In osteitis fibrosa cystica the serum phosphorus is low because its reabsorption by the renal tubules is blocked by the action of excess parathormone, and the calcium is high. In osteomalacia the calcium is low and the phosphorus tends to be high. The alkaline phosphatase level in these conditions is raised if there is any significant degree of reparative osteoblastic activity.

The alkaline phosphatase can also rise with liver disease, but one can differentiate the two reactions of the enzyme by heat deactivation and/or isoenzyme determinations. Although it may be elevated in metastatic bone disease, it frequently remains normal even in the presence of extensive metastases. On the other hand, rapid growth in adolescence or microfractures of osteoporosis may markedly elevate it, causing false concern about the possibility of bone tumor. Paget's disease produces a striking elevation of the alkaline phosphatase. Table 5-1 outlines the expected values in several conditions.

ACID PHOSPHATASE

Acid phosphatase is usually elevated in cases of carcinoma of the prostate that have extended beyond the capsule of the gland.

ENZYMES

There are no specific laboratory studies to differentiate various muscle diseases. Several serum enzymes including the serum creatine phosphokinase (CPK), serum glutamic-oxaloacetic transaminase (SGOT), lactic dehydrogenase (LDH), and aldolase provide nonspecific information of muscle necrosis. Even the sensitive CPK may not be elevated in some cases of acute polymyositis. The enzymes are not elevated in polymyalgia rheumatica and fibrositis. Many of the same enzymes are elevated in liver and heart diseases, as illustrated in Table 5-2.

Urinalysis

A routine urinalysis sometimes provides important data in patients with back pain. Pyuria

Table 5-1. Calcium and Phosphorus Metabolism in Bone Diseases

Disease	Serum Calcium	Serum Phosphorus	Alkaline Phosphatase	Urine Calcium
Hyperparathyroidism	I	D	I	I
Rickets and osteomalacia	D or N	D or N	I	D
Osteoporosis	N	N	N	N
Paget's disease	N	N or I	I	N or I
Metastatic bone neoplasm	N or I	V	N or I	V
Multiple myeloma	N or I	V	N or I	N or I

I = increased; N = normal; D = decreased; V = variable

should prompt a search for the cause and location of the urinary tract infection. In women a midstream specimen must be obtained because of possible vaginal contamination. Hematuria may be caused by a wide variety of renal and extrarenal causes. Upper urinary tract infection, calculus, trauma, and tumor are of importance when a musculoskeletal pain problem is concerned. Painless hematuria always suggests tuberculosis. Persistent pyuria with negative cultures in a patient not receiving antibiotics also suggests tuberculosis. Proteinuria in the rheumatic diseases may suggest secondary amyloidosis or nephropathy. In multiple myeloma patients one may find Bence Jones proteinuria as well; but it should be remembered that Bence Jones protein may be present in other conditions and may be absent in the presence of multiple myeloma.

Myoglobinuria, such as is found in acute muscle necrosis and in myositis, may cause urine to be dark on voiding. Dark urine may also be caused by porphyria, although in lead porphyrinuria the urine color is normal. The urine will become dark by alkalinization in the presence of alkaptonuria, a condition associated with painful arthritis and extraarticular calcification. Bile in the urine turns it dark;

Table 5-2. Serum Enzymes

Condition	CPK	SGOT	Aldolase	LDH	Alkaline Phosphatase
Polymyositis					
Acute	++++	+++	++++	+++	o
Chronic	++	+	++	+	o
Muscular dystrophy	++	+	++	+	o
Myocardial infarction[a]	+++	++	++	++	o
Liver disease	o	++	++	+	+++[b]
Polymyalgia rheumatica	o	o	o	o	o
Fibrositis	o	o	o	o	o

[a] The elevation of the various enzymes changes depending upon the time of the onset of the infarction.
[b] Variable.
+ to ++++ = relative elevation

this is present in hepatitis and is important since hepatitis patients often present muscle symptoms.

Biopsy

MUSCLE BIOPSY

Unfortunately, routine muscle biopsies are often read as showing nonspecific changes compatible with chronic inflammation and hence provide little clinical information. Only half of the patients with chronic polymyositis will have an abnormal biopsy, but the yield is much higher with acute disease. It is important to take the biopsy sample from a weak but not markedly atrophic muscle, since the latter may be largely replaced by fibrous tissue. To improve the yield of positive results one can detect abnormal muscles by electromyography, and the biopsy can then be taken from the same muscle in the opposite extremity or a different part of the same muscle, to avoid artifacts created by the needle insertions.

BONE MARROW ASPIRATION AND TREPHINE BONE BIOPSY

Bone marrow and bone biopsy material that can be studied at the same time may provide helpful data in certain cases. The marrow aspirate may show the elevated number of plasma cells in myeloma, and the stained bone spicules from the biopsy can give an impression of the metabolic condition of the bone. In cases of primary bone tumors or cysts, an open surgical biopsy can be performed prior to or at the time of surgery.

SYNOVIAL BIOPSY

Open surgical biopsies may have to be made diagnostically in joint diseases. The examination of the biopsy material may be nonspecific. Nevertheless, a biopsy may allow one to differentiate a nonspecific synovitis from specific diseases like tuberculosis, a synovioma, or pigmented villonodular synovitis. In established rheumatoid arthritis, serial biopsies have been used to follow the response to treatment especially with cytotoxic agents. Needle biopsies are seldom justified.

ARTERY BIOPSY

Temporal artery biopsy is useful in the diagnosis of polyarteritis and polymyalgia rheumatica. The material should be taken from an area of involvement as demonstrated clinically by thickening, tortuosity, and tenderness.

Synovial Fluid Examination

Examination of the synovial fluid may aid differential diagnosis. The presence of blood and pus in a specimen are of obvious significance. When pus is present, the sensitivity of pathogens to various antibiotics may be determined. However, it should be remembered that the initial response of the synovium to infection is that of a reactive synovitis up to a period of as long as 48 hours. Therefore, joint aspiration within that time may produce misleading diagnostic data and precious treatment time may be lost. This is a special danger in hematogenous pyarthrosis in infants and in gonococcal arthritis.

Miscellaneous Tests

Examination of the *cerebrospinal fluid* is important in undiagnosed back problems. Mild elevation of the spinal fluid protein occurs with a herniated disc; more pronounced elevation occurs with tumor. An albuminocytological dissociation with elevation of the spinal fluid protein but not the cell count occurs in infectious polyneuritis (Guillain-Barré syndrome) but may not take place for a period up to two weeks after onset of symptoms. *Serological tests* for syphilis in the spinal fluid should be performed. *Skin tests* are used for tuberculosis, sarcoid, and brucellosis. *Blood cultures* are necessary to isolate circulating pathogens. *Stools* should be examined in a search for ova and parasites.

APPEARANCE AND VISCOSITY

Normal synovial fluid is straw-colored and sufficiently clear to allow one to read newsprint through it. Its viscosity can be tested by dripping the fluid drop by drop from the aspirating needle or by stringing a drop between thumb and forefinger. The fluid should elongate for at least 1 inch with either method. Adding a few drops of glacial acetic acid to a test tube of synovial fluid (Ropes's test) produces a firm, nonfriable clot in normal fluid and in those effusions resulting from trauma, osteoarthritis, and systemic lupus erythematosus. In inflammatory diseases such as rheumatoid arthritis and gout, the clot breaks up easily into a flocculent amorphous haze.

MICROSCOPIC EXAMINATION

Cell Count. Normal synovial fluid has almost no leukocytes whereas an acute pyogenic infection will produce a cell count as high as 100,000 per milliliter, with polymorphonuclear cells predominating. In tuberculous arthritis there will be a lower white cell count, with lymphocytes predominating. The noninfectious, inflammatory processes will have a still lower count, with polymorphonuclear cells predominating. However, it should be borne in mind that elderly or debilitated patients may have a pyogenic infection even though the total white count is not as high as one might expect.

Crystals. The presence of sodium urate crystals in joint fluid, especially when seen within leukocytes, is pathognomonic of gout. The crystals are best seen by the use of polarized light, and their appearance is rod- or needle-shaped. They exhibit negative birefringence. In contrast, the rod-shaped or rhomboid crystals of chondrocalcinosis (calcium pyrophosphate crystals) demonstrate weak positive birefringence. Other structures seen by microscopy include cholesterol crystals, which are flat and platelike and have notched corners; steroid crystals, seen after repeated injections; cartilage fragments and other debris secondary to osteoarthritis; and metallic fragments subsequent to a prosthetic implant.

LABORATORY TESTS

Glucose. Comparison of the glucose concentration in synovial fluid with that in blood taken at the same time is a useful test. Diminished glucose in synovial fluid indicates inflammation. Glucose is markedly lowered in tuberculous arthritis but also lowered, although to a lesser degree, in rheumatoid arthritis and gout.

Culture. If gonococcal arthritis is suspected, the joint aspirate should immediately be cultured on Thayer-Martin medium. If other pyogenic organisms are suspected, standard

Table 5-3. Synovial Fluid Analysis

Criteria	Normal	Noninflammatory[a]	Inflammatory[b]	Purulent[c]
Appearance	Straw-colored, clear	Straw-colored, may be blood-tinged	Cloudy	Turbid to opaque
Mucin clot	Firm	Firm	Friable	Friable
Cell count (% polymorphs)	<200 (<25%)	200–2000 (<25%)	2000–50,000 (50%–75%)	20,000–100,000 (75%)
Glucose	⅔ of blood	⅔ of blood	Decreased	Markedly decreased
Crystals	None	None	May be present[d]	None
Culture	Negative	Negative	Negative or positive[e]	Positive or negative[e]

[a] Osteoarthritis, trauma.
[b] Rheumatoid arthritis and variants: gout, pseudogout, tuberculosis, acute gonococcal disease, and fungal infections.
[c] Acute bacterial infections.
[d] Uric acid, calcium pryophosphate, cholesterol, and steroid.
[e] Appropriate culture media necessary.

culture media should be used. In both cases, Gram's stain is a useful screening method, and when tuberculosis is suspected, an acid-fast stain should be ordered.

Table 5-3 outlines the abnormalities present with the various pathological conditions.

BIBLIOGRAPHY

Cohen, A. S. (Ed.). *Laboratory Diagnostic Procedures in the Rheumatic Diseases* (2nd ed.). Boston: Little, Brown, 1975.

Crain, D. C. The hands in arthritis. *J.A.M.A.* 170:795, 1959.

Davidsohn, I., and Henry, J. B. *Todd-Sanford Clinical Diagnosis by Laboratory Methods* (16th ed.). Philadelphia: Saunders, 1974.

Hollander, J. L., and McCarty, D. J., Jr., *Arthritis and Allied Conditions* (8th ed.). Philadelphia: Lea & Febiger, 1972.

Jackson, W. P. U., and Harris, F. Gout with hyperparathyroidism. Report of case with examination of synovial fluid. *Br. Med. J.* 2:211, 1965.

Jacoby, G. A., and Swartz, M. N. Fever of undetermined origin. *N. Engl. J. Med.* 289: 1407, 1973.

Kahn, C. B., Hollander, J. L., and Schumacher, H. R. Corticosteroid crystals in synovial fluid. *J.A.M.A.* 211:807, 1970.

McAfee, J. G., and Dunner, M. W. Differential diagnosis of radiographic changes in the hands. *Am. J. Med. Sci.* 245:592, 1963.

McCarty, D. J., Jr., and Hollander, J. L. The identification of urate crystals in gouty synovial fluid. *Ann. Intern. Med.* 54:452, 1961.

McCarty, D. J., Jr., et al. The significance of calcium phosphate crystals in the synovial fluid of arthritic patients. The "pseudo-gout syndrome." *Intern. Med. Clin. Aspects I* 56: 711, 1962.

Paul, L. W., and Juhl, J. H. *Essentials of Roentgen Diagnosis of the Skeletal System* (3rd ed.). New York: Harper & Row, 1972.

Petersdorf, R. G., and Beeson, R. B. Fever of unexplained origin. Report on 100 cases. *Medicine* 40:1, 1961.

Ropes, M. W., and Bauer, W. *Synovial Fluid Changes in Joint Disease.* Cambridge, Mass.: Harvard University Press, 1953.

Wagner, H. N., Jr., *Principles of Nuclear Medicine.* Philadelphia: Saunders, 1968.

6
Electrodiagnosis

Electrodiagnostic methods are increasingly used to assist the physician in the differential diagnosis of musculoskeletal and neuromuscular pain problems. They also play a role in prognosis, since recovery from nerve damage is often revealed earlier by electrical testing than by clinical signs.

Some of the basic neurophysiology of electrodiagnosis is included here, both for a better understanding of the use and limitations of this diagnostic method and because it is applicable to other physical modalities of treatment, discussed in the next chapter. Figure 6-1 illustrates some basic details of muscle, nerve, and the neuromuscular junction, which will be discussed in the next few pages.

ANATOMY AND PHYSIOLOGY

Skeletal Muscle–Histology

THE MUSCLE FIBER

Each skeletal muscle fiber is an elongated multinucleated cell, measuring from 1 to 4 cm in length and 10 to 100 microns in diameter. It contains the contractile elements (the myofibrils) and the cytoplasmic substance (the sarcoplasm). These components are surrounded by the sarcolemma, a double-layered, semipermeable membrane. Surrounding each muscle fiber is a thin connective tissue layer, the endomysium. Bundles of muscle fibers, numbering from 12 to 150, are surrounded by another layer of connective tissue, the perimysium; each bundle forms a unit known as a fasciculus. Surrounding the entire muscle is a layer of deep fascia known as the epimysium.

MUSCLE BANDS

Within the muscle fibers the myofibrils are arranged in groups, the groups being separated by sarcoplasm. Each group is called a Cohnheim area. Electron microscopy reveals that myofibrils contain even finer elements, the myofilaments. When viewed by light microscopy, the myofibrils are noted to be composed of alternating bands of light (A band) and dark (I band) color. The dark band itself is bisected by a thin dark line, the Z line. Extending from one Z line to the next is the structural unit of the myofibril: the sarcomere. Biochemically, within each sarcomere are the proteins myosin and actin. The energy of muscle contraction is derived from the conversion of adenosine triphosphate (ATP) to adenosine diphosphate (ADP). The actin filaments slide past the myosin filaments in a pistonlike action, resulting in shortening of the muscle. A new protein, actomyosin, is thereby formed. ATP is regenerated from ADP by reaction with creatine phosphate (creatine PO_4 + ADP $\rightleftarrows$ creatine + ATP), thus renewing

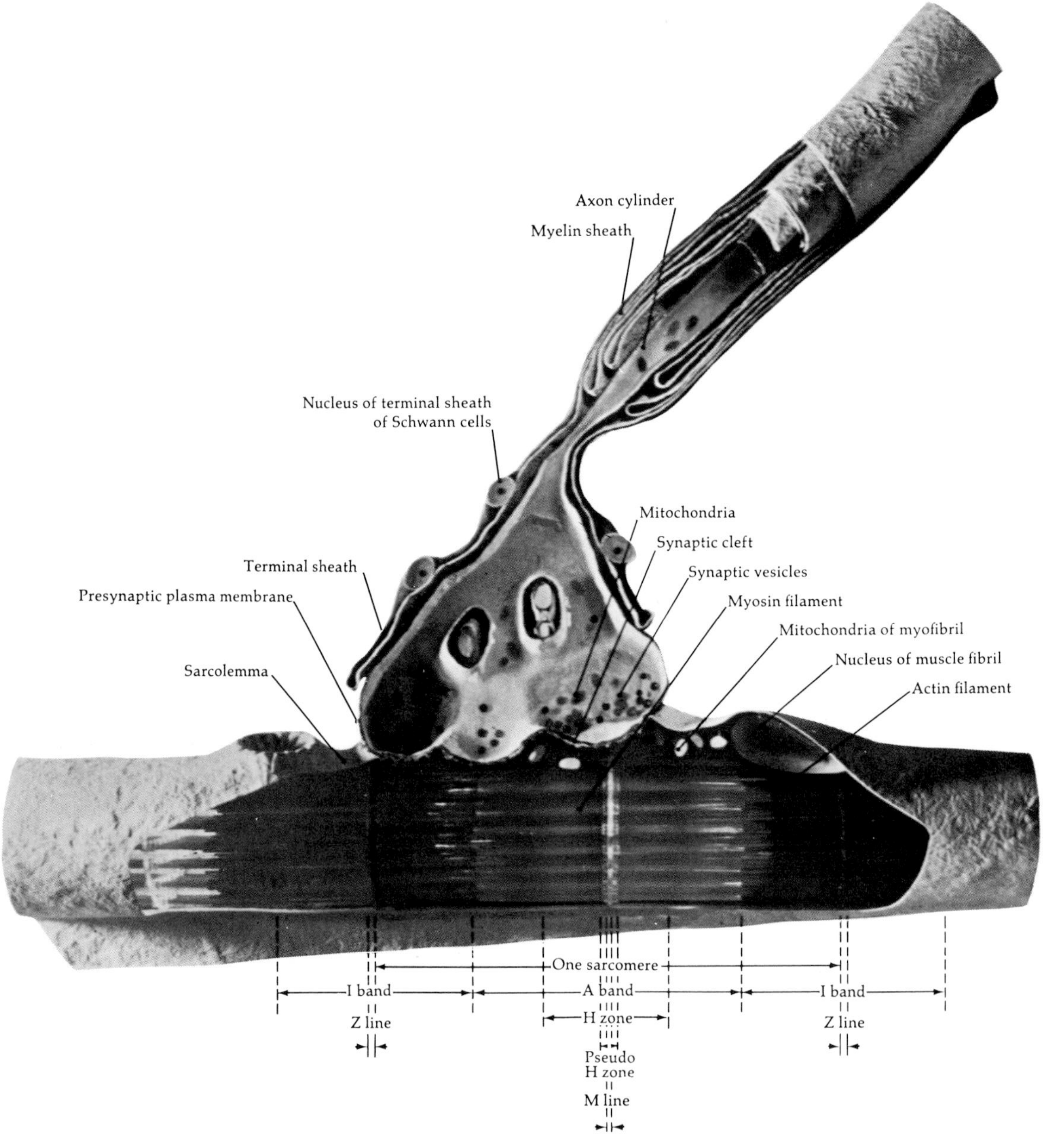

Figure 6-1. Cross-section of nerve, muscle, and neuromuscular junction illustrating some of the basic anatomy and histology. (Courtesy of Hoffman-LaRoche Inc.)

the energy cycle. Relaxation occurs when dephosphorylation of ATP ceases. As will be seen later, there is a neurophysiological basis of muscle contraction and relaxation as well as this biochemical basis.

SARCOPLASM

Sarcoplasm, which contains mitochondria and Golgi bodies, is for the most part made up of myogen, an albumin fraction of the total protein. It contains enzymes that are necessary for muscle metabolism, notably phosphorylases, aldolase, and dehydrogenase.

SARCOLEMMA

The sarcolemma is a semipermeable membrane separating the potassium-rich interior of the cell from the sodium- and chloride-rich exterior. The sarcolemma plays an important role in the transmission of the stimulus that initiates the entire contractile mechanism.

MUSCLE SPINDLES

Muscle spindles are elongated cylindrical structures measuring 0.5 to 3.0 mm in length. There are 70 to 100 of them within each muscle. The muscle spindle itself contains muscle fibers, connective tissue, blood vessels, and nerve fibers. The muscle fibers are primarily within the spindle (intrafusal) but extend outside to attach by means of a slip to a neighboring extrafusal fiber.

The nerve supply of the spindle is both motor and sensory. The motor fibers consist of myelinated alpha fibers and unmyelinated gamma fibers. The sensory fiber endings are primary or annulospiral endings, which wrap around the intrafusal muscle fibers, and secondary or flower-spray endings, which branch over the muscle fiber surfaces. Mechanical distortion of the intrafusal fibers by stretching produces an excitation of the sensory endings. This occurs if the length of the surrounding extrafusal muscle fibers is greater than the length of those fibers in the spindle. The excitation causes a sensory impulse to be relayed through the spinal cord, which in turn produces a contraction of the extrafusal muscle fibers followed by a return to their normal resting length. Thus, a biological process is set up to maintain homeostasis by a self-regulating device or servomechanism.

GOLGI TENDON ORGANS

Golgi tendon organs are receptors that lie in the fibers of the muscle tendons. They are less responsive than the muscle spindle receptors and are stimulated only by tension on the tendon, thereby serving to apprise the central nervous system of muscle overload. Signals from the receptors are transmitted to the cord via type A alpha fibers.

MUSCLE MOVEMENTS

A muscle principally responsible for a particular movement is the *agonist* or *prime mover*. The muscle whose function is the opposite and whose relaxation or deceleration is necessary for effective function of the agonist is the *antagonist*. This reciprocal action permits smooth movement, which is assisted by *fixator muscles*. These fix the base, so that the prime mover can act. For example, the wrist dorsiflexors fix the wrist joints while the fingers flex. The combined functioning of all these muscles is called a *synergic action*.

MUSCLE CONTRACTION

When a muscle contracts, its tension increases, work is performed, but shortening does not necessarily occur. *Isometric contractions* are those in which the muscle ends do not approximate each other. The contractions in which muscle ends do move closer to each other are *isotonic contractions;* because of this approximation, they are also considered *concentric contractions*, as opposed to *eccentric contractions*, in which the muscle develops tension as it is lengthening. The lengthening contraction acts as a decelerator. An example is the use of the biceps brachii to lower the body slowly after a chin-up exercise.

MUSCLE TONE

Muscle tone (or tonus) has both a clinical and a physiological context. On the basis of Sherrington's concept of postural tonus reflexes, it was thought for a long time that continuous neuromuscular activity occurs to maintain bodily posture, this activity being responsible for muscle tone. However, electromyographic studies have shown that muscles involved in the maintenance of the upright posture may go for prolonged periods of time without alpha motoneuron activity—that is, without recordable electrical activity. Change of position leading to loss of balance does produce an immediate firing of the motor units, presumably activated by the gamma-loop system. This has led to the newer concept stated by Basmajian that muscle tone consists both of "passive elasticity or turgor of muscular [and fibrous] tissues and the active [though not continuous] contraction of muscle in response to the reaction of the nervous system to stimuli."

Peripheral Nerves—Histology

THE NERVE CELL

The individual nerve cell or neuron consists of a cell body, dendrites, and an axon. Surrounding the axon is a curled layer of phospholipids called the myelin sheath, while surrounding this in turn is a nucleated sheath called the neurilemma or sheath of Schwann. It is believed that the myelin sheath is elaborated by the plasma membrane of the neurilemma cells. At regular intervals the neurilemma indents to come into contact with the axon, and the indentations are known as the nodes of Ranvier. Outside the neurilemma, and not forming part of the nerve itself, is the sheath of Henle, or the endoneurium. Groups of nerve fibers are called fascicles and are enclosed by a connective tissue sheath called the perineurium; another connective tissue sheath, the epineurium, encloses the entire nerve.

MOTOR UNIT

The motor function of the neuromuscular system is based upon motor units. The motor unit is the smallest unit under central nervous system control and consists of an anterior horn cell, its axon, and a variable number of muscle fibers innervated by the neuron through the motor end-plate (Figure 6-2). The number of muscle fibers varies with the quality of function required. Extraocular muscles have as few as 10 muscle fibers per motor unit while muscles with gross functions such as each of the quadriceps muscles have a thousand or more. It has been shown that the fibers of a motor unit have their own territory—that is, they are confined to an area within a diameter of 4 to 6 mm. This territory, however, comprises several muscle bundles, so that there is an overlapping of motor units within each muscle bundle. Each motor unit is further divided into 50 to 100 subunits.

EFFERENT (MOTOR) NERVES

An efferent nerve fiber has its cell body in the anterior horn of the spinal cord. Two types of efferent fibers are present in peripheral nerves, alpha and gamma. Alpha nerve fibers are large

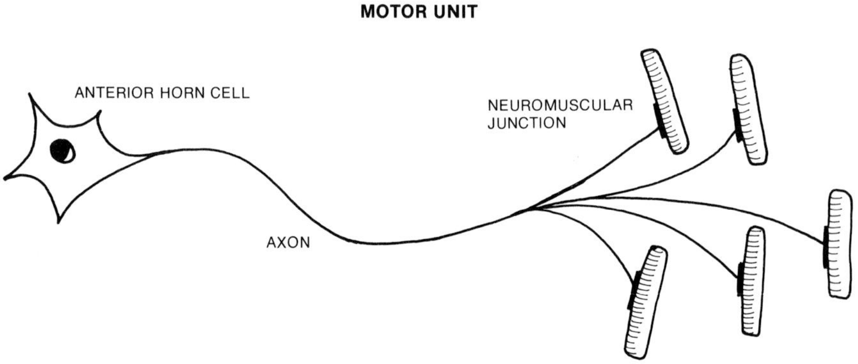

Figure 6-2. Schematic representation of a motor unit showing the basic components: anterior horn cell, axon, and muscle fibers.

in diameter (1 to 20 microns), are heavily myelinated, and are fast conductors of nerve impulses. Alpha nerve fibers branch repeatedly when they reach the muscle, innervating individual muscle fibers through the motor end-plate. Gamma nerve fibers are smaller in diameter and branch out to innervate the intrafusal fibers of the muscle spindle.

AFFERENT (SENSORY) NERVES

Cell bodies for sensory nerves are located not within the spinal cord but in the spinal ganglia adjacent to the cord. Fibers are both myelinated and unmyelinated. Each cell is unipolar, with division into a central and a peripheral branch. The central branch transmits impulses to the spinal cord. The peripheral branch receives impulses from the environment via specialized nerve endings, the somatic receptors. Somatic receptors are divided into those which are affected by changes in the external environment (the exteroceptors) and those which are affected by changes in the internal environment of the muscles, muscle spindles, and tendons (the interoceptors).

Receptors in the skin subserve two broad classifications of sensibility. The more primitive type (protopathic) responds to pain, crude touch and pressure, and gross temperature changes while the others (epicritic) respond to light touch, vibration, two-point discrimination, and minor temperature changes. Protopathic fibers are small and poorly myelinated and conduct slowly; epicritic fibers are large and heavily myelinated and conduct rapidly.

MOTOR END-PLATE

Contact between nerve and muscles occurs at a structure called the motor end-plate. One end-plate is present for each muscle fiber at its midpoint. As the nerve nears the muscle, the endoneurium (sheath of Henle) becomes continuous with the endomysium and the neurilemma becomes continuous with the sarcolemma. The axon passes beneath the sarcolemma and breaks off into clublike terminals (telodendria) embedded in the sarcoplasm, the whole structure making up the end-plate.

THE MOTOR POINT

The motor point is the surface projection of the place in the muscle where the motor nerve dips into it. Large muscles may have more than one motor point. Denervation of the muscle was thought to bring about a distal movement of the motor point. The motor point has been demonstrated to be an anatomical entity, however, and not merely a physiological property of the muscle fiber. Thus, with denervation, the motor point cannot shift or descend as had been previously described. Rather, uniform excitability of the entire muscle occurs, but the portion of muscle near the tendon contracts more obviously than the rest of the muscle, accounting for the apparent shift or "descent." Lack of a recognizable motor point is diagnostic of complete denervation of muscle.

PHYSIOLOGY OF MUSCLE CONTRACTION

Resting Potential. If microelectrodes are inserted into a cell, a difference in potential between the interior of the cell and its external environment is noted in the order of magnitude of 50 to 100 millivolts, with the interior negative relative to the exterior. This is the resting potential of the cell.

Depolarization. Upon receipt of a stimulus of adequate intensity, alterations of the cell membrane permeability occur, with ion migration and partial redistribution of ion concentrations. This process—depolarization—produces a change of potential at the site with a resultant flow of current, which rapidly traverses the entire membrane. A trailing area of recovery produces repolarization or a wave in the opposite direction. This results in a biphasic or triphasic wave known as an *action potential.* The minimal stimulus necessary to produce an action potential is known as a *threshold stimulus.* A stimulus of lesser magnitude is called a *subthreshold stimulus* and produces no response. Following a threshold stimulus with a resultant depolarization and production of an action potential, there occurs a period of 1Σ (1/1000 second) duration in which no stimulus of any intensity can produce an impulse.

This is the *absolute refractory period*, and it is followed by another period in which a response can be obtained only with a stimulus of increased intensity. This is called the *relative refractory period* (15Σ). If two stimuli are delivered, the latter occurring before the first muscle twitch has been completed, the response is greater than the sum of the two, a phenomenon known as *summation*. If enough stimuli are delivered within a period of time, relaxation does not take place at all and a tetanic contraction occurs.

All-or-None Law. With reference to a motor unit, a stimulus above threshold in intensity will produce an impulse no different from one produced by a stimulus of threshold intensity. This is the all-or-none law and as used here applies to a single axon rather than to the whole nerve. Because an external stimulus cannot be applied with equal intensity to all of the nerve fibers and also because the size of fibers within the nerve varies, the intensity of current necessary to produce a response varies from nerve fiber to nerve fiber. As intensity increases, more axons are recruited until maximal stimulation of the whole nerve has been reached.

Neuromuscular Transmission. When an impulse arrives at the motor end-plate, acetylcholine is elaborated locally. This alters the permeability of the end-plate sarcolemma, setting up an end-plate potential, which in turn sets up a propagated impulse in the adjacent muscle sarcolemma and an action potential results. Recovery of this mechanism is dependent upon the destruction of the chemical mediator acetylcholine, a function performed by the enzyme cholinesterase.

PATHOLOGY

Nerve Injury

CLASSIFICATION OF TRAUMATIC NERVE INJURIES

Many different classifications of traumatic nerve injuries have been proposed, based on clinical, pathological, or electrodiagnostic findings. No one classification has been completely satisfactory and therefore uniformly accepted. We favor the one proposed by Seddon, which is the simplest and most applicable clinically. An electrodiagnostic classification based on Seddon's work is on p. 103. It should be borne in mind that each of the three categories in this classification can vary greatly in extent depending on the nature of the injury and the proximity of the nerve to the trauma.

NEURAPRAXIA

Neurapraxia is the first category. It is also known as physiological block, conduction block, or reversible nerve block. Compression in the form of trauma or edema is applied to a segment of nerve. Either ischemia or mechanical deformation of the axonal substance occurs, but axonal continuity is preserved and wallerian degeneration does not take place. The block is restricted to the segment damaged, and, although no volitional signal can pass through it, the nerve responds to a stimulus applied above and below the site of the block. For instance, should the radial nerve be compressed at the middle third of the humerus, stimulation of the nerve at the level of the brachial plexus or axilla should produce a response in the triceps since the innervation here is above the level of the block, while stimulation in the forearm should produce a response in the extensors since the innervation here is below the level of the block. Stimulation above the level of the block would not, of course, produce a response in the extensor muscles. The degree of block may vary from extremely light compression with effects lasting minutes or hours, as occurs on prolonged crossing of the legs, to loss of function for up to six weeks, as in a crutch palsy.

AXONOTMESIS

Axonotmesis is the second category of nerve injury. Here the continuity of the axons and myelin has been disrupted, but the sheath of Schwann remains intact. The number of axon fibers affected and their degree of disruption may vary greatly. The nerve trunk loses its excitability within 72 hours, but the smaller intramuscular nerve fibers do not lose theirs for seven to ten days. In three to five days the axon begins to break up into small, irregular segments, and it gradually undergoes complete disintegration. The myelin sheath becomes fragmented, breaks into small droplets, and finally disappears, as a result of the phagocytic action of proliferated Schwann cells and histiocytes of the endoneurium. Retrograde changes take place for only a few internodes. However, changes do occur in the cell body, the degree of change being dependent on the extent of the injury and its proximity to the cell body. Changes in the cell body include swelling, dissolution and disappearance of Nissl bodies, and displacement of the nucleus toward the periphery. The process of wallerian degeneration is the sum of all these changes and is completed by the fifth to eighth week. All that remains is the tube of Schwann cells surrounded by endoneurium that is completely emptied of debris.

NEUROTMESIS

Neurotmesis, the third category, is a severance of the nerve trunk with loss of anatomical continuity. Similar changes of wallerian degeneration take place, but the retrograde effects are more severe.

Nerve Repair

STAGES OF REGENERATION

Sunderland describes several steps in the regenerative process. First is the recovery of the neurons from the retrograde effects described above and the onset of regeneration commencing at the severed end of the axon. The axon tip grows to the site of injury and then must traverse the zone of injury. The axon continues its growth down the endoneurial tube below the site of injury, and then reattachment of the end of the nerve to the muscle takes place. The time elapsed up to this point is known as the *neurotization time*. Functional maturation follows, with restoration of the original anatomical and physiological relationships. Last, recovery of a sufficient number of fibers to produce an adequate mechanical contraction occurs. The time required for these processes is the *maturation time*. With neurotmesis, there is a proliferation of cells of the endoneurium and of fibroblasts to bridge the gap of the injury.

AXONAL SPROUTING

Resistance met by the regenerating axon at the site of scar tissue results in the formation of multiple sprouts at the axon tip. The sprouts may enter a single endoneurial tube, or, by branching, the single axon may enter more than one endoneurial tube. Depending on the outcome, such branching is either advantageous or disadvantageous. A desirable effect might be the innervation of more than the normal number of muscle fibers by a single axon with resultant improvement in function. An undesirable outcome would result if indiscriminate budding took axons into foreign tubes, resulting in impaired motor function.

TIME SEQUENCE FOR REGENERATION

The duration of the regeneration process depends in part on the period of initial delay, during which time neuronal regeneration and axon growth near the site of injury take place, determined by the severity of injury and the proximity to the cell body. The time required for axonal fibers to traverse the zone of injury is known as the *scar delay*. Failure of the axonal sprouts to find a distal tube will lead to neuroma formation, while failure to find the correct tube will lead to axonal sprouting and to alteration of fine motor function.

AXON GROWTH TIME

Axon growth time is the time it takes the axoplasm to traverse the neural tube to the muscle. This depends upon several factors, but the two most important appear to be the degree of the lesion (slower regeneration takes place with a severed nerve) and the level of the lesion (the closer the axon tip to the cell body, the faster the growth rate). The rate of recovery slows as it proceeds distally. This rate, therefore, cannot be regarded as constant for any phase of the recovery process, which may at least partially account for the wide range of outcomes reported by different investigators. Another variable is introduced by differences in methods employed clinically for testing regeneration. The recovery rate falls between 1 and 2 mm of linear nerve regeneration per day.

Changes in Other Tissues

BONE

Changes in the bone structure as a result of denervation are noted wherever denervation occurs, but the effect is most marked in the small bones of the hands and feet. The cortex becomes thinned, and the trabecular structure fades; this is osteoporosis. The effect is greatest in the vicinity of the joint. In growing bone, osteoporosis of disuse over a long period of time is complicated by a decrease in the size of the bone.

MUSCLE

Individual fiber atrophy is noted, with a reduction of the amount of contractile substance in each fiber. All the connective tissue elements proliferate, so that the muscle fibers constitute a smaller percentage of the muscle. However, the normal fascicular pattern is maintained.

PHYSICS–ELECTRONICS

Characteristics of Current and Current Flow

CURRENT FLOW

In any electrical circuit, current flow follows Ohm's law: The intensity of current (amperes–I) is directly proportional to the electromotive force (volts–E) and inversely proportional to the resistance–R ($I = E/R$. The flow of current passes from the cathode (negative pole) to the anode (positive pole) and occurs only when the circuit is completed–that is, on the "make." When an electrical current is passed through a muscle, muscle contraction occurs on both the "make" and the "break" of the circuit but normally much more briskly on the "make."

DIRECT CURRENT

Direct (galvanic) current is a steady unidirectional flow of current through tissues connected between the positive and negative terminals of a source. Direct current may be applied either continuously or in an interrupted manner. The effects of continuous current are primarily chemical, and no contraction of muscle takes place. As a result of resistance to electrical flow, heat is generated in the tissues. The principal physiochemical effects of a continuous direct current are vasodilation, transfer of ions, and raised threshold of excitability and conductivity of muscle and nerve. When the direct current is interrupted, however, a muscle contraction occurs. This phenomenon forms the basis for several types of electrodiagnostic testing. Interrupted current is used as a stimulating current.

PROPERTIES OF STIMULATING CURRENT

A stimulating current must have three properties if a response is to be obtained. It must be of sufficient *intensity* to exceed the threshold of the tissue being stimulated. Its *rate of rise* must be rapid enough to prevent accommodation of the tissues to the stimulus. Its *duration* must be long enough to produce a response.

ALTERNATING CURRENT

Alternating current is a flow of current that changes its direction of flow rhythmically. The average current flow in the positive direction equals the average current flow in the negative direction, and the sum of flow in each direction completes one cycle. The number of cycles per second is termed the *frequency*, measured as Hertz (hz). Low-volt therapy utilizes frequencies of from 1 hz to 3000 hz. The commonly used waveforms are the sine wave, the square wave, and the pulse wave (Figure 6-3). In addition, it is possible to increase or decrease the peak current inten-

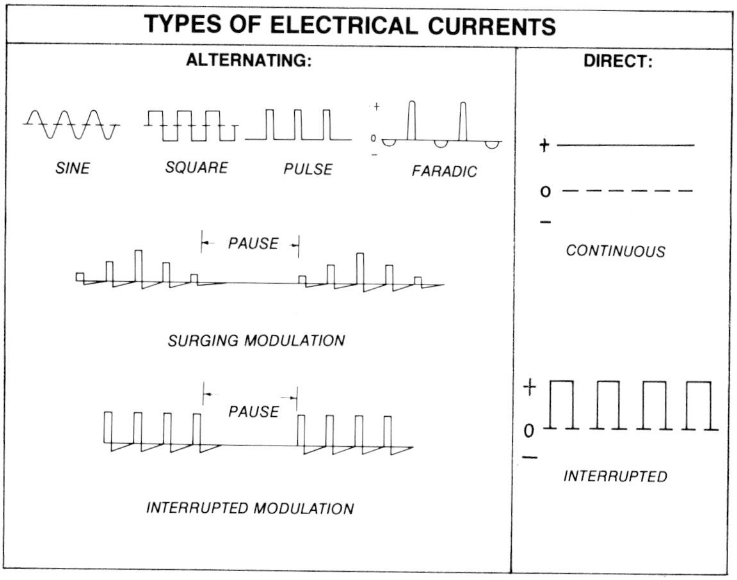

Figure 6-3. Commonly used waveforms.

sity. This adjustment is known as *modulation.* If it is done slowly, it is called *surging modulation;* if it is done abruptly, it is called *interrupted modulation.*

FARADIC CURRENT

The faradic current (Figure 6-3) is a type of alternating current produced originally by an induction coil but now by electronic means. This is an asymmetrical, interrupted current, the impulse of which lasts about 0.001 second, a duration brief enough to stimulate nerve tissue but not muscle tissue. This property is put to use in performing the classic electrodiagnostic testing as described on p. 109.

PHYSIOLOGICAL EFFECTS

The basic physiological effect of the application of an alternating current is muscle contraction. No chemical effect is produced by this current since the average flow in one direction is equal to the average flow in the other and thus there is no net ion transfer. If an alternating current of 1 hz is applied to normal muscle, a single twitchlike contraction occurs. When the frequency is increased to above 20 hz, the contractions merge and a tetanic contraction results.

ELECTROMYOGRAPHY

Electromyography, coupled with motor and sensory nerve conduction measurements, has become the predominant method of electrodiagnostic testing. Electromyographic tests have demonstrated increased effectiveness in diagnosis, localization of a nerve injury, and prognosis after nerve injuries when compared with earlier methods of electrodiagnostic testing. However, these are ancillary clinical tests and cannot be substituted for good clinical judgment. Their effectiveness may be enhanced by coupling them, when necessary, with chronaxy and strength duration curves (see p. 109 for definition), nerve excitability testing, or galvanic-faradic testing.

The usefulness of electromyography is limited to the diagnosis and management of lower motor neuron and muscle disorders. It is based on the principle that an electrical current precedes contraction of the muscle, and that this current can be detected by an electrode, magnified by a preamplifier and amplifier, recorded visually on a cathode-ray oscilloscope, and registered audibly on an audio-amplifier. Electromyography is therefore analogous to electrocardiography since it depends on the recording of the electrical processes which accompany muscle contraction.

Electromyographic Examination

ELECTRODES

Two types of electrodes are commonly used in clinical testing. The *coaxial electrode* is a 25-gauge hypodermic needle into which is inserted a second solid insulated core serving as the negative electrode. The *monopolar electrode* is a solid 27-gauge Teflon-coated needle with just the bare tip exposed. Its use requires a second negative electrode. Both electrodes require the use of a ground electrode to complete the circuit. The coaxial needle picks up electrical activity from a smaller area than the monopolar needle, which may or may not be desirable. The use of the coaxial needle electrode is less comfortable when an extensive examination is undertaken. Surface electrodes are never used in clinical electromyography but have a role to play in kinesiological research.

PROCEDURAL CONSIDERATIONS

Since these clinical examinations are not performed as a "routine" but are modified to suit the circumstances, they should be preceded by an appropriate history and physical examination. An explanation of the nature of the test, as well as distracting conversation during its performance, does much to make the examination tolerable to the patient besides promoting his maximal cooperation, an essential component of an accurate examination. Adequate sampling of the muscles must be carried out, especially when equivocal changes are noted or when there is a discrepancy between the electrical responses and the clinical opinion.

Permanent records can be made on magnetic tape, on film, or, in newer equipment, on recording paper but are not essential for the routine clinical examination. Shielding of the room is not necessary with modern equipment. To avoid possible complications such as the transmission of serum hepatitis, sterility of electrodes must be ensured.

Electrical Patterns

When a normal muscle is at rest, no electrical activity is present and a flat baseline is observed. Figure 6-4 illustrates the basic patterns of electrical activity seen on the oscilloscope. Each has a characteristic sound as well.

VOLUNTARY MOTOR UNIT POTENTIALS

On volition, biphasic or triphasic potentials measuring from 100 to 2000 microvolts and lasting from 2 to 10 milliseconds appear on the screen. These potentials produce moderately deep and thumping sounds over the audio system of the electromyograph. Each action potential represents the firing of a single motor unit and is called a *voluntary motor unit potential*. When many voluntary motor unit potentials appear on stronger volition, they fill the oscilloscope, blending into each other and creating an *interference pattern*. Failure to create such a pattern indicates either the loss of many motor units or lack of cooperation on the part of the patient.

POLYPHASIC MOTOR UNIT POTENTIALS

Polyphasic potentials probably represent the loss of synchronous discharge of the motor unit. Less than 5 percent of these appear normally, and any greater percentage is evidence of abnormality. The amplitude may vary widely, from 50 to 5000 microvolts. Polyphasic potentials with low amplitude are referred to as *nascent motor units* and occur during the process of reinnervation. Their small size is a reflection of the relatively small number of motor fibers reinnervated. Polyphasic potentials at the upper end of the amplitude scale are referred to as *giant motor units* and occur either in conditions of chronic denervation or in the process of reinnervation. It is thought that the increased size of these potentials results from hypertrophy of remaining motor units or abnormal reinnervation due to axonal sprouting. The duration of these potentials is usually prolonged, and therefore their sound is low-pitched and chugging, like that of an idling motorboat.

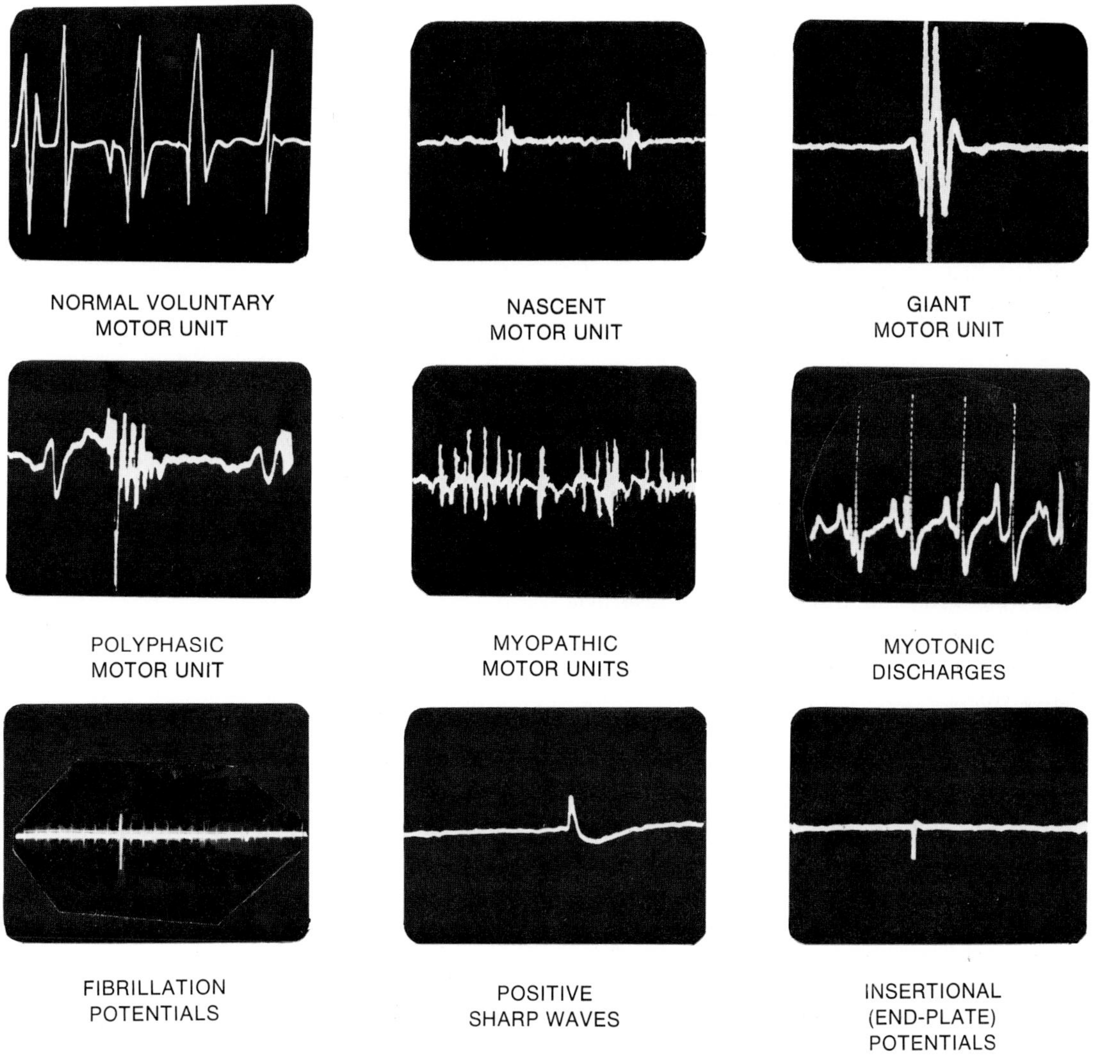

Figure 6-4. Characteristic electromyographic waveforms.

FASCICULATION POTENTIALS

Fasciculation potentials are similar in appearance and sound to the voluntary motor units but differ from them in that they occur when the muscle is at rest rather than in action. Fasciculation potentials represent the spontaneous firings of motor units. It is not possible to determine from the electromyographic examination alone whether these fasciculations are a normal phenomenon, as in fatigue or stress, or a reflection of a pathological process, as occurs in anterior horn cell disorders or nerve root lesions, unless they are accompanied by other changes characteristic of a disease process.

MYOPATHIC POTENTIALS

Voluntary motor units which are poorly organized, of low amplitude, and of brief duration, and which may be accompanied by many polyphasic motor unit potentials, are called *myopathic potentials.* They represent disintegration of the motor unit and occur in primary muscle disorders.

FIBRILLATION POTENTIALS

When a nerve has been damaged and adequate time has elapsed for wallerian degeneration to take place (10 to 14 days), spontaneous potentials not under voluntary control appear. These are brief (1 to 2 milliseconds in duration), low-amplitude (10 to 200 microvolts), biphasic waves with a high-pitched clicking or tapping sound. Called *fibrillation potentials,* they represent the firing of a single muscle fiber, separated from its motor unit and exhibiting its properties of intrinsic excitability and contractility. Although a fibrillation potential is the classic sign of nerve degeneration, it also occurs in a variety of primary muscle disorders, particularly those inflammatory processes in which alteration of membrane excitability has taken place.

POSITIVE SHARP WAVES

The *positive sharp wave* is a biphasic potential of variable amplitude with the positive peak sharply pointed. Its exact etiology has not been established but probably has to do with mechanical deformation of an irritable muscle fiber by the probing electrode. Like the fibrillation potential, it reflects nerve degeneration, but it too occurs in a variety of primary muscle disorders. The presence of one or both of these potentials is required for a diagnosis of nerve degeneration to be made with certainty, although nerve irritability can be diagnosed by a constellation of other findings.

MYOTONIC BURSTS (DISCHARGES)

This striking potential appears on insertion of the needle in conditions of myotonia and consists of trains of high-pitched waves of brief duration which wax and wane, producing sounds that have been compared to those of a dive bomber. The discharges may last for many seconds and reoccur each time the electrode is moved. Potentials of similar sound that last for a shorter period of time and cease abruptly are known as *high-frequency discharges* (pseudomyotonic bursts). They occur in several conditions, most commonly in the inflammatory disorders of muscles but also in motor nerve damage.

INSERTIONAL POTENTIALS

In normal muscle, the insertion of the needle produces a very brief discharge of electrical potentials. If these persist, and appear as small, brief, high-pitched biphasic waves with an initial negative deflection, they are *end-plate potentials.* Increased pain usually accompanies the end-plate potential. All other forms of increased insertional activity are abnormal.

ARTIFACTS

Certain extraneous potentials may be artifacts, such as 60-cycle source interference, diathermy equipment, or loose electrodes. Once these have been identified, they are usually easily corrected.

Sites of Pathology

Both neuropathic and myopathic conditions lend themselves to electromyographic study. It is the first task of the electromyographer to place the disorder in one of these categories. Beyond that, however, it is important to further localize the disorder to one of six anatomical sites of the motor unit (Figure 6-5). These are: (1) anterior horn cell, (2) root, (3) plexus, (4) peripheral nerve, (5) myoneural junction, and (6) muscle.

ANTERIOR HORN CELL

Conditions involving the *anterior horn cell* include amyotrophic lateral sclerosis and the other spinal muscular atrophies, poliomyelitis, and a late manifestation of syringomyelia. Abnormalities seen by electromyography in these disorders include fasciculation potentials and fibrillation potentials. In chronic denervation states such as poliomyelitis extremely large motor units can be seen. They represent either hypertrophy of the remaining motor units, axonal sprouting of the remaining motor units

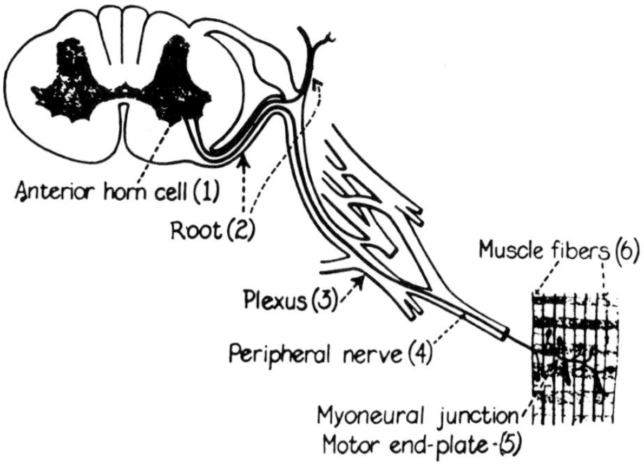

Figure 6-5. Six anatomical sites along motor unit pathway where specific pathological processes may produce abnormalities electromyographically. (Adapted from Y. T. Oester and A. Rodriguez, in S. Licht [Ed.]. *Electromyography and Electrodiagnosis.* New Haven, Conn.: E. Licht, 1961.)

to include more muscle fibers, or regenerated motor units with a higher than normal innervation ratio. The sometimes difficult differentiation of amyotrophic lateral sclerosis from cervical spondylosis can frequently be helped by electromyographic examination. In the upper extremities, cervical spondylosis shows involvement of one or perhaps two roots whereas amyotrophic lateral sclerosis demonstrates abnormal potentials in a random distribution. An early manifestation of amyotrophic lateral sclerosis is the presence of denervation potentials in the lower extremities as well. This is obviously an impossibility if the cause is a cervical root lesion. A related finding of denervation potentials in the tongue, due to bulbar involvement, is sometimes present in amyotrophic lateral sclerosis; never in a cervical root lesion.

ROOT

At the root level, electromyography is a valuable aid in the diagnosis of both cervical and lumbar lesions. The high level of accuracy of the examination coupled with its lack of morbidity makes it invaluable when diagnosis becomes a problem in such cases. However, its limitations must be understood. Such an examination is difficult to interpret, painful to the patient, and rarely necessary in thoracic root lesions. The electromyogram, in contrast to the myelogram or discogram, diagnoses motor nerve root compression and not mechanical deformation. It must be remembered that root compression does not necessarily arise from herniated nuclear material. Either epidural or intrathecal masses (e.g., tumors or vascular anomalies), fractures or callus formation, abscesses, foreign bodies (e.g., shell fragments), osteoarthritic spurs, and swelling of a facet joint capsule may produce root compression. In patients whose only problem is pain due to sensory root compression the electromyogram will be negative.

Adequate time must have elapsed for wallerian degeneration to take place before characteristic changes develop. Sampling of the paravertebral musculature should be included as part of the examination because these muscles are innervated by the posterior primary rami whereas the extremities are innervated by the anterior primary rami; both may be compressed in the same manner (Figure 6-6). The yield of abnormal findings in the paravertebral muscles is much greater in the lumbar spine than in the cervical spine, possibly because of the greater overlap of myotomes in the cervical spine or because the posterior rami between the fifth and eighth cervical levels have been demonstrated to be significantly smaller than elsewhere in the spinal cord.

Multiple studies have shown the accuracy of electromyography in diagnosing root lesions to be in the order of 80 percent. Comparison with the accuracy of myelography is very favorable, and the two combined provide an ex-

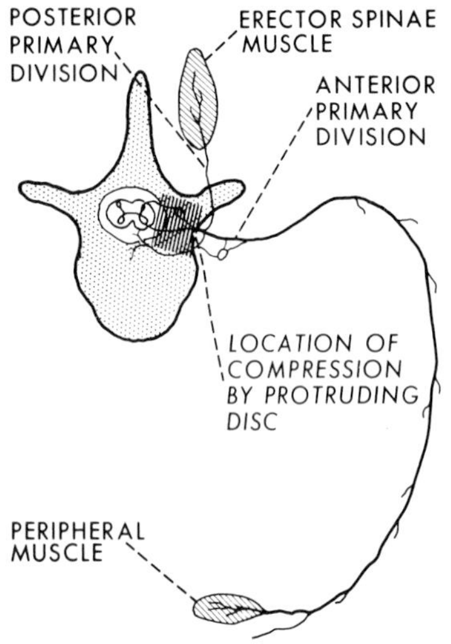

Figure 6-6. Division of nerve root into anterior primary division (to extremity muscle) and posterior primary division (to back muscle). Compression of nerve root produces abnormalities of both divisions.

tremely high level of accuracy, approaching 95 percent.

Careful sampling of the muscles of the whole extremity is important since the level of the root lesion needs to be determined by deduction. Table 6-1 shows that if the quadriceps muscle is normal, the tibialis anterior, extensor digitorum longus, and gluteus medius show profuse denervation, the peroneus longus and gastrocnemius show little or no denervation, and the interossei show no denervation at all, it can be surmised that a lesion of the fifth lumbar nerve root is present. The reason is that abnormalities are present throughout the fifth lumbar myotome—that is, those muscles innervated by a single spinal root. Some overlapping occurs because most muscles are innervated by more than one root. Abnormalities in the paravertebral musculature confirm the diagnosis of nerve root compression but, because of overlapping of myotomes, do not necessarily establish the level.

A similar approach is applicable to the upper extremities. For example, the combination of minimal abnormality in the deltoid, maximal abnormality in the biceps and brachioradialis, minimal abnormality in the triceps and forearm extensor muscles, and no abnormality in the intrinsic muscles of the hand would place the lesion at the level of the sixth cervical root (see Table 6-2).

PLEXUS

Electromyographic studies, performed in conjunction with nerve conduction measurements, help to establish the correct diagnosis where avulsion of the brachial plexus or nerve root has occurred. In nerve root avulsions the lesion takes place proximal to the spinal ganglion; therefore, wallerian degeneration of the motor fibers but not of the sensory fibers occurs, since the latter are in continuity with their cell bodies. Thus denervation potentials will be found in muscles whose nerves have degenerated, and abnormal motor nerve conduction velocity will also be observed. However, sensory nerve conduction studies yield normal results since these fibers have not degenerated.

Table 6-1. Root Supply of Selected Lower Extremity Muscles

Muscle	L2	L3	L4	L5	S1	S2
Quadriceps	+	++	++			
Tibialis anterior			++	+		
Extensor hallucis longus			+	++	+	
Extensor digitorum longus			+	++	+	
Gluteus medius			+	++	+	
Extensor digitorum brevis			+	++	+	
Peroneus longus			+	+(+)	+	
Medial hamstrings			+	++	+	
Lateral hamstrings				+	++	+
Lateral head—gastrocnemius				+	++	+
Medial head—gastrocnemius				+	+	++
First dorsal interosseous					+	++

++ = major supply; + = minor supply

Table 6-2. Root Supply of Selected Upper Extremity Muscles

Muscle	C5	C6	C7	C8	T1
Deltoid	++	+			
Supraspinatus	++	+			
Infraspinatus	++	+			
Biceps	+	++			
Brachioradialis	+	++			
Triceps		+	++	+	
Extensor digitorum communis		+	++	+	
Opponens pollicis			+	++	+
Abductor digiti quinti				++	++
First dorsal interosseous				++	++

++ = major supply; + = minor supply

Nevertheless, since these fibers are no longer in contact with the spinal cord, there will be anesthesia of the part despite normal peripheral sensory conduction studies.

Stretch injuries of the plexus frequently masquerade clinically as isolated peripheral nerve lesions. However, electromyographic sampling of the entire extremity may disclose abnormal potentials in muscles not innervated by the peripheral nerve thought to be injured, thus indicating that only a more proximal (i.e., plexus) lesion could account for denervation of muscles innervated by more than one peripheral nerve.

Prognosis for recovery after brachial plexus injury can be deduced from serial electromyographic sampling, since electrical changes usually precede clinical evidence of healing. This is especially important for reassuring the patient (or parents, in the case of birth palsies) and helps the surgeon decide when to intervene.

PERIPHERAL NERVE

The detection of abnormalities of the peripheral nerve is one of the more important applications of clinical electromyography. This study should always be combined with nerve conduction measurements. In addition to damage caused by trauma or compression, the various neuropathies, infectious polyneuritis (Guillain-Barré syndrome), and progressive peroneal muscular atrophy (Charcot-Marie-Tooth disease) are all to be considered. In the nontraumatic conditions, denervation potentials may be scarce or absent, and slowed conduction times alone suggest the diagnosis. This is the case when the lesion involves primarily the myelin sheath, as it does in diabetes and alcoholic polyneuropathy. When the lesion involves primarily the axon, as it does with various drug-induced neuropathies, extensive changes seen by electromyography including a decreased recruitment pattern, increased polyphasic potentials, and denervation potentials are seen. A mixed picture occurs in many neuropathies, including advanced states of the diseases mentioned above, and a mixed electrical picture is likewise seen.

In addition to determining the extent of damage and helping to localize the site, serial electromyographic studies are useful in evaluating the process of regeneration.

Electrical Classification of Nerve Damage. Electrical classification of the degree of nerve damage is useful but frequently difficult. A simple method is shown in Table 6-3, where abnormalities are graded from 1+ to 4+ depending upon how many needle insertions produce abnormalities and how profuse the abnormalities are. A more complex classification is the electrical analogue of Seddon's classification to which reference was made on p. 94. This classification is most helpful in determining the extent of damage but cannot be quantitated since the electromyographer's judgment becomes an important factor.

Neurapraxia (Physiological Block). Neurapraxia produces an absence of voluntary motor units unaccompanied by denervation potentials—merely electrical silence. Since compression seldom involves all the axons equally, mild cases show scattered motor units but no interference pattern while severe cases

Table 6-3. Electromyographic Classification of Nerve Damage

Degree of Abnormality	Insertional Activity	Denervation Potentials	Voluntary Motor Units
1 +	Increased irritability Prolonged firing of voluntary motor units Scattered denervation potentials	Scattered–not present on all insertions	Present
2 +	Increased irritability Mixed picture of voluntary motor units and denervation potentials	Present in different areas of needle insertion	Decreased–may be poorly formed
3 +	Increased irritability Predominantly denervation potentials	Present on most needle insertions	Rare
4 +	Increased irritability Showers of denervation potentials on all insertions	Profusely present on all needle insertions	Absent

show no voluntary motor units but some scattered denervation potentials. An externally applied electrical stimulus is unable to pass through the point of the block. In all cases, however, since wallerian degeneration does not take place, there should be a response to an external stimulus applied to muscles innervated above the level of the block. When the stimulus is applied below the level of the block, a response should also be obtained.

Axonotmesis. Axonotmesis may be partial or complete; denervation potentials, however, should be present in all cases, and motor units are present only minimally or not at all. In mild cases it is possible to transmit an electrical stimulus, although a stimulus of greater intensity than that used for normal nerves may be required.

Neurotmesis. In neurotmesis no trace of voluntary motor units is seen. Fibrillation potentials and positive sharp waves are profuse and readily found. They are seen in all muscles innervated by the involved nerve.

Regeneration. By means of serial testing, regeneration can usually be demonstrated with great accuracy. The earliest sign is a decrease in fibrillation potentials. Not infrequently positive sharp waves become the dominant denervation potential seen. As the motor units are reinnervated, they initially appear as nascent polyphasic units, gradually being replaced by normal voluntary motor units. Because of aberrent regenerative processes as previously described (p. 95), giant motor units appear.

MYONEURAL JUNCTION

Abnormalities of the myoneural junction include myasthenia gravis, the myasthenic syndrome of Eaton and Lambert occurring in association with carcinomas, and the myasthenic state secondary to hyperthyroidism. In a normal individual, motor unit potentials produced either by volition or by repetitive stimulation can be maintained almost indefinitely at the same amplitude. In myasthenia gravis, the response to repetitive stimulation is a rapid decrease in the amplitude of the action potentials so that their amplitude is soon only a fraction of their former size. This reaction does not happen throughout the body but only in those muscles whose end-plates are clinically involved. When a short-acting anticholinesterase preparation such as Tensilon is injected intravenously, in a dosage of 8 mg after a trial dosage of 2 mg, the amplitude of the potentials increases again, returning toward normal. A similar picture of decreased amplitude on repetitive supramaximal stimulation occurs in the myasthenic state secondary to hyperthyroidism, but there is no response to the injection of Tensilon. Still a different response occurs in the myasthenic syndrome of Eaton and Lambert. In the Eaton-Lambert syndrome, on volition or by repetitive stimulation, there is a brief initial increase in the size of the motor units, followed by a gradual decline. A single supramaximal stimulation will produce a marked reduction in the size of the action potential, a finding which does not occur in myasthenia gravis. A poor or absent response to the Tensilon test is characteristic.

MUSCLE

Primary abnormalities of muscle include the muscular dystrophies, myotonias, inflammatory disorders of muscles such as polymyositis and dermatomyositis, and secondary myopathies occurring in association with steroid therapy, endocrine disorders, and cancer.

The muscular dystrophies are frequently difficult to diagnose electromyographically in the early stages of the disorder. As muscle fibers are replaced by fibrous tissue, the motor units become smaller and both their amplitude and their duration decrease. Considerable variation in the size of the motor units can be observed, particularly in the early stages. As the disease progresses, it is necessary to recruit many more units to produce the same effect as that produced by just a few normal motor units. Therefore, a signal to the patient to produce a small contraction against mild resistance fills the screen with dystrophic potentials. In the latter stages of the disease process, fibrous tissue becomes interposed between the muscle and its nerve, and denervation potentials in the form of fibrillations appear.

The myotonias, myotonic dystrophy, myotonia congenita (Thomsen's disease), and paramyotonia can be easily diagnosed electromyographically. One cannot differentiate electromyographically between these conditions. Each needle insertion, or tapping of the muscle after insertion of the needle, is characterized by the prolonged trains of "dive-bomber potentials," or myotonic bursts. The prolongation of the bursts differentiates them from the brief high-frequency discharges that occur in conditions of rapid peripheral nerve degeneration or muscle necrosis.

Myopathic potentials are of short duration (1 to 2 milliseconds) and low amplitude. It is sometimes difficult to establish the presence of a myopathy with equipment currently available for clinical electromyography because the range of amplitude and duration in normal potentials is wide and many of the potentials seen may be of borderline magnitude. A useful role of electromyography, however, is mapping out likely sites for biopsies. The individual performing the biopsy should be cautioned to take a section of muscle untouched by needling since the needled area is changed histologically as a result of local tissue reactions to the needle trauma.

Polymyositis and dermatomyositis exhibit a constellation of abnormal electrical potentials. Increased insertional activity in the form of high-frequency discharges, fibrillations, and sharp waves or increased trains of normal motor units may be present. Many of the motor units may appear as myopathic potentials and an increased number of polyphasic potentials may be found.

NERVE CONDUCTION STUDIES

Nerve Conduction Velocity Measurements

Evaluation of the conducting efficiency of the peripheral nerve has added a new dimension to electrodiagnostic testing. Four different determinations can be made with this test: motor nerve conduction velocity, motor latency, sensory nerve conduction velocity, and sensory latency.

Conduction Velocity

Conduction velocity is the speed at which motor and sensory impulses traverse a given segment of nerve, reported in meters per second (refer to p. 106 for procedural details). The speed is proportional to the diameter of the axon, larger axons conducting faster than smaller ones. Proximal segments conduct faster than distal segments. Newborn infants' nerves have slower velocities, reaching normal adult values in two to five years. Any decrease in the local tissue temperature slows the conduction velocity, and there is a slowing of conduction in the elderly.

Latency

An externally applied electrical stimulus which initiates the electrical activity of the nerve is called a stimulus or shock artifact. Latency is the time in milliseconds from the application of the stimulus artifact until the action potential appears on the oscilloscope. Included in the motor latency are several unmeasurable events such as the utilization time (time required to produce rheobasic stimulation) and the end-plate potential. These events are eliminated when a two-point stimulation study is performed. Intensity of the stimulus should be increased until the maximal potential is obtained. It is important to observe the contours of the action potential in addition to the time lapse. With damage to some of the muscle fibers, a temporal dispersion of the action potential takes place. Since the latency includes other potentials such as end-plate potentials, it does not represent purely the conduction of the nerve fiber. However, in terminal portions of the nerve, latency is the only evaluation that can be performed. In some nerves or segments of nerves where only one site can be stimulated for reasons of anatomical inaccessibility, latency measurement must replace normal conduction velocity.

Motor Nerve Conduction

Any stimulus, if strong enough, can be propagated nonspecifically through the soft tissues of an extremity, but effective specific motor nerve stimulation can be applied only to nerves that lie close to the surface. The study is performed as follows: A pickup electrode (surface or needle) is placed over the muscle serving as the end point, and an electrical shock is applied to the nerve through the overlying skin by means of a stimulating electrode. The

Table 6-4. Normal Conduction Values

Nerve	Motor Latency	Sensory Latency	Motor Conduction Latency	Sensory Conduction Velocity
Median	<4.5 milliseconds (at wrist)	<4.0 milliseconds (at wrist)	45–70 meters/second (elbow to wrist)	50–70 meters/second
Ulnar	<4.0 milliseconds (at wrist)	<3.5 milliseconds (at wrist)	45–70 meters/second (elbow to wrist)	50–70 meters/second
Radial	–	<3.5 milliseconds (at lower third of radius)	50–70 meters/second	–
Facial	<4.0 milliseconds (angle of jaw to orbicularis oris)	–	–	–
Femoral	<8.0 milliseconds (inguinal ligament to vastus medialis)	–	–	–
Common peroneal	<7.0 milliseconds (at ankle)	–	40–65 meters/second (knee to ankle)	–
Tibial	<7.0 milliseconds (at ankle)	–	40–65 meters/second (knee to ankle)	–
Sural	–	–	–	>40 meters/second

Note: There is slight variation in values from laboratory to laboratory.

latency (T_1) is recorded, and then a more distal point along the same nerve is similarly stimulated. This latency (T_2) is also recorded. Subtracting the shorter latency from the longer yields the time taken for transmission of the impulse between the two nerve points stimulated. The distance, in centimeters, between the two points of stimulation is recorded, and a simple formula–velocity is the distance (centimeters) divided by time (milliseconds), multiplied by 10 to convert from centimeters to meters

$$\left(V = \frac{D}{T_1 - T_2} \times 10 \right)$$

–establishes the conduction velocity in meters per second. Normal values for various nerves are shown in Table 6-4. The above technique is illustrated by Figure 6-7. Although the upper and lower ranges for conduction are listed, values that fall at the lower end of the scale should be regarded with suspicion and, when possible, comparison should be made with the same nerve in the unaffected limb. There are no abnormalities known to be associated with increased conduction velocity. Motor nerve conduction velocities are more markedly reduced in conditions affecting primarily the axon of the peripheral nerve than in conditions involving the anterior horn cell or root. Diseases affecting the myelin sheath show greater slowing of conduction than those affecting the axoplasm. Striking reduction occurs in infectious polyneuritis and Charcot-Marie-Tooth disease.

Sensory Nerve Conduction

Sensory conduction velocity measurements differ from motor nerve conduction testing in that the action potential of the nerve itself rather than of a muscle serves as the observable

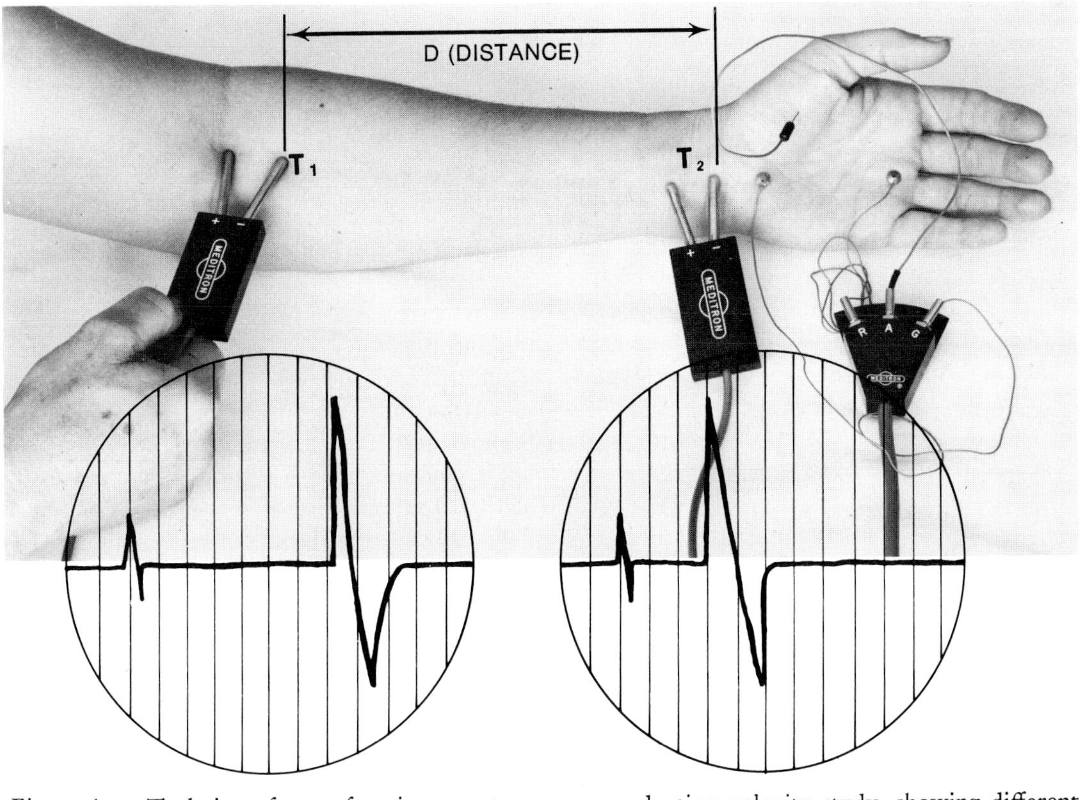

Figure 6-7. Technique for performing a motor nerve conduction velocity study, showing different oscilloscope patterns when stimulator is applied to elbow (T_1) or wrist (T_2).

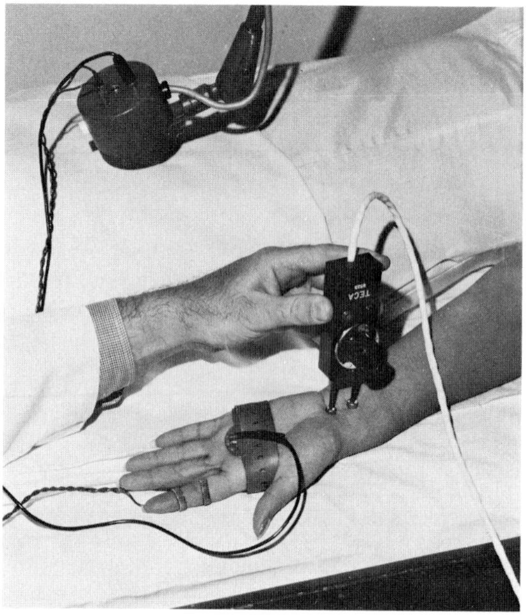

Figure 6-8. Performance of a sensory nerve study by means of the antidromic technique. Equipment is from TECA Corp., White Plains, New York.

end point. Sensory nerve action potentials are of much smaller magnitude, and the sensitivity of the equipment needs to be adjusted accordingly. When stimulated externally at some point along its course, a nerve will conduct an impulse both orthodromically (propagating the impulse along the axon in the normal direction or distal to proximal) and antidromically (propagating the impulse in the direction opposite to normal or proximal to distal). Equipment can be so modified that the tests may be performed either way depending on the preference of the examiner, with no significant difference in results (Figure 6-8). The performance of the examination is the same as for the motor nerve conduction velocity, and the normal values are listed in Table 6-4. Diabetic neuropathies frequently produce a striking decrease in sensory nerve conduction velocity.

Motor and Sensory Latencies

Motor and sensory latencies are most helpful in evaluation of the distal portion of the peripheral nerve. Prolonged terminal latencies occur in the entrapment syndromes such as carpal tunnel or tarsal tunnel syndromes and to a lesser degree in the peripheral neuropathies. The normal terminal latencies as well as latencies of nerve trunks are listed in Table 6-4. Distortion of the action potential with multiple phases and prolonged duration (temporal dispersion) is seen with entrapment syndromes. Absence of the nerve action potential in a sensory latency test, if satisfactory technique is assured, is evidence of abnormality in the sensory fibers of the nerve.

OTHER TESTS

Nerve Excitability Testing

The principle of the nerve excitability examination is similar to that of the nerve conduction time test: A stimulus is applied to a nerve trunk, and a response is observed. However, instead of measuring the latency before the appearance of an action potential on the oscilloscope, one must record the minimal amount of current in milliamperes necessary to produce a visible muscle contraction (threshold contraction) and compare it with the finding on the opposite uninvolved side. This test provides an evaluation of the status of the nerve trunk rather than of the intramuscular nerves. Since degeneration causes the nerve trunk to lose its excitability within 72 hours whereas the intramuscular nerves maintain their excitability for seven to ten days, an earlier evaluation of the status of the nerve is possible. Stimulation should be performed distal to the site of the lesion with a stimulator capable of delivering an impulse with a duration of 0.001 (1/1000) second. The examination should be performed in a room with good light so that minute responses can be observed. For clinical purposes a response within 2 milliamperes of intensity of current of the opposite uninvolved side may be considered normal. An increased intensity of current, greater than 2 milliamperes of difference between the two sides, is necessary to produce a response indicative of partial nerve degeneration, while no response at all indicates total nerve degeneration.

HYSTERIA AND MALINGERING

In peripheral nerves, hysteria may be differentiated from a neurapraxia by direct electrical stimulation of the nerve above the supposed site of the lesion. If no response is obtained, or if a response is obtained only with increased intensity of current, a neurapraxia exists. This technique is useful clinically in the early diagnosis of hysteria or malingering, the early diagnosis of nerve injury, and the estimation of the severity of nerve damage.

BELL'S PALSY

Nerve excitability testing has been found to be of particular worth in evaluating the status of the facial nerve in Bell's palsy. As compared with other diagnostic tests (except for the nerve latency test, which also measures the integrity of the nerve trunk), it makes earlier evaluation possible and permits both adjustment of the treatment program and reassurance to the patient. It is easy to perform and produces minimal patient discomfort. Since a small number of patients whose test indicates merely a neurapraxia subsequently develop nerve degeneration, the test should be repeated biweekly for the first two to three weeks. Where any doubt exists, it is best to use this test in conjunction with other methods of electrodiagnosis.

Classic Electrodiagnostic Testing

REACTION OF DEGENERATION

Classic electrodiagnostic testing, devised by Erb, is the traditional way to test for nerve degeneration. Direct and alternating currents are used over the motor point of a muscle. Stimulation of a normal muscle at its motor point, with an alternating current, will produce a brisk contraction. When a reaction of degeneration has taken place and the muscle is stimulated 10 to 14 days following injury to the nerve, such stimulation elicits no contraction since the duration of flow is not long enough to stimulate muscle fibers directly. When interrupted galvanic current is applied to a normal muscle, a brisk contraction takes place at the "make" and "break" of the current. However, when denervation has occurred, a greater amount of current is required, the response is limited to a sluggish or vermiform contraction, and the motor point is difficult to find or is absent. Normally, stimulation with direct current produces the best contraction when the cathode or negative electrode is closed (i.e., on the make of the current). But in denervated muscle the anodal closing current usually produces a better response than the cathodal closing current. It should be noted that this change is not always present in denervation. The opening currents are affected in the same way. Nerve damage, therefore, produces a change in the stimulus and a change in the muscle response. The reaction of degeneration is expressed as:

- Absence of response to faradic current
- Increased intensity of galvanic current required to produce contraction of muscle at the make and break of the current
- Sluggish, vermiform contraction in response to galvanic current
- Inability to localize the motor point

A partial denervation will demonstrate these changes to a lesser degree.

Classic electrodiagnostic testing is advantageous because it is simple to perform. It provides a qualitative rather than a quantitative result and is conveniently used as a screening test. It has no prognostic value since return of voluntary function usually precedes by several days the response of denervated muscle to faradic current.

CHRONAXY AND STRENGTH DURATION CURVE

In order to derive the chronaxy and the strength duration curve, the rheobase must first be determined.

The *rheobase* is the minimum intensity of current of prolonged duration (clinically taken to be 100 milliseconds) necessary to pro-

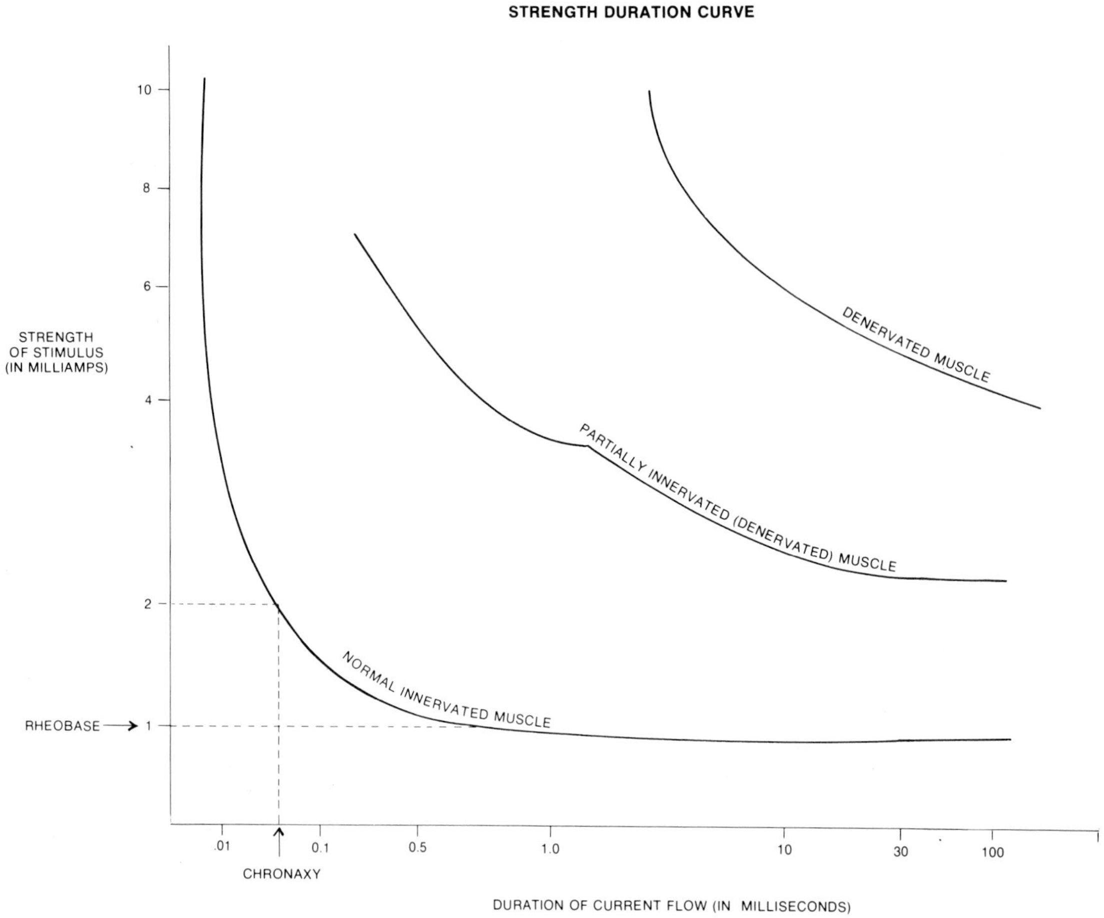

Figure 6-9. Strength duration curves showing normal muscle, denervated muscle, and partially innervated (or partially denervated) muscle. From the strength duration curve the chronaxy can be plotted.

duce a threshold contraction of a muscle. It is expressed in milliamperes or volts. If this point is plotted on a graph on which the vertical axis is the intensity (strength) and the horizontal axis is the duration (time), a strength duration curve is obtained and from this the chronaxy can be determined (Figure 6-9).

The *chronaxy* is defined as the minimum time, expressed in milliseconds, required for a stimulus of twice the rheobase strength to produce a threshold contraction. In the presence of complete denervation, chronaxy values may be elevated 50 to 200 times those of normally innervated muscle. However, in the presence of partial denervation, the value observed may be normal, slightly elevated, or grossly abnormal, depending on the accessibility of fibers to the stimulating electrode.

For a normally innervated muscle the strength duration curve is low and flat, representing the relative ease of stimulating innervated muscle. A totally denervated muscle shows a shift of the curve to the right, representing the increased intensity of current necessary to stimulate muscle isolated from its nerve supply. A partially denervated muscle or one that is evidencing reinnervation shows a discontinuity in the curve, representing partial nerve response and partial muscle response (Figure 6-9). Serial strength duration curves are of value in that they may show the shift from the right to the left of the curve, which is characteristic of recovery. The shortcoming of these electrical tests lies in their inability to localize the site of nerve injury. Their usefulness lies in the fact that they can be performed without sophisticated apparatus and by a paramedical person if an electromyographer is not available.

BIBLIOGRAPHY

Basmajian, J. V. *Muscles Alive: Their Functions Revealed by Electromyography* (3rd ed.). Baltimore: Williams & Wilkins, 1974.

Bauwens, P. Electromyographic definition of the site of and nature of peripheral new lesions. *Ann. Phys. Med.* 5:149, 1960.

Beardwell, A. The spatial organization of motor units and the origin of different types of potential. *Ann. Phys. Med.* 9:139, 1967.

Brady, L. P., Parker, L. B., and Vaughen, J. An evaluation of the electromyogram in the diagnosis of the lumbar-disc lesion. *J. Bone Joint Surg.* 51A:539, 1969.

Dawson, G. D., and Scott, J. W. The recording of nerve action potentials through skin in man. *J. Neurol. Neurosurg. Psychiatry* 12:259, 1949.

Eaton, L. M., and Lambert, E. H. Electromyography and electric stimulation of nerves in diseases of motor unit; observation on myasthenic syndrome associated with malignant tumors. *J.A.M.A.* 163:1117, 1957.

Gilliatt, R. W. Electrodiagnosis and electromyography in clinical practice. *Br. Med. J.* 2:1073, 1962.

Goodgold, J., and Eberstein, A. *Electrodiagnosis of Neuromuscular Diseases.* Baltimore: Williams & Wilkins, 1972.

Knutsson, B. Comparative value of electromyographic, myelographic and clinical-neurological examination in diagnosis of lumbar root compression syndrome. *Acta Orthop. Scand.* (Supp.) Vol. 49, 1961.

Lambert, E. H. Electromyography and Electrical Stimulation of Peripheral Nerve and Muscle. In Mayo Clinic and Mayo Foundation, *Clinical Examination in Neurology.* Philadelphia: Saunders, 1957. Pp. 287–317.

Licht, S. *Electromyography and Electrodiagnosis* (3rd ed.). New Haven, Conn.: E. Licht, 1961.

Melvin, J. L., Schuchmann, J. A., and Lanese R. R. Diagnostic specificity of motor and sensory nerve conduction variables in the carpal tunnel syndrome. *Arch. Phys. Med. Rehabil.* 54:69, 1973.

Seddon, H. J. Three types of nerve injury. *Brain* 66:237, 1943.

Shea, P. A., Woods, W. W., and Werden, D. H. Electromyography in diagnosis of nerve root compression syndrome. *Arch. Neurol. Psychiatry* 64:93. 1950.

Sunderland, S. *Nerve and Nerve Injuries.* Baltimore: Williams & Wilkins, 1968.

Tenckhoff, H. A., Boen, F. S. T., Jebsen, R. H., and Spiegler, J. H. Polyneuropathy in chronic renal insufficiency. *J.A.M.A.* 192:1121, 1965.

Upton, A. R. M., and McComas, A. J. The double crush in nerve-entrapment syndromes. *Lancet* 2:359, 1973.

Zohn, D. A. Advances in clinical electrodiagnosis. *Med. Ann. D.C.* 34:9, 1965.

Zohn, D. A., and Duke, C. J. Bell's palsy: Management based on prognosis. *G.P.* 36:99, 1967.

II
Treatment

7

Modalities of Physical Treatment

It is not necessary for physicians to know in detail the physics or the biophysical or biochemical properties of the modalities used in physical therapy for them properly to prescribe physical treatment for the relief of their patients' symptoms. However, it is essential for anyone using physical treatment to know the principles which should be followed in its prescription, including its limitations and its contraindications.

The number of available modalities may come as a surprise to many who are accustomed to thinking of physical therapy only in terms of some form of heat and massage followed by some sort of exercise program.

Physical therapy modalities include the following:

- Manual therapy
  - Exercise
  - Massage
  - Manipulation
  - Relaxation
- Mechanical therapy
  - Traction
  - Compression
- Heat
  - Superficial heat
    - Dry—infrared
    - Wet—Hydrocollator
    - Low-temperature fever therapy
  - Deep heat
    - Short wave diathermy
    - Microwave diathermy
    - Ultrasound
  - Induced-fever therapy
- Cold
  - Vapocoolant spray
  - Ice packs
  - Ice massage
  - Hypothermia
- Ultrasound
  - For its properties other than heat
  - Phonophoresis
- Electrotherapy
  - Alternating current—faradism; sinusoidal current; transcutaneous nerve stimulation
  - Direct current—anodal (medical) galvanism; interrupted galvanism; iontophoresis
- Biofeedback training
- Heliotherapy—sunlight
- Actinotherapy—ultraviolet light
- Hydrotherapy
- Supports—braces, strapping, and assistive devices
- Occupational therapy

OBJECTIVES OF PHYSICAL THERAPY

These modalities, when properly prescribed, promote a climate in the host to favor healing by: (1) relief of pain, (2) production of relaxation, (3) alteration of metabolism, (4) prevention of morbidity, and (5) restoration of function which may have been lost during the healing phase of treatment.

Predictability of Results of Physical Therapy

The effect of physical therapy modalities should be predictable. If the predicted effect is not achieved in a predicted time, either the diagnosis of the condition being treated is wrong or the prescription is wrong or both. The continuing use of any modality of physical therapy without measurable improvement in a patient's condition is indefensible. So, too, is treatment by trial and error in the hope that some modality may eventually help the patient. Rather, every effort should be made to establish the correct diagnosis, which is a prerequisite to proper and effective treatment with any physical modality, as it is, indeed, in all other fields of medical therapeutics. The empirical use of machines by untrained people is simply a pretense at giving physical treatment.

If a modality of physical therapy is prescribed for any reason other than for the reasons of principle—namely, (1) to prevent morbidity, (2) to maintain as normal a physiological state as possible in parts only secondarily affected by the primary pathological condition, (3) to effect rest from function in the anatomical structure in which the primary pathological condition is situated and, (4) when healing has occurred, safely to restore any lost function—then physical therapy is being abused.

The proper prescription of each modality will be discussed in the order which may be most familiar to the reader.

HEAT

Whatever the intricacies of the physics of heat, in therapy, for the most part, heat is heat, and the way it is employed is usually predicated on convenience of application and availability of method. Its efficacy may depend on how deeply the heat penetrates the tissue, but its properties are the same.

Therapeutic Properties of Heat

EFFECT ON CIRCULATION

Initially, the application of heat may produce transient vasoconstriction, but this is rapidly succeeded by vasodilation and hyperemia. An improved blood supply both to and from the part being treated results. However, if the application of heat is prolonged, congestion occurs, and this is a morbid state. So a beneficial effect can readily turn into a morbid effect; indeed, if prolonged beyond the therapeutic range, the use of heat may actually produce tissue change which, when extreme, results in "cooking" or frank burning.

EFFECT ON METABOLISM

The metabolism of heated parts is increased. If the removal of the products of increased metabolism is impaired because of poor circulation, morbidity results unless some other modality is used to disperse the accumulation of catabolites.

EFFECT ON THE SKIN

Heat promotes sweating, which may itself help to get rid of waste products resulting from disease or increased metabolism. Misused therapeutic heat produces burns.

EFFECT ON NERVES

Heat increases the threshold of sensory nerve endings and hence is pain-relieving unless the primary condition is neuritis, in which case heat may aggravate pain. However, if the sensory nerve supply to the heated part is deficient, burns more easily occur.

EFFECT ON PSYCHE

Warmth is relaxing to a patient, and psychic relaxation promotes physical relaxation.

REFLEX HEATING

With arterial deficiency in the legs it is possible to produce reflex dilation by using short

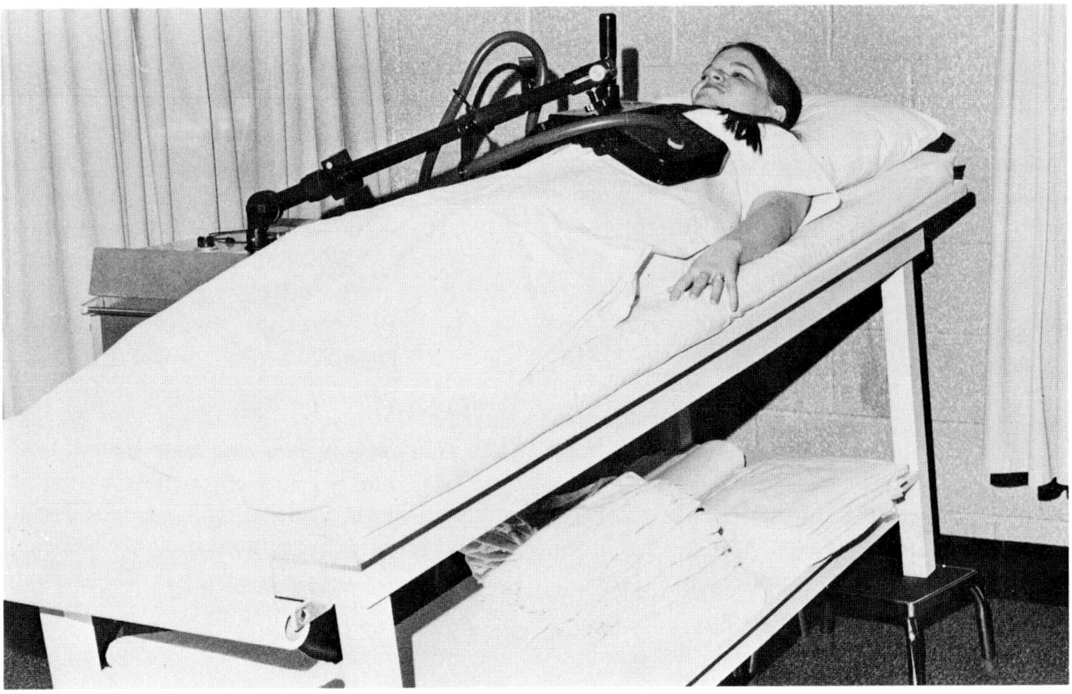

Figure 7-1. One method of using short wave diathermy in heating the lumbar sympathetics for reflex arterial dilation in the legs. Note elevation of head of table.

wave diathermy to produce deep heating of the lumbar sympathetic nerves. A method of effecting this reflex heating is illustrated in Figure 7-1. Heat directly applied to the legs in such cases could precipitate gangrene. Reflex heating avoids this danger.

Contraindications to Heat

The contraindications to the use of heat are:

- Deficient vascularity due to organic disease of the blood vessels
- Deficient sensation
- Malignant neoplasms in the area being treated
- Bleeding
- Active tuberculosis
- Pregnancy (over the uterus)
- Overuse

The use of deep heat by short wave diathermy is contraindicated in tissue in which metal is implanted (because metal concentrates heat, and burning at the interface would occur) and in the presence of clinical osteoporosis since it promotes further osteoporosis.

Special Uses of Heat

Since the advent of antibiotics physicians have tended to forget the special uses of heat in certain conditions in which it may still be beneficial. In osteomyelitis, when drainage has been achieved, diathermy promotes healing in a way that cannot be due solely to its local heating property. In cases of gonococcal arthritis, prostatic or pelvic diathermy indirectly promotes cure of the disease and relief of the arthritis. In cases of sinusitis, diathermy promotes healing. Reflex heating is of value in peripheral vascular disease or any vascular deficiency syndrome. Diathermy is effective in treating nonspecific muscular chest wall pain, as it is in the treatment of unresolved pneumonia and lung abscesses.

Hyperpyrexia

Hyperpyrexia using a moist air bath or low-degree fever therapy (patient's temperature to 102° F) is useful in treating diffuse fibrositis. It is interesting to note that during the symptomatic phase the patient's sweat, when tested with litmus paper, is found to be highly acid, and that as improvement occurs the sweat becomes neutral in reaction. Gouty patients respond to low-temperature fever therapy. At high degrees (up to 106° F) fever therapy may still have a place in the treatment of the fourth stage of syphilis and in recalcitrant gonorrheal arthritis. It has also been used with benefit in malignant cases of diseases such as rheumatoid arthritis and multiple sclerosis. In these cases fever may be induced by body heating in a moist air cabinet or by protein shock or induced malaria.

ULTRASOUND

Because all energy is eventually converted by the body into heat, one of the effects of ultrasound in therapy is deep heating. The advantage of ultrasound over diathermy is that the heat from ultrasound is not concentrated by metal implants as it is with diathermy. Otherwise, its only heating property advantage is deeper penetration. Thus, a deep-seated joint like the hip can be adequately heated.

Ultrasound Properties

PHYSICAL PROPERTIES

Ultrasound increases the permeability of semipermeable membranes and is useful in dispersing fluid accumulations within inaccessible synovial joints, especially in the spine. For the same reason it is useful in the treatment of inflamed and swollen nerve roots.

MECHANICAL PROPERTIES

Ultrasound by its micromassage effect softens scar tissue. Thus it is useful in fibrositis, with painful scars, and in such conditions as early Dupuytren's contracture, Peyronie's disease, and plantar warts.

CHEMICAL PROPERTIES

The use of ultrasound for the local introduction of drugs through the skin is a relatively new application of this modality. Hydrocortisone can be introduced into deep tissue, including nerves, by means of 10 percent hydrocortisone cream acting as a coupling agent. Salicylates and local anesthetic agents can also be introduced locally through the skin for pain relief. This therapeutic use of ultrasound is called *phonophoresis.*

Diagnostic Use of Ultrasound

The echogram makes use of ultrasound to detect brain tumors and other masses such as a subdural hematoma that shift the midline. Recently, sound is being used to assess healing of fractures and for diagnosing cardiac and renal lesions.

Dangers of Ultrasound

Overdosage of ultrasound destroys tissue but should be harmless with continuous movement of the sound head and in its therapeutic range of 0.5 to 1.5 watts per square centimeter of sound head. When ultrasound is applied under water, the dosage may be increased to 2.5 watts per square centimeter.

Pulsed Ultrasound

Some patients complain of a deep aching pain while others complain of a sensation of uncomfortable warmth, even when the proper treatment technique is used. Both of these side-effects are probably due to an accumulation of heat in the tissues. Application of the ultrasound on a pulsed rather than a continuous basis prevents accumulation of heat and at the same time permits usage for the physical, chemical, and mechanical effects mentioned above. Pulsed ultrasound is not applicable when the primary purpose of the ultrasound is to produce a deep heating effect.

Ultrasound in Conjunction with Muscle Stimulation

Ultrasound may be combined with muscle stimulation. The mechanical pumping of muscle by electrical stimulation promotes the dissipation of the products of increased metabolism or edema fluid away from the muscle being treated, thus restoring its physiological state to as near normal as possible. However, the muscle stimulation must not be metronomic or it promotes fatigue and causes increased morbidity. The relaxation phase of muscle stimulation must be long enough to satisfy the relaxation and refractory periods of muscle. By its mechanical action ultrasound probably helps to prevent or reverse symptoms arising from adherence in interstitial tissue and softens microscopic fibrous tissue adhesions and scars.

EXERCISE THERAPY

Exercise therapy as it is commonly prescribed shares with massage the status of being the most abused modality of physical treatment. Active range of motion exercise is absolutely contraindicated in any part of the musculoskeletal system in which there is an active pathological process. Rest from function (voluntary movement using the muscles) together with adequate passive movement of any anatomical structure involved in a pathological condition is a prerequisite to healing with minimal morbidity. In prescribing either passive or active exercise, one must develop a sense of dosage of each. Too little movement can be just as harmful as too much.

If it is correct to prescribe rest from function to promote healing in an extremity joint or muscle, it is illogical to prescribe an active functional exercise program for the treatment of a similar pathological condition in the back, which is just another part of the musculoskeletal system. *Movement* is one thing; it may be passive, assistive, or active. *Function* is another and denotes the addition of normal use to movement.

Rationale for and against Exercise Therapy

When a patient has a musculoskeletal problem, his basic symptom is loss of function usually accompanied by pain. Loss of function almost invariably means that some movement is lost. If a lost movement could be restored by exercise (i.e., by muscle action), the patient would have no need to consult his doctor, so the prescription of exercise to treat pain arising from a joint whose movement is impaired is unsound. The prescription of exercise to treat aching which may be due to muscle weakness or chronic ligamentous strain from poor body mechanics is sound. In this case a properly designed exercise program may be a very appropriate part of the overall management of a musculoskeletal problem. In the same vein, the prescription of pain medication and muscle relaxants for musculoskeletal disability can only be a part of a well-planned treatment program. Muscle spasm is seldom a primary condition but is a secondary reflex reaction of muscle to splint and prevent painful movement.

Principles of Exercise Therapy

Therapeutic exercise, however, certainly has a place, and an important one, in the treatment of musculoskeletal pathology but only when its prescription follows the basic principles of physical therapy. If, on examining a patient by the use of passive movement, the physician is unable to move the joint which the patient says he cannot move, the examiner can be sure that the prescription of any exercise will only aggravate the problem and create morbidity. If, on examination, there is impaired movement of a joint, then muscles, however normal, cannot move the joint. Equally, a normal joint cannot be moved normally by abnormal muscle.

Movement Therapy vs. Exercise Therapy

In considering exercise therapy, one must recognize that there are two parts to this modality. The first is movement therapy, which may well play an important role in the treatment of an acute pathological condition; the second is true exercise therapy, which is usually restricted to the restorative phase of treatment after healing of the primary pathological condition has occurred.

Types of Movement Therapy

Each phase of therapy is subdivided into passive and active, and the active phase is subdivided into assisted, resisted, and free. Progressive resistive exercise should not be used in treating musculoskeletal pathology which produces pain. This form of exercise is designed primarily to hypertrophy normal muscle or muscle atrophied solely from disuse. It may be used, however, in two situations: (1) to overdevelop a normal muscle that may have to act as a substitute for a muscle that cannot recover from its pathological state and (2) to develop muscles that are going to have to substitute for deficient ligaments following injury.

Prescription of Movement Therapy

Each phase of exercise therapy must initially be directed at either the joint or the muscle. Only in the restorative phase of treatment should exercise be directed at both to reestablish function.

Movement in the Presence of Joint Pathology

When there is primary pathology within a joint structure, movement of the joint should be maintained passively within the limit of pain. One degree of movement is better than none, 5 degrees is better than 1, 10 degrees is better than 5, and so on. However, too much movement is no better and may be worse than too little movement. Each produces morbidity. That is why the therapist must develop a sense of dosage of movement.

While the joint pathology is healing, the second principle of the proper use of physical therapy must be remembered, namely, the maintenance of as normal a physiological state as possible in the structures uninvolved by the primary pathology—in this case the muscles. The normal pumping action of muscle has to be maintained to minimize atrophy, to keep circulation of blood and lymph as normal as possible, and to preserve the normal neuromuscular mechanisms. To achieve all this, isometric exercises should be prescribed. If a patient cannot be taught to perform them properly, faradic muscle stimulation can be used instead.

Isometric Muscle Exercises

By definition, isometric muscle exercise moves and exercises muscle without moving the joint on which the muscle acts. The movements have to be taught to a patient properly; in other words, they have to be performed in a manner that takes into consideration the normal physiological cycle of muscle action: relaxation period, refractory period, contraction period. If a second contraction of muscle is attempted before the relaxation and refractory periods are completed, fatigue results, and this creates morbidity instead of maintaining the normal physiological state. The performance of any exercise with metronomic rhythm defeats its purpose. The proper rhythm of an isometric (or, for that matter, isotonic) exercise is contract-hold-relax-rest. For a rule of thumb, if the contraction is held for a count of six, the rest period should be for not less than a count of three.

The progress of this movement therapy is graduated and judged by the patient's clinical improvement. Unfortunately, too much movement may have to be done once before one knows the correct amount. This "too much" does not in reality do much harm so long as it is recognized and the dose of movement is adjusted accordingly. This is the reason why movement therapy should be given by a well-trained physical therapist who knows the signs of "too much," which are fatigue, resting pain, increased weakness, and maybe spasm.

Movement in the Presence of Muscle Pathology

When there is primary pathology within a muscle, movement of the muscle should be maintained passively within the limit of pain. Stretching of a damaged muscle (particularly in the presence of paresis or paralysis) must be avoided, and, if necessary, the muscle must be supported to prevent stretching. To move a muscle without function is virtually impossible except by the use of graduated surging electrical muscle stimulation. The principles of muscle physiology and the normal cycle of muscle action must be scrupulously observed, and movement must be within the limit of pain. While the muscle pathology is healing, the second principle of the proper use of physical therapy should be reviewed: maintenance of as normal a physiological state as possible in the structures uninvolved by the primary pathology—in this case the uninvolved muscles and the joints. To achieve this objective the uninvolved muscles are treated by electrical stimulation (and/or isometric exercises) of greater intensity than that used in treating the pathological muscle, and maximum joint movement is maintained without pulling on and stretching the damaged muscle. At least the full normal range of joint-play movements must be maintained by the therapist. In this context, joint-play movements that do not produce any range of voluntary movement while being performed can be likened to isometric muscle movements used to maintain normal muscle physiology without muscle function. Here joint-play movements maintain as normal a physiological state within the joint as possible without joint function.

Exercise Therapy in the Restoration Phase

The transition from passive movement and isometric movement to active movement and isotonic movement should occur when the primary pathological condition is healed. But this is not a transition to function. Though early return to function is the major rehabilitation aim of therapy, too early return to function may delay recovery and even be the cause of a relapse.

Recovery Phase

The recovery phase from the pathological state and any unavoidable morbid changes which have accompanied it is better thought of as the phase of reeducation, or the restoration of strength and function. To describe this process in a prescription we use the term *manual muscle reeducation.* Prior restoration of freedom of joint movement is assumed. Part of this has usually involved the restoration of lost joint-play movements by manipulation. By the use of manual techniques a well-trained therapist knows when a patient should progress through graduated assisted movement to graduated manually resisted movements. The progress is determined by clinical signs of improvement in strength and range of movement and absence of pain. However, in this phase of treatment some of the principles of therapy are changed and some discomfort experienced by the patient is permissible but only to the extent that there are no accompanying signs of injury—for instance, joint swelling and increased pain—indicating that too much has been attempted. Such permissible discomfort should not last longer than an hour or so, should not require medication for its relief, and should never keep the patient awake at night. In a joint, the appearance of an increase of synovial fluid indicates excessive therapy. In muscles, decrease in strength or earlier fatigability indicates excessive zeal on the part of the therapist or the patient or both.

Isotonic Exercises

Isotonic exercises are used in the final phases of muscle reeducation and training. In this form of exercise the muscle contracts with progressively heavier resistance and with associated movement of the joint. As opposed to the isometric exercise, muscle shortens with contraction. The exercise program may be designed to develop strength, in which case few repetitions with a heavy load are employed, or to develop endurance, in which case many repetitions with a light load are employed. Development of strength occurs only if sufficient resistance is applied to stress the muscle.

The use of exercise machines is no substitute for the careful supervision of the exercise program by a therapist; also, excessive use of machines by an overenthusiastic patient can result in harm since a paretic or atrophied muscle can be further weakened by overuse. Nevertheless machines do play a useful role in muscle reeducation, offering a good objective guide to progress and providing more resistance than can be provided manually. A further advantage of some exercise machines is that stretching can be done at the same time that the muscle is strengthened. This also permits strengthening in ranges of muscle action that are not normally used in daily activities, which is of importance to the amateur athlete who will only irregularly use the extreme ranges and who is likely to harm himself because of sudden excessive use of joints that have lost full range of motion and strength.

A variation of the usual isotonic exercise is the isokinetic exercise in which, by use of a machine, the speed of contraction is kept constant throughout the full range of motion. This converts acceleration into torque and allows the muscle to exert maximal force throughout its entire range of motion. Although the machine is a good dynamometer for testing purposes, it has not proven to be more effective than standard isotonic exercises in strengthening muscle except in neurological conditions where strength can be developed in a time sequence more fitting normal gait.

Return to Function

The time at which return to function is attempted is largely a measure of clinical experience. However, a useful rule of thumb is that no attempt should be made to resume function until on a manual muscle test the muscles have reached the grade of "good" (see Chapter 3, p. 40). Even then the return to function should be graduated by a little more each day or so, and strapping (or in some cases bracing) of either joint or muscle to protect it in the initial stages of return to function may be a wise precaution against overuse. It is not necessary to achieve a full anatomical range of joint movement before resuming function; a useful range of motion is all that is required. However, the joint-play movements in that functional range must be free if pain on resumed function is to be avoided.

If a patient fails to advance to the grade of "good" on manual muscle testing and reaches a plateau of improvement at a lower grade, supportive devices may have to be used or deliberate muscle substitution may have to be developed. But in everyday problems resulting from simple trauma to the musculoskeletal system, failure to restore normal pain-free function usually means that some correctable facet of normal functioning of the component parts of a joint has been overlooked.

Motivation

A word should be said about motivation. Other medical treatments may be done *to* a patient but treatment of most musculoskeletal ills can only be done *with* a patient. Therefore, the active cooperation of the patient—his motivation—plays an essential role in the treatment. However, it should be recognized that all too often lack of motivation is imputed where lack of skill should be blamed. A therapist may be as lacking in motivation as a patient, or more. When a therapist early reports lack of "patient motivation," this judgment should be suspect.

Exercise Routines

There are some exercise routines that physicians prescribe somewhat empirically by name or by the name of the originator without knowing their rationale or how to perform them. Examples are Williams's (flexion) exercises, Kraus's exercise regimen for low back problems, Buerger's exercises for arterial insufficiency states, Frenkel's exercises for ataxia and incoordination, Bobath's method of patterning, proprioceptive neuromuscular facilitation exercises for spastic conditions, rebound exercises, rhythmic stabilization exercises, Jacobson's relaxation exercises, DeLorme's progressive resistive exercises, Guthrie Smith's exercises using slings and springs, Codman's circumduction exercises (Figure 7-2), and various "breathing exercises" for various respiratory conditions. There are many more. When an exercise program is prescribed by a name, probably no two therapists would interpret the prescription in the same way or teach patients the same thing.

It is not cogent to this work to describe exercise routines in detail. It is better that the

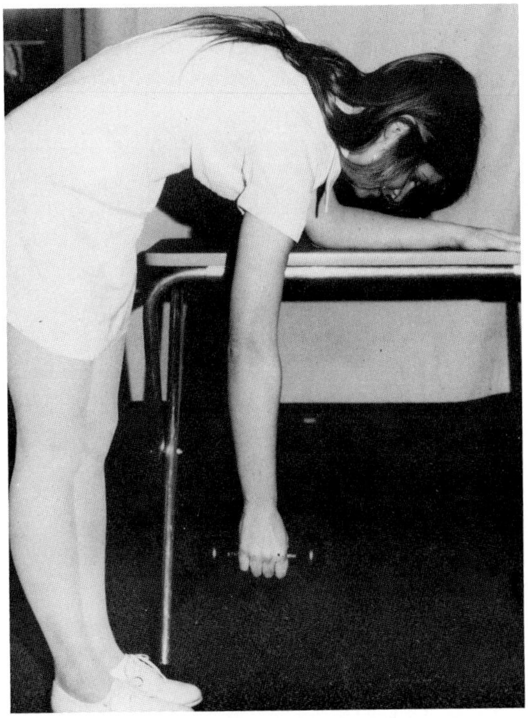

Figure 7-2. The proper way of positioning a patient to perform Codman's circumduction shoulder exercises. Holding a weight helps to stretch the capsule. The weight should not be heavy enough to produce pain in the shoulder.

physician give the therapist the patient's diagnosis, goals, and contraindications of treatment in ordering exercise therapy and that the details of the program be left to the therapist.

WILLIAMS'S EXERCISES

Williams's exercises were designed originally specifically for treating a prolapsed disc problem. They were not designed for treating chronic low back pain that persists month in and month out. Since their function is to strengthen the abdominal muscles and stretch the tight structures in the low back, they are probably more successful in treating facet joint dysfunction problems. If these flexion routines are used with rebound, limited by the therapist, from the maximum flexed position, they are remarkably effective in restoring lost joint play in facet joints. A sacral tilt added by the therapist as the limit of flexion is attained by the patient may be required to "free" the lumbosacral facets.

Most lumbar facet joint dysfunctions occur in flexion as the patient starts to resume the upright position. One of the objects of the Williams "exercises," if not the most important one, is to get the patient beyond the point of flexion at which the "locking of the facets" occurs, trusting that on passive rebound (i.e., when there is almost complete muscle relaxation) the locked facet will move through its impaired range and freely move again. If relaxation of muscle is a prerequisite for success of this treatment regimen, it can scarcely be called an exercise program, as most physicians think it is.

ABDOMINAL STRENGTHENING EXERCISES

Abdominal strengthening exercises used as a part of a program of treatment for "backache" should be undertaken because the abdominal muscles are the anterior supporting muscles of the lumbar spine. They should be done with the knees and hips flexed to avoid substituting use of the hip flexors for the abdominal muscles. Back exercises performed symmetrically—that is, with avoidance of twisting—are deferred until the restorative phase of treatment following the resolution of most back problems.

REBOUND EXERCISES

Rebound exercises are difficult to describe either in words or by pictures. Their performance is an art in which pressure, release of pressure, and timing are each important. The student must learn by watching, doing, and having them done on him. They should be practiced on a normal subject before being incorporated into a therapeutic situation.

Rebound exercises are useful in the restoration phase of treatment when one is restoring flexion to stiff finger joints or to a stiff knee joint and in overcoming recalcitrant joint dysfunction in a facet joint between the fourth and fifth lumbar vertebrae or at the lumbosacral junction. For example, consider facet joint dysfunction at the lumbosacral junction. The patient, supine, pulls up both knees, with the legs spread apart, until he has attained full hip joint flexion. The knees are in relaxed flexion. The patient is then instructed, or even assisted by the therapist, to rock the pelvis by exerting an additional pull on his knees, thereby further flexing his hips, and then quickly to let go of his knees. The patient's legs automatically rebound, and the therapist catches them in his hands, which he is holding ready about 6 inches away.

When trying to attain added flexion to, say, stiff metacarpophalangeal joints, the therapist cradles the patient's stiff fingers with the palm of his hand (his right hand if he is dealing with a patient's left hand; his left hand if he is dealing with a patient's right hand), with the patient's stiff joints at their maximum pain-free angle of (limited) flexion. With his other hand the therapist stabilizes the patient's palm, wrist, and forearm. The patient is instructed to extend his fingers as hard as he can, and the therapist resists this extension movement firmly. The patient is thus doing an isometric exercise of his finger (and wrist) extensor muscles. As soon as the therapist feels that the patient has produced his maximum extension effort, he tells the patient to let go. The patient's fingers immediately rebound into flexion, and a few extra degrees of flexion of the stiff joint(s) are achieved. On no account does the therapist allow the hand he has been using as a resistive force to follow the patient's rebounding fingers to force them into additional flexion.

When performing rebound on a patient with a stiff knee, the therapist has the patient lie with the bad leg on the edge of the treatment table. He allows the patient to let the leg dangle over the edge, supporting the patient's foot at the heel in the palm of his hand which is farthest from the patient. The knee is thus supported at its maximum pain-free range of flexion. The therapist rests his other forearm on the treatment table under the patient's femoral condyles. The patient is now encouraged to do isotonic quadriceps contractions through this predetermined pain-free arc of movement while the therapist holds his hand ready to catch the patient's heel again at the same place each time when he releases his quadriceps contraction. In other words, the patient, having straightened his knee, is encouraged to let go to allow a completely relaxed, gravity-assisted return through the pain-free arc of limited flexion. When the patient performs this exercise with confidence, the therapist, without telling the patient, and while the patient is holding his knee extended, drops his hand one inch downward and closer to the patient's pelvis. Thus on relaxed release of the extended knee the patient's lower leg rebounds through an extra small arc of flexion, achieving greater range of flexion at the knee. This maneuver is done only once at each therapy session. The therapist *must* catch the patient's heel and may have to regain the patient's confidence.

RHYTHMIC STABILIZATION EXERCISES

Rhythmic stabilization exercises are used to achieve added range of movement in a stiff joint and are based on the physiological principle that full muscle relaxation occurs after contraction. The object of the exercise is to obtain simultaneous relaxation of the agonist and antagonist muscles, which move the joint. Then the joint can be moved a few degrees painlessly during that relaxation period.

Rhythmic stabilization exercise is particularly useful in the restoration phase of treatment when one is dealing with a stiff shoulder, a stiff elbow, or a stiff neck. With a stiff shoulder, for instance, with the patient recumbent, the arm is carried to the limit of any restricted movement—say, forward flexion. The therapist grasps the upper end of the patient's upper arm. The patient's elbow is flexed and relaxed. The therapist then in an alternate and rhythmic fashion prevents the patient from pulling the arm down and pushing the arm up. That is, the patient maintains the shoulder in a static position by using alternately the agonist muscles of shoulder forward flexion and the antagonist muscles that tend to produce shoulder extension, as the therapist pushes and pulls against them. After four to six movements by the therapist (resisted by the patient) in each direction the patient is told immediately to relax, and the therapist can passively move the joint through an additional range of pain-free movement.

FRACTIONAL MANIPULATION

Fractional manipulation is used for treating stiff joints that normally have a large range of voluntary movement: the shoulder, the hip, the knee, and the fingers. Remembering that manipulation by definition means restoring lost joint-play movements, the therapist, when manipulating one of these joints, restores the lost joint play at the joint's angle of limited movement. In the joints mentioned there is likely to be only a partial return of functional movement following manipulation (i.e., a fraction of the lost range is restored). This increase of pain-free range is consolidated and may be improved on by rebound techniques (see p. 123) or rhythmic stabilization exercises (see p. 124); by heat, spray, and stretch techniques (see p. 127); by passive and active exercises within the limit of pain; and by using the joint within the limit of pain. When it appears that a plateau of improvement is reached, manipulation is again undertaken, and new gains are consolidated by therapy. Usually, no more than three manipulative sessions are required in treating the joints (if anesthesia is used) for the patient to achieve full, pain-free, functional movement. A therapist may use fractional manipulation once every two or three days depending on the patient's reaction to each treatment session. The patient may experience three or four hours of discomfort after the fractional manipulation, but the local joint discomfort should not interfere with his adjunctive therapy program. If it does, then either there has been an error in technique or too much force has been used by the manipulator.

WEIGHTLESS EXERCISES

Where weightlessness is required to carry out a movement program of treatment and a pool is not available, the Guthrie Smith sling and spring apparatus is invaluable (Figure 7-3). Its chief use is to teach relaxation and coordination exercises of a weightless character. It is also used for skilled reeducation of weakened muscles, particularly in old polio patients who have had recent decompensation due to a less-

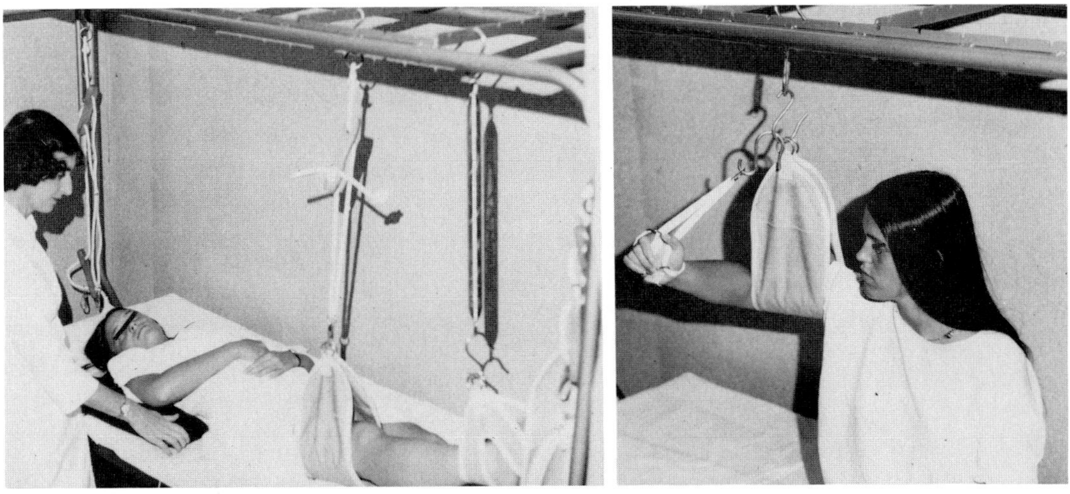

A *B*

Figure 7-3. *A*, Guthrie Smith sling and spring apparatus used for exercising a hip passively and weightlessly. *B*, The same apparatus being used for shoulder exercise. The pelvis, back, and shoulder can be exercised similarly.

ened level of activity associated with retirement or debilitating illness. It is used as an assist for joint mobilizations, in the aftercare of fractures or arthroplasties, and to exercise patients in heavy plaster casts following back surgery or other orthopedic procedures. It is helpful in rheumatoid and other painful arthritic conditions and in conditioning of geriatric patients. Exercise under water has a similar weightlessness effect and is discussed on p. 149.

JACOBSON'S RELAXATION EXERCISES

The regimen is based on the appreciation of the normal physiological cycle of muscle function. For a muscle to be able to contract normally, it is necessary for the contraction to be preceded by complete relaxation. Whenever muscle is in spasm, pathophysiological changes take place in it unless relaxation is attained. Psychological tension or serious pathological bone and joint conditions may cause spasm, which perpetuates pain.

In Jacobson's regimen relaxation, a prerequisite to the eventual relief of pain, is produced by the gentle initiation of a minimal contraction in every muscle in the body, which must be followed by a maximum relaxation in each muscle. At the end of the "exercise" not only is the whole body relaxed but so is the mind. Sleep often follows the proper application of this treatment. The regimen is particularly valuable in the treatment of patients suffering from rheumatoid arthritis and tension states associated with other chronic musculoskeletal pain syndromes. Moreover, it preconditions a patient who requires manipulative therapy and is useful in conjunction with vapocoolant therapy in the treatment of myofascial trigger points.

COLD

To understand the proper use of cold, which should produce predictable results in treatment, it is necessary to recall not only the physical properties of cold but also some neurophysiological facts.

Cold is used for five reasons in medicine:

To stop bleeding and reduce the morbidity associated with it immediately following acute injury
To relieve pain and spasm
To relieve spasticity
To produce hypothermia
To prevent tissue changes in the first aid treatment of second degree burns and to relieve pain from the burn

In physical treatment the main use of cold is for relief of pain. Its use in the treatment of spasticity produces only transient improvement in spastic neurological diseases, and the results do not justify its cumbersome application. However, in treating a recovering hemiplegic, icing spastic muscles is a useful adjunct to the physical restoration program.

The physiological effects of cold are vasoconstriction, local anesthesia, production of a superficial biochemical reaction (a histamine-like response within the skin), and counterirritation. Therapeutically, cold is applied by the use of ice packs, vapocoolant spray and muscle stretch techniques (Figure 7-4) (see Chapter 8, p. 158), ice massage, or ethyl chloride (for its local anesthetic property).

Vapocoolant Spray

The principles underlying the successful use of the vapocoolant spray for the relief of pain depend only on whether muscle spasm is the cause of pain or an irritable trigger point is causing the spasm which is causing the pain.

When painful muscle spasm is being treated, the jet stream is directed initially at the origin of the irritated muscle. Then it is swept down the length of the muscle to the muscle's point of insertion. Three or four sweeps of the spray in parallel lines in one direction only are usually sufficient to make the muscle "let go." The number of sweeps depends on the size of the muscle being treated. It is only necessary to chill the skin representation of the muscle. If the muscle itself becomes chilled the spasm becomes worse. The jet stream should hit the skin at an angle of about 35° from the horizontal. The rate of sweep is about 4 inches per second.

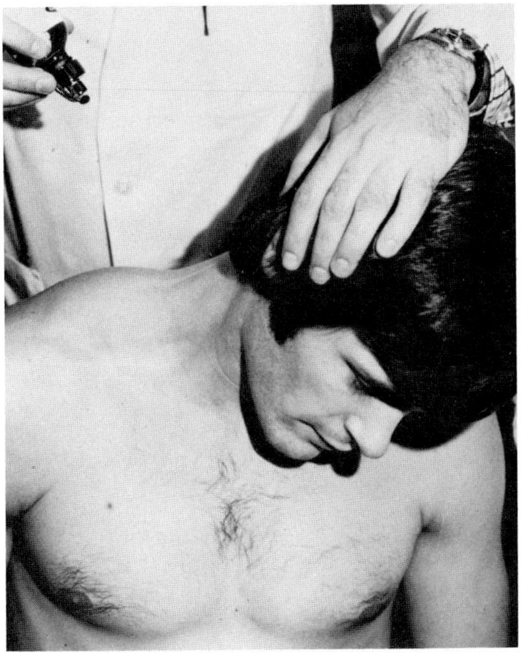

A

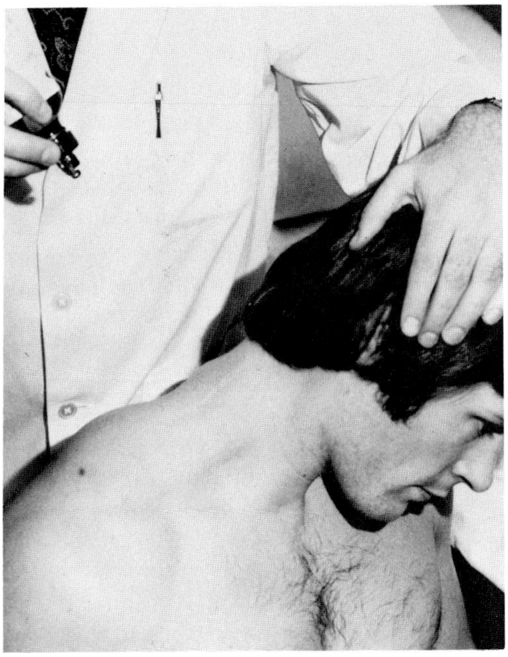

B

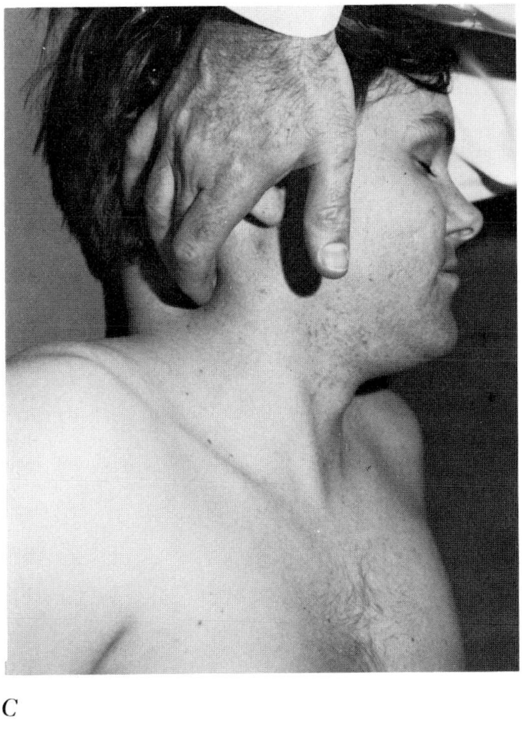

C

Figure 7-4. Use of the vapocoolant spray (Fluori-methane) with stretch for an irritable trigger point in *A*, the right splenius capitis and *B*, the shawl area of the right trapezius. *C*, How to turn the head away from the right sternomastoid muscle with flexion of the neck for spraying and stretching it. As symptoms subside, the head is swung around to the opposite side and the neck is fully extended to obtain a full stretch of the muscles.

At the completion of each sweep of the jet stream the muscle is gently stretched. When there has been sufficient cooling, it should be possible to stretch the muscle to its normal resting length. The normal resting length of muscle is its pain-free state. Overstretching causes spasm; overchilling causes spasm. To achieve success in treatment it is necessary that the vapocoolant spray is correctly applied.

When treating pain arising from an irritable trigger point the principles of application of the spray are the same. In this case, however, the jet stream is directed initially at the irritable trigger point and the sweeps, again in one direction, in parallel lines, aimed at the skin at an angle of about 35° from the horizontal, and progressing at about 4 inches per second, are directed over the skin areas of the referred patterns of pain. The muscle is again stretched gently at the end of each sweep until its normal resting length is achieved. The stretching of the involved muscle to its normal resting length is probably the most important part of treatment. The stretching must be done passively, i.e., by the therapist.

To maintain relief of pain, gentle heat followed by gentle stretching exercises are prescribed. Bad habits of muscle use must be unlearned. These habits may result from bad posture, overuse in occupational situations, fatigue, disuse, or unaccustomed activity. Several sessions of therapy over three to four weeks may be necessary for successful results. Sometimes one treatment session may be enough.

The use of the vapocoolant spray and stretch techniques fails to bring a patient relief from pain if the underlying cause of painful spasm is overlooked and has not been eradicated. It fails if a satellite trigger point is mistakenly treated for the parent one. However, satellite trigger points must be treated as well as the parent one. It also fails if the spasm is residual to a fracture which has healed with loss of bone length. This is because the muscle can only be stretched to an adapted normal resting length. It fails too following, for instance, a surgical fusion procedure because the action of the muscle is changed. It fails if the spasm is a manifestation of radicular or neuritic pain because stretching the muscle also stretches the involved nerve, further irritating it.

We stress that just because pain relief is achieved by use of vapocoolant treatment does not necessarily mean that the cause of the pain has been discovered.

Ice Massage

Ice massage may be used in lieu of the vapocoolant spray. With this technique an ice cube is held in a face cloth that is also used to mop up the water from the melted ice to keep the skin dry. If muscle spasm is being treated the affected muscle is stroked from origin to insertion in parallel lines; if an irritable trigger point is being treated the stroking extends outward from the trigger point over the referred pattern of pain. Passive stretching of treated muscle to its normal resting length is an essential part of the treatment; failure to include this will in all likelihood fail to produce relief of pain.

The patient's response is usually in four phases: cold, burning, aching, and numbness. As soon as the skin is cold it becomes intensely red, a histamine-like reaction contrary to the expected skin reaction of blanching. During the next phases the patient is conscious of a tender mass over the site of maximum pain that is not palpable to the examining fingers. When this sensation disappears, the pain the patient complained of is relieved and the treated part is cold and numb. Appropriate stretching can then begin.

INJECTION THERAPY

We recognize that injection therapy is not a true modality of physical therapy, but it has so many advantages in physical treatment that this section would be incomplete without discussion of it.

Injection therapy means infiltration with some local anesthetic, with or without corticosteroid solutions added, for the relief of pain. It acts by "breaking the pain cycle" of contused or apparently locally inflamed areas in muscle (fibromyositis), tendinitis and traumatic tenosynovitis, bursitis, and traumatized ligaments or capsules. Also considered are epidural injections, intraarticular injections, and nerve blocks.

Injection of Trigger Points

For the most part, we believe that when the injection of soft tissue pain points is effective it is because a trigger point has been injected. When this treatment fails, on the other hand, it is mainly because a site of referred pain has been injected or because bleeding sets up further irritation of the trigger point. For injection of trigger points (Figure 7-5) we use a solution of one percent procaine or lidocaine without epinephrine, in physiological saline solution. In patients who have sensitivity to procaine, physiological saline solution alone may be used because it contains a preservative, benzyl alcohol, which itself has a mild local anesthetic effect. The effect of the injection may or may not be due to the local anesthetic property; perhaps it is due to the substitution of the pain of the needle and the injection, which produces a substitutive sensation sufficiently strong to block the noxious sensation

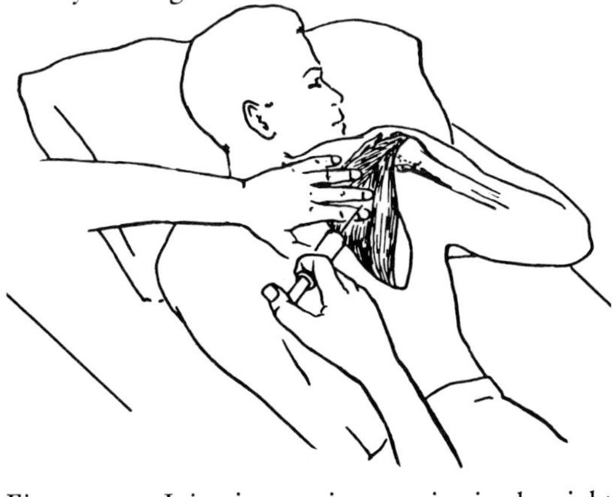

Figure 7-5. Injecting a trigger point in the right infraspinatus muscle. Note that the index and middle fingers of the physician's left hand stretch the skin and produce pressure on either side of the needle to prevent bleeding. (From J. Travell, The myofascial genesis of pain, *Postgraduate Medicine* 11:432, 1952. Copyright McGraw-Hill, Inc.

from the trigger point that has set up a noxious conditioned reflex state. The injection allows the restitution of the normal reflex pattern so that the affected muscle can resume its normal resting length. This is prerequisite to its pain-free state. The effect achieved by injection therapy of trigger points (and by the vapocoolant spray) has an interesting similarity to the effect claimed by those who use acupuncture for pain relief. It is becoming apparent that both of these techniques, and perhaps others as well, obtain their result by means of a reflex inhibition of the pain impulse, either at the level of the substantia gelatinosa of the spinal cord or higher, at the level of the thalamus. To use the terminology of Melzack and Wall, "the gate is closed." However, there is a curare-like property of procaine in injection therapy which cannot be ignored; its effect has yet to be studied.

In other words, the success of substitution therapy—whether achieved by vapocoolant spray, ice, or injection—is the restoration of the normal resting length of muscle. Thus, the most important factor in this form of therapy is manually to stretch the muscle being treated. Failure to do so means failure of the treatment.

Use of any steroid preparation in this type of injection therapy is not always necessary but does seem to prolong the effect of the injection in some patients. Hemostasis is, however, essential to successful therapy because blood is irritating to the tissues and can be a source of a new trigger point or can reactivate the trigger point being treated. The other requirements of successful treatment, as in all forms of injection therapy, are good surgical aseptic preparation of the skin and the use of clean apparatus, preferably disposable syringes and needles. When using any form of injection therapy the physician should have resuscitative apparatus and appropriate medications readily available.

Intraarticular Injection Therapy

STEROID INJECTION THERAPY

That there is a use for the intraarticular injection of steroids is undoubted, but we believe this should be limited to the treatment of painful joints in the collagen vascular diseases. However, joint pain in other conditions that is recalcitrant to other, simpler forms of therapy may be improved by it. Steroids should not be used in intraarticular injection therapy as the sole method of treatment but only in conjunction with other modalities of physical treatment following the basic principles of physical therapy previously stated.

USE OF LACTIC ACID AND PROCAINE

In osteoarthritic joints that have pain and limitation of movement, a thickened and possibly adherent capsule is associated with joint dysfunction, sometimes excessive synovial fluid from synovitis, secondary muscle spasm, and either primary or secondary trigger points in the muscles that exhibit protective spasm.

Considering the joint problem alone, a contracted thickened capsule militates against success of all other therapy. Thus the most important part of treatment must be the loosening and stretching of the capsule. This is the basis, then, for intraarticular injection therapy in such cases. Fluid being incompressible, the injection of fluid into the capsule is a logical approach to treatment. Lactocaine–lactic acid 0.2 percent, with procaine 1 percent in solution buffered to pH 5.2–has proved to be a nonreactive solution. To stretch the capsule of a hip or knee about 8 ml of solution is required; for a shoulder about 6 ml is adequate. While the capsule is free of pain because of local anesthetic and stretched because of the fluid, the joint articular surfaces are also cushioned by the fluid; manipulative therapy to restore lost joint play can be undertaken without causing pain.

INTRAARTICULAR INJECTION OF AIR

It is sometimes said that the use of air in an intraarticular injection for relief of pain is a flight into fantasy. There may be no known scientific reason why it is effective, but it is, especially when the air replaces the aspirated blood of hemarthrosis. Injection of air into the joint is contraindicated in the presence of an associated fracture. It has been used successfully in the relief of pain in acute gonococcal arthritis. This use of air is based purely on clinical experience and is mentioned because of the remarkable pain relief.

INTRAARTICULAR INJECTION OF ANTIBIOTICS

The instillation of antibiotics into an infected joint may be a useful adjunct to the administration of systemic antibiotic therapy once the sensitivity of the culpable organism has been established.* If renal and ototoxic antibiotics must be given intraarticularly, the risks of renal failure and deafness must be considered.

LOCAL ANESTHETIC INFILTRATION OF PAINFUL LIGAMENTS

Traumatized ligaments may be infiltrated with local anesthetic agents, when simple sprain is the underlying cause of pain, to allow the early return to function. If there are ligament fiber tears, infiltration of the ligament is just as effective, but further treatment by support of the joint must be undertaken while healing is taking place and before full function is resumed.

In ligament rupture, infiltration with a local anesthetic is also pain-relieving but is no substitute for specific treatment to promote healing, whether immobilization in a cast or surgical repair. However, the injection of a local anesthetic may be essential to allow the taking of stress x-ray pictures of the joint to demonstrate the diagnosis of ligament rupture; this is often a more serious disability than a fracture.

Another use of local anesthetic infiltration is its introduction into the hematoma at a fracture site to allow reduction of the fracture without the use of general anesthesia.

DIAGNOSTIC USE OF LOCAL ANESTHESIA

Injection of a local anesthetic often aids differential diagnosis. For instance, in the shoulder it is sometimes difficult to determine whether pain is arising from the long head of the biceps tendon or its sheath. Blocking the

* In Hollander's *Arthritis,* intraarticular antibiotics are described as having little if any value over parenteral antibiotic treatment.

suprascapular nerve modifies or abolishes pain arising from the tendon sheath and leaves tendon pain unchanged. In the back it is sometimes difficult to differentiate pain arising from a sacroiliac joint and that originating at the lumbosacral junction. Injection of the posterior ligaments of the sacroiliac joint relieves pain arising from it and leaves lumbosacral junction pain unchanged. Further, infiltration of an interspinous ligament at an intervertebral junction anywhere in the back helps to determine the level from which pain is arising. If the ligament at the right level is injected, pain is modified. If the wrong level is injected, the pain remains unchanged. In the foot, muscle and joint pain cannot be differentiated from vascular and causalgic pain by blocking the posterior tibial nerve behind the medial malleolus, but it may be an efficient way of obtaining a sympathetic, as well as a peripheral, nerve block in breaking a pain cycle. The use of local anesthetic block of the stellate ganglion in the neck and the lumbar sympathetic chain in the low back differentiates vascular pain and causalgic pain in the upper and lower extremities, respectively, from other types of musculoskeletal pain.

EPIDURAL INJECTIONS

Epidural injections have proved to be a useful adjunct in the management of patients with sciatica. In general, best results are obtained in patients who have had sciatica without either severe back pain or severe neurological deficit. Even when limited to the above category, however, results tend to be extremely variable, from permanent relief of pain to transient relief of pain to no response at all. It may well be that those who do not respond well have a severe underlying mechanical cause, such as a ruptured disc, whereas those who respond well have edema of a nerve root or synovitis of a facet joint as the primary cause of pain, but this supposition has not been established.

Injections are usually given epidurally at the suspected level of pathology but may be done through the caudal canal (see below), particularly if there has been a prior laminectomy. Lidocaine and cortisone, mixed to a volume of 30 ml, is the drug combination, and anesthesia without motor loss is an indication of a successful injection. One may take advantage of the anesthesia to manipulate the joints if evidence of joint dysfunction is present. Some physicians advocate the injection of the local anesthetic-steroid mixture intrathecally at the suspected level of pathology, especially when treating a patient who has had back surgery.

Before the introduction of the use of steroids in medicine, injection of 2 percent fresh, sterile saline into the epidural space was used with very satisfactory results. We advocate this today. It is a safe, innocuous office procedure.

The injection is made through the sacrococcygeal hiatus. Up to 60 m of 2 percent warmed saline is used. The success of this method is based on the physical rather than the medicinal properties of the injected fluid. Fluid being incompressible, if on injection into a space it meets an obstruction in the space (for example, a prolapsed disc, adhesions, a tumor), the obstruction must be compressed. In this case the symptoms (usually the radiating pain) are reproduced or aggravated. The injection then becomes a diagnostic tool. If small adhesions are present around a nerve root they may well be torn. Then the injection becomes a therapeutic tool. If the cause of the radicular pain is from nerve root edema, the edema fluid is withdrawn into the injected hypertonic saline solution and relief of the pain symptoms is immediately achieved. The injection is a therapeutic test. Conceivably, edema in the contents of the intervertebral foramina is similarly affected, thereby relieving pressure on a nerve root. The injection, again, becomes a therapeutic tool. Whatever the mode of action, a successful injection is by its immediacy very dramatic and satisfying to the patient.

Of the various local injections used by different physicians for the relief of radicular pain (epidural injections of steroid, anesthetic combinations, the radiopaque medium used in discograms) the common physical property of the injected solutions is their hypertonicity. Our explanation of the reason for success of our method described in this section suggests that the use of 2 percent saline is a most reasonable treatment and, being innocuous, should be tried first. Repeated injections at seven- to ten-day intervals have been found necessary, but not more than three times.

PROLOTHERAPY

For some years the use of sclerosing solutions in ligaments for the relief of pain in unstable joints has been much in vogue. That the therapy works in some cases there is little doubt. That it works because of ligament tissue proliferation, which tightens the ligaments, is a matter of conjecture even though proliferation has been demonstrated, at least in animals.

A more likely explanation of the pain-free condition of joints following prolotherapy would seem to stem from the teaching of prolotherapists that the injection must instill the solution as close as possible to the origin of the ligaments of the joint being treated. This happens to be the location of the maximum congregation of sensory nerve endings, and probably the sclerosing solution effectively "kills" them off and thus prevents the appreciation of pain. The solution, containing glucose and phenol, is available commercially. Its injection is often very painful.

BEHAVIORAL CONDITIONING AND BIOFEEDBACK

Modification of behavior has long been known in the laboratory and recently has been applied to clinical situations. Specific behaviors or operants can be conditioned by positive reinforcement (the response is strengthened) or by negative reinforcement (the response is reduced). Behavior not followed by either positive or negative reinforcement tends to decrease in frequency, a process known as *extinction.*

With regard to musculoskeletal pain problems, conditioning programs are based on the assumption that chronic pain responses are learned behaviors. In a clinical setting, positive and negative reinforcement and extinction techniques are carried out by the treatment staff, or by informational feedback from a machine. In the latter case, an electromyograph is used to obtain information about the patient's physical condition by way of electrodes attached to him. It supplies him with visual and auditory clues regarding the contractile status of the muscles. From the clues he learns control of the affected muscles and how to relax them. Negative reinforcement in this setting is in the form of electrical shocks when he fails to produce the appropriate response. Conditions for which the technique has been used include tension headache, neck injuries of a chronic nature, spasmodic torticollis, and chronic low back pain. A nonpainful musculoskeletal condition in which this technique has proved useful is that of muscle alienation (see p. 134).

It should be understood that the techniques of conditioning and biofeedback are in their infancy and that they are not applicable to all patients, since positive motivation, patience, and comprehension are required. Further, even if behavioral patterns are successfully modified, the long-term results of such modifications have yet to be assessed.

ELECTROTHERAPY

Three types of current are used in electrotherapy: direct current–galvanism; sinusoidal current; and alternating current–faradism.

Galvanism

Galvanic treatment has three uses:

Muscle stimulation. In normal and denervated muscle the direct current produces a muscle contraction at the make and break of the current, and this property is used in the treatment of denervated muscle.

Iontophoresis (ionization)—the introduction of drugs locally through the skin or into the skin.

Medical (anodal) galvanism.

INTERRUPTED GALVANISM

In any lower motoneuron involvement in which recovery may be reasonably expected, any treatment that helps prevent morbid changes is obviously beneficial. The morbid changes are muscle atrophy, which is followed by proliferation of fibrous tissue, collapse of the vascular tree, stagnation of the lymph system, and osteoporosis. Joint dysfunction and skin changes also occur.

It has been shown that the use of intermittent galvanism delays the onset of muscle atrophy for about six weeks and retards its progress subsequently. However, it has also been shown that treatment for 2 minutes, 10 times a day, is more effective than a single treatment session for 20 minutes a day. Ideally the patient should be provided with a home stimulator and instructed in its use. Part of the treatment must be to educate the patient to accept the fact that nerve regenerates very slowly—at the approximate rate of about 1 mm a day.

Simple muscle stimulation by the use of intermittent galvanism is not enough. Paralyzed muscle must not be stretched if it is to recover, and prevention of stretching is necessary if eventual recovery is to be optimal. The galvanic twitch is inadequate to maintain the return circulation of blood, to prevent lymphatic stagnation, or to maintain elasticity or normal turgor of the skin. Massage, with the benefit of proper positioning to overcome gravity, is necessary in this situation. Nor does electrical muscle stimulation prevent joint dysfunction and loss of movement. Care of the joints by passive movement, therefore, is also prerequisite for successful treatment.

IONTOPHORESIS

The use of galvanism in iontophoresis is based on the fact that when a current passes through a salt solution there is ionization of the salt, the positive ion being repelled by the anode (the positive electrode) and the negative ion being repelled by the cathode (the negative electrode).

Some clinicians have looked askance at iontophoresis, regarding it as akin to witchcraft. However, recent work in which ions have been tagged with radioactive substances shows that iontophoresis is effective in the introduction of elements, as has been known clinically for decades.

Commonly used in iontophoretic therapy are zinc$^+$, copper$^+$, procaine$^+$ (or other local anesthetic), histamine$^+$, carbachol or Mecholyl$^+$, and hyaluronidase (Wydase), which does not itself ionize but attaches to the positive ion of a salt and is introduced when the anode is the active electrode. Zinc forms a coagulum with protein and acts as a living dressing over, for instance, an indolent ulcer. It also stimulates the production of granulation tissue. Copper is fungicidal. Procaine has its usual local anesthetic action. It is preferable to inject the local anesthetic first and ionize over the injected area to disperse it. Histamine and carbachol are used for their local vasodilation effect and a deep counterirritant effect. Wydase lessens the viscosity of tissue fluid and so aids absorption of blood or excess synovial fluid within a joint or interstitial edema fluid. The anodal current, in addition, has an anesthetic effect which relieves pain (see Medical Galvanism, p. 134).

The halogens, chloride (Cl^-) and iodide (I^-), have the property of softening scar tissue. When they are used, the cathode is the active electrode.

Electrotherapy texts say that the current used in iontophoresis is 1½ to 3½ milliamperes per square inch of active electrode. In practice, the current is turned up until the patient feels a mild prickling sensation. Too high a current burns, and if the borders of the electrodes become dry, burning also occurs around the electrode. This is called the edge effect. As treatment progresses, the resistance of the skin decreases and the current increases.

Herein lies a danger. A constant check of the current must be maintained. Deficient skin sensation precludes the use of this therapy. The dispersive inactive electrode should be larger than the active one.

MEDICAL GALVANISM

Medical galvanism, or anodal galvanism, is the use of the galvanic current, with the anode active, to relieve pain and to lessen tissue swelling. The latter result is presumably achieved by promoting an osmotic effect by ionizing the salts in the tissue fluids. The pain-relieving effect is certainly a clinical fact but its cause is obscure; possibly pain relief is due to an iontophoretic effect—that is, a humoral effect at the nerve endings or a physical effect induced by changed polarization of the nerve endings or both.

TRANSCUTANEOUS NERVE STIMULATION

Based on the work of Melzack and Wall cited elsewhere (see p. 159) there has been a renewed interest in the use of electrical currents for the relief of chronic pain, particularly pain of neurogenic origin. The physiologic basis for the treatment is that stimulation of larger myelinated afferent fibers tends to block, at the level of the spinal cord, the passage of painful impulses carried by smaller diameter afferent fibers.

Transcutaneous nerve treatment is accomplished by stimulating the dorsal root or the peripheral nerve with a pulsed alternating current wave of 0.1 millisecond duration, a frequency of approximately 100 hz, and a variable voltage to the point of tingling or paresthesias but not pain. If only partial or temporary relief is obtained the patient may be supplied with a portable stimulator which he can carry with him and use as necessary. It is advisable to use this treatment in conjunction with other modalities of physical therapy, particularly when treating causalgia. These modalities include heat or cold, ultrasound, and hydrotherapy.

ELECTRODIAGNOSIS

Galvanism and the nature of muscle response to it are used in electrical diagnosis, as is discussed fully on page 109.

ELECTROCAUTERY

Galvanism is also used as a cautery in skin surgery, as in, for instance, the removal of warts and the treatment of corns and plantar warts.

Sinusoidal Current

The sinusoidal current is the 60-cycle current as provided commercially. It is an alternating current of smooth waveform.

MUSCLE STIMULATION

The alternating current is used for muscle stimulation when the muscle has a normal nerve supply. This use produces normal muscle contraction and assists a patient to activate a muscle without using it to move a joint (i.e., to induce an isometric type of exercise). Also, by this means an alienated muscle can be brought into action in a situation in which another muscle has taken over, by substitution, the alienated muscle's function. Alienation of the intrinsic foot muscles is almost invariably present in patients suffering from foot pain. However, the induced faradic current is easier to control when it is surged, and the physiological cycle of contraction–relaxation–refractory period is more easily imitated by its use. The muscle stimulation current employed in conjunction with ultrasound (p. 118) is the sinusoidal current.

Faradism

MUSCLE STIMULATION

The faradic current is an induced alternating current. It is less comfortable and rougher than the sinusoidal current (the common 60-cycle current), but by surging it produces a more natural muscle contraction. Its use is the same as that of the sinusoidal current.

ELECTRICAL DIAGNOSIS

The faradic current is used in electrical diagnosis, and the absence of muscle reaction to it is a basis for the reaction of degeneration test (R.D. test) described on page 109. The return of muscle response to faradic stimulation is an indication of regeneration of the nerve.

SUBTONAL FARADISM

Subtonal faradism is done without surging the current. The electrodes are placed over the opposite ends of long bones or at either end of the spine. This use of faradism temporarily relieves bone pain. The current is just below the intensity required to stimulate a contraction in a nearby muscle. The reason for its clinical effect is not known.

MASSAGE

Massage is not just rubbing. It is a highly skilled modality of physical treatment. That it has fallen into disfavor in the United States because it has been abused in so-called massage parlors and because of spurious lay claims that it assists in weight reduction and is a substitute for exercise is a poor reason for abandoning it.

The classic terms used to designate the various types of massage are *effleurage*, *pétrissage*, and *tapotement*. However, we are using descriptive American terms.

Stroking

Light stroking massage properly used produces relaxation both psychologically and physically by its counterirritant effect, as in the use of the vapocoolant spray, but in this case light touch is the substitute sensation instead of the touch-cold sensation previously discussed. Massage lightly given also appears to heighten the threshold of skin sensory nerve endings, thereby decreasing pain and increasing relaxation. It has a vascular effect, moreover, which has to be a reflex phenomenon. An overdosage of light stroking may be irritating rather than sedative to some patients, especially those in the older age groups; these patients show an early change of response from sedation to irritation. After almost all physical therapy treatment, stroking massage should be given to produce relaxation.

Heavy stroking is reserved for its mechanical effect on the vascular and lymphatic systems (see Edema Massage, p. 137).

Kneading

Kneading massage includes firm kneading of muscle, skin rolling, and stretching massage; in prescribing kneading one must specify the type. These forms of massage have a mechanical effect of helping the return of venous blood, lymph, and catabolites into the mainstream of the circulation. Positioning to overcome the effects of gravity is an important adjunct to the use of kneading massage. Muscle stretching which helps to overcome spasm may also form a part of kneading. The stretch may be either in the length of the muscle or across it.

SKIN ROLLING AND CUPPING MASSAGE

With chronic muscle spasm and joint pathology, the skin becomes adherent to the underlying superficial fascia, loses its movement and its elasticity, and is painful, especially over the back and the iliotibial bands. The same condition occurs all over the chest wall secondarily to chronic lung disorders such as emphysema. It is also found in the palms of the hands of patients suffering from rheumatoid arthritis and, less obviously, in the soles of their feet. The only way to overcome the pain of this condition, which we call panniculofibrositis, is by skin rolling or, if that is too painful, by cupping massage. Skin rolling is performed by picking up the skin and superficial fascia between the thumbs and fingers and rolling the skin backward over the advancing thumbs, which keep up their pressure behind the roll all the time (Figure 7-6). No lubricant is used. If a suction cup has to be used (a breast pump works very well), a liberal application of pe-

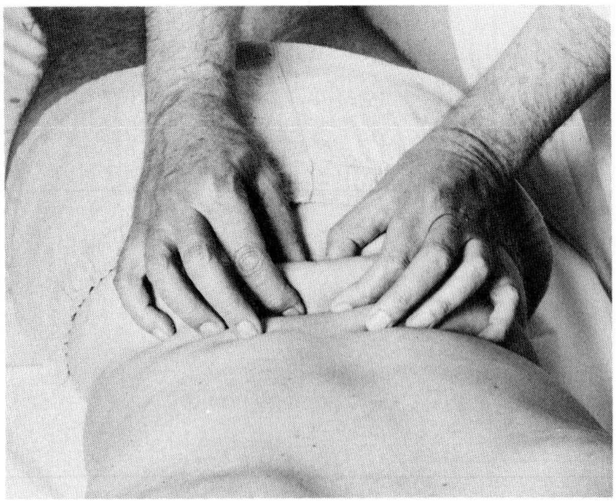

Figure 7-6. The maneuver of skin rolling in therapy.

troleum jelly over the skin is necessary to maintain suction. The cup is rhythmically moved over the skin across the painful area, and the suction is adjusted to the patient's tolerance. Figure 7-7 illustrates the method of using a breast pump for cupping massage.

Friction Massage

Friction massage means any method of treatment of soft tissues (muscles, fascia, interstitial tissue, scars) using intermittent pressures. The effect of this form of treatment may be local and mechanical or, by producing a form of counterirritation, it may affect the deep circulation.

LOCAL AND MECHANICAL EFFECTS

Softening of scar tissue may be facilitated by the use of friction massage and is particularly effective in the following conditions: (a) in the wrist and hand when the effects of trauma and surgical procedures are being treated; (b) in the knee when the quadriceps mechanism is plastered down as it so often is following skeletal traction and surgical procedures; (c) at the ankles following prolonged immobilization in plaster; (d) in the foot following prolonged disuse; (e) in the shoulder girdle following surgical procedures or in treating a frozen shoulder; (f) as part of the treatment of indolent ulcers to prevent adherent scarring around the edge of a healing ulcer which prevents marginal granulations and further healing; and (g) in treating pathological fibrosis in plantar fasciitis, Dupuytren's contracture, and tightness of the iliotibial bands. Overdosage of friction massage increases muscle irritability and aggravates rather than relieves the condition being treated.

Friction massage may be performed by the use of a vibrator through the skin. The initial effect is sedative and relaxation results. Overuse of a vibrator is irritative and reverses the therapeutic effect of properly performed friction massage.

Friction massage used to soften scar tissue in the skin is performed by picking up the skin between the thumb and fingers and subjecting it to intermittent pressure.

Friction massage used for its effect on muscle, interstitial tissue, and fascia is performed by using the ball of the thumb or the heel of the hand to dig into the deep tissues and subjecting them to small circular intermittent pressures which are frictions.

No lubricant is used in friction massage for indolent ulcers. The friction between the therapist's hand and the patient's skin adds to the

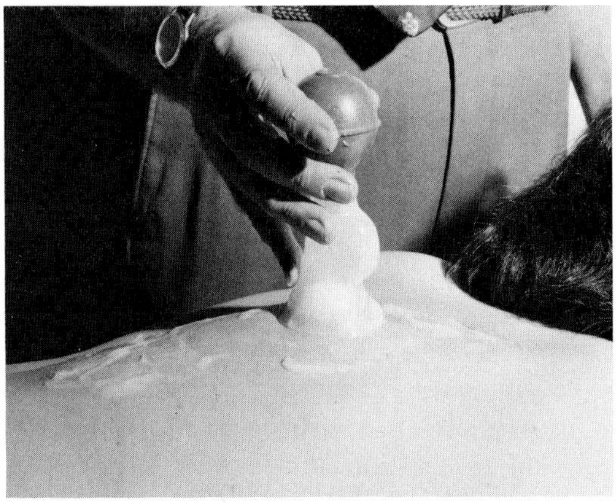

Figure 7-7. Use of breast pump for cupping massage. The back is liberally covered with petroleum jelly to facilitate moving the cup over the skin. The cup is held slightly raised and the intensity of the suction is varied according to the patient's tolerance by compressing the bulb of the pump as necessary.

effectiveness of the modality. A lubricant such as lanolin or cocoa butter is given the patient when he continues his program at home. Friction massage has to produce some degree of irritability as in spasm. When friction massage is used, stroking massage should be used to restore relaxation at the end of a treatment session.

Pressure

Sustained pressure up to about a minute over a trigger point appears to block noxious impulses from muscle and produces relief of pain and relaxation. Pressure over a fibrositic nodule has the same sort of effect.

Edema Massage

Venous and lymphatic stasis often presents a major problem in the restorative phase of physical treatment, and massage is a useful modality to try to overcome it. When using massage to get rid of edema, the therapist must start at the proximal end of the limb and progress to its distal end. Elevation of the limb is an important adjunct to drainage of edema; in the lower extremity it is important to elevate the pelvis as well as the leg.

Tonic Massage

Slapping or percussion of muscle is called *tonic massage*. This is used to "tone up" normal muscle and has no place in the treatment of conditions giving rise to musculoskeletal pain.

Any form of massage other than light stroking tends to cause irritation of muscle. Following the use of all the other types of massage, therefore, light stroking must be given to produce relaxation once more.

ASSISTIVE DEVICES, SPLINTS, AND SUPPORTS

In this section we make no pretense at covering the subject of the use of assistive devices except insofar as they serve as an adjunctive treatment for pain. When used for this reason, supports, braces, and dynamic or static splints of all kinds must be considered a modality of physical therapy.

Crutches

Crutches relieve the joints and muscles of the lower extremities from weight-bearing and painful function while the pathological condition producing pain is active. In the restorative phase of treatment they assist in the gradual return to function.

There is more to the use of crutches than most physicians realize. For their proper and safe use they must be fitted to the patient and the patient must be trained in their use. This program cannot be achieved at a moment's notice; indeed, some elderly patients need several days of training before they can safely use crutches.

If crutches are necessary, nothing is to be gained by using one rather than two except trouble. To use them safely and with full benefit the patient must have strong shoulder depressor muscles, triceps muscles, and wrist muscles. If axillary underarm crutches are used, the patient must be taught not to rest his weight on the axillary bars because of the danger of pressure neuritis, most commonly of the radial nerve. Since the radial nerve involvement is usually motor and not sensory, there is no warning pain before a sudden wrist drop. In fitting a patient one should always be sure that there is a three-fingerbreadth clearance between the top of the crutch and the axilla.

When the abandonment of crutches is contemplated, the physician would do well to consider whether recovery from the original pain-producing condition is sufficiently complete for the patient to behave as though he again has a normal joint or a normal muscle. If there is any doubt, crutches should continue to be used as a prophylactic form of treatment. The abandonment of crutches often becomes a rehabilitation goal of the physician for his patient; the same goal may be unrealistically cherished by the patient himself. Too often the underlying motivation is vanity rather than the patient's best long-term interest. For example, prosthetic replacement of hips would probably be far more successful and more enduring if patients with such replacements were discouraged from believing that, because the hip is pain-free, they once again have a normal hip and no longer require any assistive device.

In active younger people the use of Lofstrand forearm crutches is advocated. They allow the patient to be more versatile. The hand rest in whatever crutch is used should be high enough to maintain a 30-degree angle of flexion at the elbow at rest.

Canes

If a single cane is to be used for the partial relief of weight-bearing and compressive muscular and mechanical forces on a leg, or for stability, it should be held in the hand on the opposite side and advanced with the painful or uncoordinated leg on ambulation. Safety is promoted if the cane is fitted with a large rubber crutch tip. A three- or four-pronged cane is safer than a simple stick for elderly people who have poor balance. A single cane can never properly free an injured limb from weight-bearing, although it does relieve to some degree the stress acting upon the injured joint. The use of a cane may promote the development of poor gait and often initiates pain-producing stresses in the low back. Maximum arm leverage is obtained, as with crutches, when the elbow of the arm using the cane is flexed to 30 degrees at rest.

If it is necessary for a patient to use two canes, he probably should be using crutches. We cannot see how two canes can ever be considered more desirable than crutches, and they are certainly less safe and less effective.

Walkers

For the aged and those untrainable in the use of crutches, the walker is the best assistive device for ambulation for all painful conditions of the lower extremity joints, but especially the hips. In pain problems, the static rather than the hinged walker is preferred. It is virtually impossible for the older patient to use the walker without any weight-bearing on one leg; such a requirement would force the patient to hop. He should be allowed at least to steady himself by touching the foot of the affected leg to the ground. The important thing is that the "legs" on the walker be high enough to keep the patient upright. Maintaining the upright posture, rather than the usual stooped position, is the major feature in the relief of pain in the hip on assisted ambulation.

Wheelchairs

If adequate relief from weight-bearing cannot be achieved by the use of crutches or a walker, a wheelchair is in order. The only important considerations in its prescription for painful musculoskeletal conditions are that the front wheels be 8-inch ones (smaller wheels make the chair difficult to propel), that the seat be adequately cushioned, and that the back is not too high.

Braces

The prescription of braces for pain conditions in the extremities is limited if only because the expense does not justify their short-term use. In conditions such as Legg-Perthes disease, however, braces are mandatory. In overcoming myostatic contractures at the knee, serial plaster casts or a long leg brace with a turnbuckle at the knee will be invaluable. Knee cages are of limited value in supporting a painful knee weakened by injury or surgery but, when used, should have side bars. The ischial weight-bearing brace or a long leg brace incorporated into a quadrilateral socket is helpful for patients with hip pain from degenerative hip disease or aseptic necrosis if, for any reason, more radical corrective procedures are contraindicated.

Stabilization of the ankle by bracing is called for primarily when muscles are weakened. However, it does find uses in the more severe arthritic conditions and painful ankle joint diseases.

Splints

Splints made of plaster or plastic may be necessary to rest the wrist and hand from function

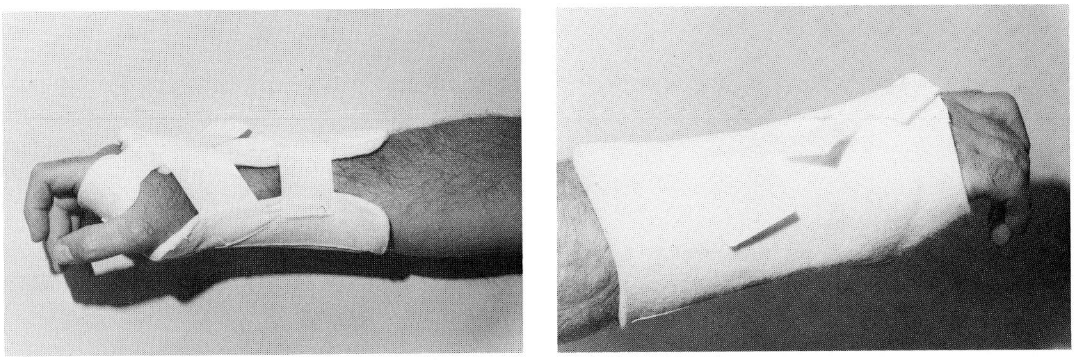

A *B*

Figure 7-8. *A*, Side view of a soft cock-up splint used in the conservative management of the carpal tunnel syndrome. *B*, The splint is fastened dorsally by Velcro straps. (These splints are available commerically.)

during the initial phases of treatment for sprains and strains and tenosynovitis. When conservative treatment of a carpal tunnel syndrome is contemplated, a cock-up splint for the wrist is most useful (Figure 7-8). In the lower extremity, plaster posterior shells may be adequate to support the foot when ligament sprains about the ankle are present or when muscles affected with paresis are being treated. But casting is required in the treatment of ligament sprains at the knee or ligament tears at either the ankle or the knee.

Back Braces, Corsets, Jackets, and Belts; Collars and Neck Braces

Back braces and corsets are often little but external consciences in the treatment of conditions characterized by low back pain. The Knight spinal brace (chair-back brace) does prevent extension and lateral bending but really does not support the back.

In painful low back conditions a flexion plaster jacket is very useful in relieving muscle spasm and pain. After it has been worn for a few days, a better clinical examination can be undertaken. Then, when a diagnosis has been made, a more definitive treatment can be started. If it is removed carefully, it can be worn in between treatments for a few days.

In painful conditions higher in the back, a plaster bed—a complete posterior plaster support from head to toe—is invaluable for the same reason.

The Goldthwait brace or Taylor brace is useful in the treatment of chronic pain conditions in the thoracic or high lumbar spines. It is particularly useful in the treatment of compression fractures of vertebral bodies.

The soft felt or foam rubber collar or even a folded Turkish towel wrapped around the neck is helpful in sparing the cervical spine joints from their weight-bearing function. For best results the collar should be worn "backward," that is, with the opening in the front and the high part in the rear to maintain flattening of the cervical spine and to prevent neck extension. But if steady support of the head on the neck is required the four-poster brace has to be used.

Corsets are of use in chronic pain conditions of the back but, though more comfortable to wear, they are less efficient than braces. In the recuperative phase of low back pain, from whatever cause, corsets serve as a useful reminder to be careful, and they provide some support.

The sacroiliac belt, ideally a 2-inch webbing belt with a 3-prong buckle, worn tightly around the pelvis with its upper edge below the anterior superior iliac spines and the belt horizontal to the floor, does restrict movement of the sacroiliac joints and the facet joints at the lower two lumbar junctions. Incorporated into a corset, it has a definite stabilizing effect on the low back. To this degree, it aids rest from function in treatment of pain conditions in the joints at these levels and relieves the erector spinae muscles of stress and resultant spasm. No pressure pads should be incorporated. Pressure over muscle produces atrophy and may compound pain problems by weakening the natural muscle support of the back.

Plastic and Leather Jackets

In the case of patients for whom surgical fusion might be indicated for the relief of pain but who, for some reason, refuse or perhaps cannot tolerate the procedure, plastic or leather jackets are of definite benefit in relieving intractable, chronic pain. They must be accurately molded.

STRAPPING AND BANDAGING

Strapping is particularly useful in the first aid treatment of ankle pain, elbow pain, chest wall pain, low back pain, and shoulder pain. Bandaging with elastic bandages does little more than support the morale of the patient.

In using adhesive tape, one should recognize that it is the tension and not the adherence of the strapping that is effective. Thus the length of adhesive is covered with gauze bandage except for the terminal inch or so at each end where adherence is needed. Tincture of benzoin should always be used to protect the skin when adhesive strapping is used.

Strapping for the Low Back

For supporting the low back when pathology is limited to the soft tissues, a method of strapping is used for which we claim no originality.

Overlapping lengths of 2-inch adhesive tape are applied. They must be long enough to encircle the back from in front of the midaxillary line on each side. The lowest tape should adhere anteriorly just below the anterior superior iliac spines and is the first one to be applied, like a pelvic band.

The tapes are covered, except at their ends, with 2-inch gauze bandage to keep them from adhering to the skin except at their ends. It is the circumferential tightness and not the area of adherence that provides the efficient support. About eight overlapping tapes will give support to a height of about 8 inches. A second pelvic band tape is cut, and to it are stuck the bases of about 12 cotton-tipped applicator sticks. This is applied over the lowest part of the original strapping. Further overlapping 2-inch strips of adhesive tape, unprotected by gauze bandage, are applied over the sticks and covering the first layer of tapes.

The whole application is made with the patient lying prone and relaxed. The tapes are started from alternate sides, and the buttocks and flanks of the trunk are pulled up and toward the spine from the side away from that on which each tape is initially anchored. This excellent support will last three or four days, which is usually long enough for its purpose, and heat can be applied through it (Figure 7-9).

A many-tailed scultetus bandage properly applied with the lowest segment wrapped around the pelvis like a sacroiliac belt serves as a very satisfactory temporary low back support.

Supporting Knee Bandages

A "wool and firm bandage" is used for supporting the knee. A full width and half the thickness of top-grade cotton wadding is wrapped around the knee, centered over the patella. A 3-inch gauze bandage is wrapped twice around the knee, again centered over the patella, tightly enough to prevent the wool from slipping and rotating. Another wraparound is then made, but this time the upper edge of the bandage passes over the center of the first wraparound over the patella. It is carried around the back of the knee, however, at the same level as the original wraparounds. Now there are one and a half widths of bandage anteriorly opposed to a single width posteriorly. The next wraparound is made so that the upper border of the bandage passes over

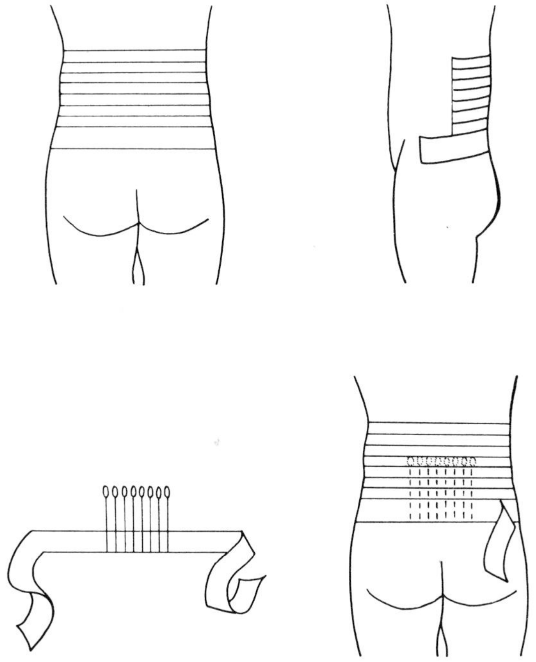

Figure 7-9. Strapping the low back. The lowest strap has its upper edge just below the anterior superior iliac spines, like a pelvic band. Successive strips of tape are started from alternate sides of the body, and the tissue from the other side is pulled across to the midline gently but firmly.

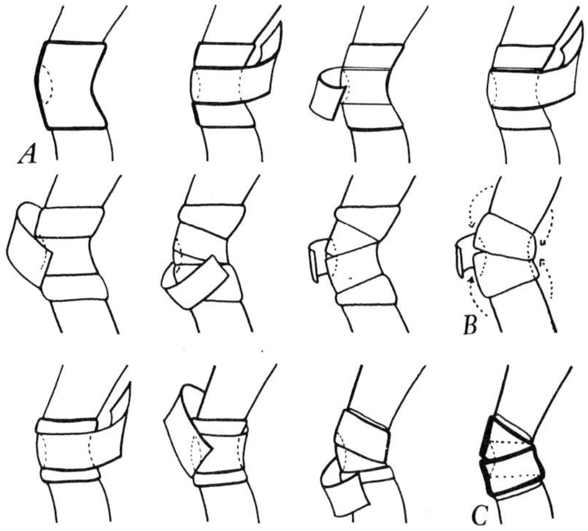

Figure 7-10. Applying the "wool and firm bandage" to support the knee. *A,* The sheet of cotton wool (cotton wadding) placed around the leg and centered over the knee cap. *B,* The point at which the cotton wool above and below the bandage is turned down from above and up from below. Thereafter the bandage is pulled as tight as possible as each successive wind is made. *C,* A final wrap of banadage (dotted line) passes over the knee cap.

the center of the patella. It is again carried around the back of the knee at the same level as before. Now there are two widths of bandage anteriorly and still only one width posteriorly. The tautness of the wrapping makes it firm enough to prevent movement of the wool.

The wool still extends above and below the bandage. The section from above is now turned down over the knee cap in front; the section from below is turned up over the knee cap in front. The sections overlap somewhat behind the knee.

Now the same pattern of wrapping is repeated, but this time the bandage is pulled as tight as possible on each wraparound and a final additional wraparound is made over the patella.

Large diaper safety pins are now inserted over the collateral ligaments of the knee through all the layers of bandage on each side. Figure 7-10 shows the main features of applying this support.

Useful Strapping around the Ankle

In the initial phase of treating a sprained ankle the application of a simple figure-eight strapping is supportive and pain-relieving. It is applied so that there is a pull in inversion or eversion under the instep (Figure 7-11), depending on whether the sprain involves the medial or lateral ligaments.

Another very efficient way of achieving initial comfort is to soak a gauze bandage in hot wax and then wrap it around the foot and ankle. The heat, though transient, is comforting, and when the wax cools it hardens to provide relieving support.

In more severe sprains, a plaster boot may be necessary for treatment initially. When the plaster is removed, the application of an Unna boot is successful in supporting the vascular

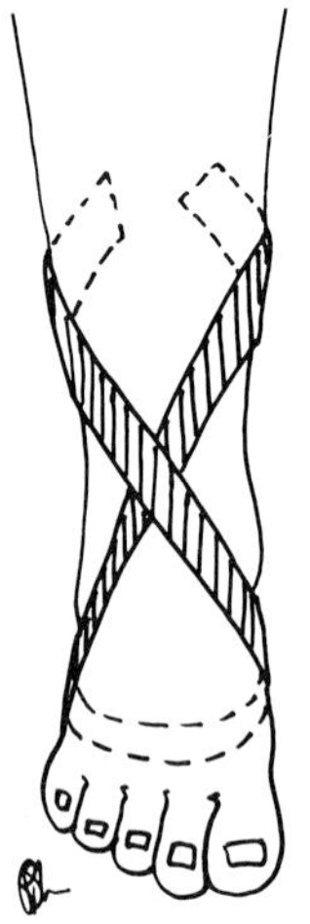

Figure 7-11. Simple figure eight strapping for a sprained ankle. In inversion sprains the strapping is started on the lateral aspect of the lower leg so that the final pull of attachment medially produces an eversion stress at the ankle. In eversion sprains the strapping starts medially so that the final pull produces an inversion stress at the ankle.

tree, which may have lost its tone, to prevent dependent edema.

Useful Strapping for the Foot

For relieving pain of sprain at the insertion of the peroneus brevis in the foot, Goldthwait's method of strapping is useful. This is illustrated in Figure 7-12.

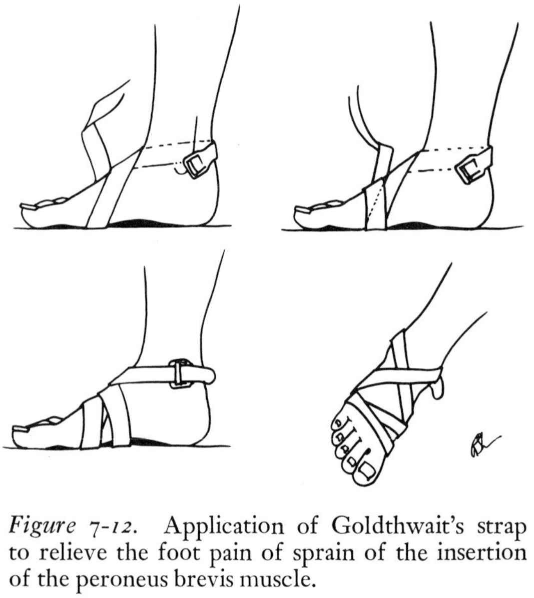

Figure 7-12. Application of Goldthwait's strap to relieve the foot pain of sprain of the insertion of the peroneus brevis muscle.

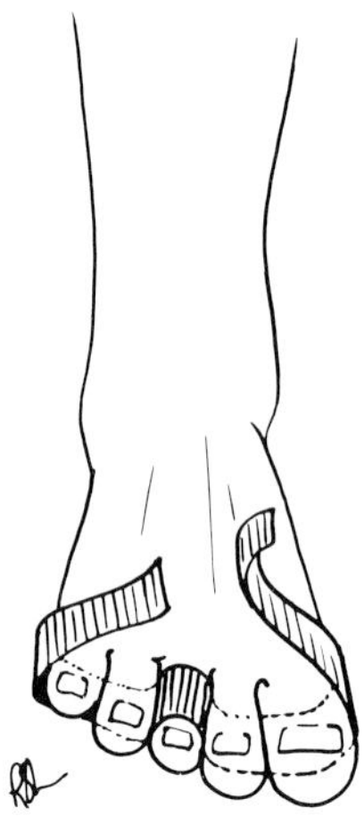

Figure 7-13. Method of applying nonadhesive tape to correct the condition of an overlapping toe.

The overlapping toe can also be treated by strapping, as is illustrated in Figure 7-13. This helps to prevent (or treat) the formation of a painful corn over the proximal interphalangeal joint. It may eventually correct the condition if the patient wears wide enough shoes.

Strapping for Extensor Muscle Tendon at the Elbow

The rationale and indications for the use of extensor muscle strapping are discused on page 179. (See Figure 7-14).

Strapping for Musculoskeletal Chest Pain

Again, the reader is reminded that it is not the adherence of tape upon which the success of strapping depends but the tightness of the tape. All the adhesive surface of the tape, therefore, is covered by gauze bandage except for the terminal inch or so at each end. Two-inch adhesive tape is applied in overlapping strips as described for strapping of the low back. The important thing is that the strap must be started on the chest wall posteriorly on the opposite side of the spine from the side of the chest wall pain and carried anteriorly beyond the midline of the front of the chest to be fixed again on the side of the chest wall opposite to the side of the pain. The tapes must be applied at the end of expiration and pulled tightly into place.

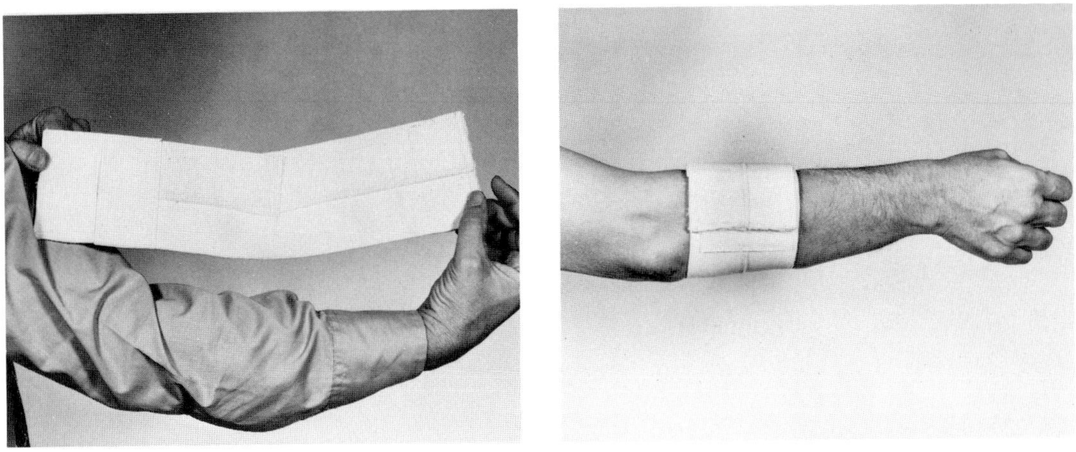

Figure 7-14. Extensor muscle strap for the forearm. One 2-inch strap is often sufficient.

Strapping for Shin Splints

A discussion of shin splints and this type of strapping (Figure 7-15) appears on page 185.

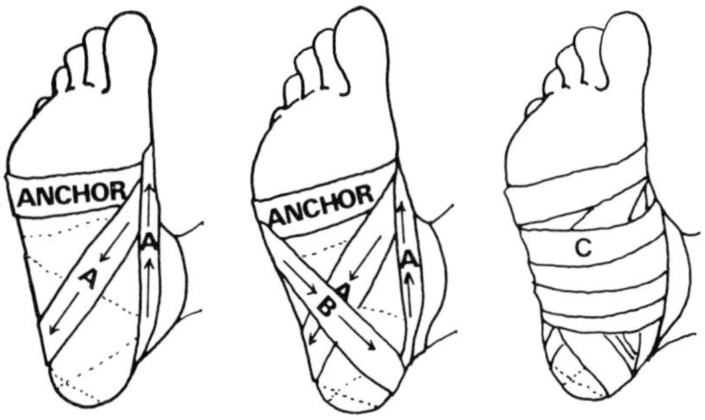

Figure 7-15. The foot should be cleansed, sprayed with Tru-Stik Tape Adherent, and wrapped with J-Wrap Underwrap. *A,* Place an anchor strip of 1-inch adhesive tape around the ball of the foot, making certain that the action of the toes is not constricted. Start the second strip at the point of the anchor strip closest to the big toe, continuing up at a sharp angle, crossing the center of the longitudinal arch, encircling the heel, and then descending along the side of the foot to the starting point. *B,* Start the second strip by the little toe, bringing the tape across the longitudinal arch, around the heel, and down the side of the foot to the starting point. *C,* Overlap each strip three times. Anchor lightly around the ball of the foot and place additional anchors over the cross of the strapping for added strength. (Reprinted by permission of the copyright owner, © Johnson & Johnson, 1973.)

Slings

We deplore the use of the conventional triangular sling, which is invariably knotted over the first and second thoracic vertebrae. This not only is pain-producing but results in a stooped forward posture. We recommend a cross-back sling made of muslin bandage. The application is illustrated in Figure 7-16. This sling avoids pressure at the cervicothoracic junction and discomfort on the shawl area of the trapezius. It does support a paralyzed shoulder.

Finally, it is well to stress the potential dangers of supports. The prescription of a support for a weakened muscle may further weaken the muscle; it may also impair movement of and create morbidity in the joint(s) it supports. "Corrective" shoes and "arch supports" never correct anything. The feet correct ill-fitting shoes and arch supports and suffer pain as they are traumatized in so doing. Pressure pads added to supports, if they compress a muscle, produce atrophy of the muscle. The use of sponge rubber or elastic in supports allows for error in their manufacture and fitting and is undesirable.

OCCUPATIONAL THERAPY

Occupational therapy must be considered an integral part of physical treatment. In the

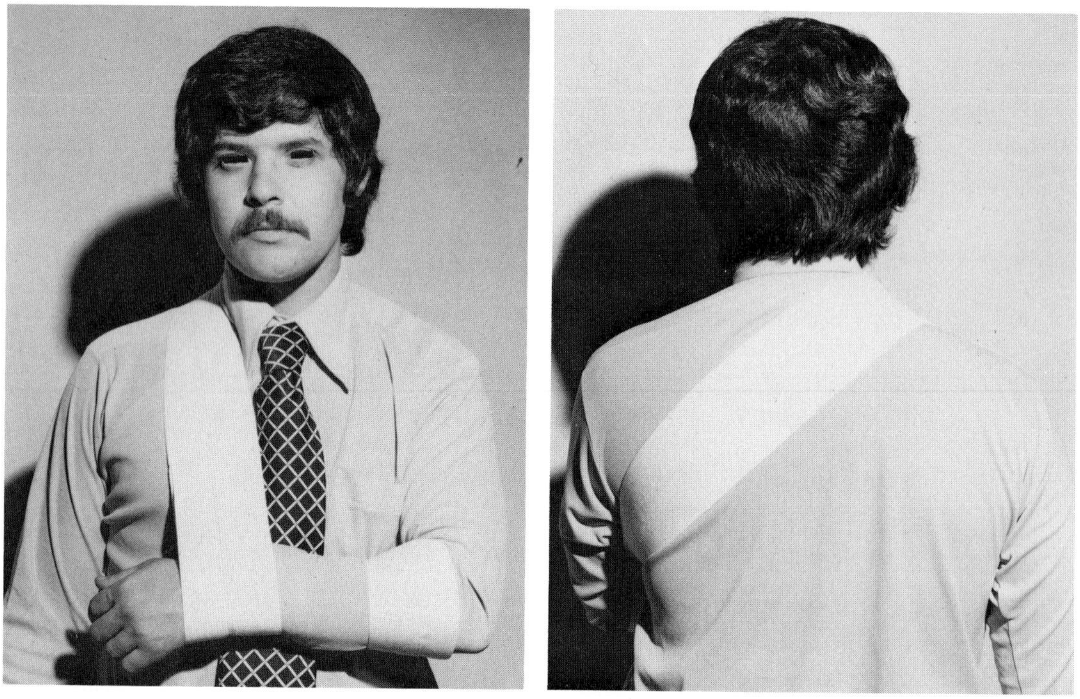

Figure 7-16. The cross-back sling that most adequately supports the shoulder.

context of treatment for pain, occupational therapy properly used is particularly helpful for developing reciprocal muscle action or retraining patients in it and for the resumption of whole-limb coordinated activity in the recovery phase of treatment. It also plays a role in the process of muscle reeducation for denervated muscle or muscles that have been immobilized for extended periods, and in the treatment of alienated muscles. A most useful role for occupational therapy is in the treatment of hand problems after reconstructive surgery and for burns, tendon transplants, and other hand and arm injuries.

In rehabilitation it should be the function of occupational therapy to retrain a patient in the activities of daily living, the activities of self-care, and the activities required for vocational independence. The prescription of "diversional therapy" is usually an abuse of both the term *therapy* and the professional status of the therapist.

Unless physical therapy and occupational therapy are integrated, there arise conflicts in aims which impair the effectiveness of both.

TRACTION

Traction is commonly used to stretch muscle which is in spasm (or contracted) and/or to produce immobilization. When used intermittently it maintains some sort of joint movement without function. The comments here are limited to cervical spine traction and pelvic traction as employed in physical therapy. Skeletal traction has specific orthopedic functions that are not within the scope of this book. Long axis extension in the range of joint-play movements is probably the basis of success in using intermittent traction in physical therapy.

Spinal Traction

CONSTANT MECHANICAL TRACTION

Mechanically performed traction is used empirically to overcome muscle spasm without searching for the underlying cause of the spasm. Were it used to restore the resting length of muscle, this would be a more logical method of treatment; but even then, it would be restricted to muscles acting in one plane—longitudinally—and is never effective if the pull of the traction is against the spinal curve. In other words, flattening of the spinal curve is prerequisite to stretching any muscle or muscle group. Traction should be adapted to the patient rather than the patient to the traction; that is, the posture of the patient should not be enforced but adapted to his comfort, and the traction should be discontinued when the patient cannot tolerate it. In effect, constant traction serves to keep a patient at rest while nature heals.

INTERMITTENT MECHANICAL TRACTION

Whatever the reason for prescribing intermittent mechanical traction, its real effect must primarily be to move the spinal joints, which can be achieved only if the spinal curve is flattened. The movement, moreover, can only be in the long axis of the flattened spinal segments. Thus one is, in fact, performing a range of joint-play movement (see Joint Manipulation, page 149). Secondarily, limited muscle stretching is achieved, which may help to overcome spasm (Figure 7-17), because intermittent traction produces only one of many movements of joint play. In the neck particularly, opening of the facet joints helps to prevent loss of elasticity of the capsule and formation of fibrous adhesions around the nerve roots and relieves osteophytic impingement (Figure 7-18). We are not convinced that anyone really knows the effects of intermittent traction, but, besides producing a joint-play movement of the apophyseal joints, it probably opens up the neural foramina, which is a beneficial result in the presence of radicular inflammation.

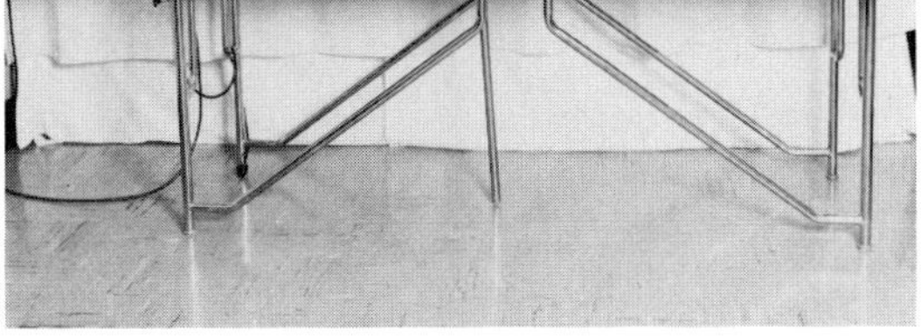

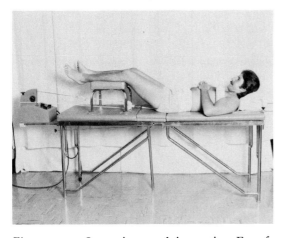

Figure 7-17. Intermittent pelvic traction. For efficient traction the hips and knees must be flexed to flatten the lumbar spine. (Courtesy of Tru-Eze Manufacturing Co.)

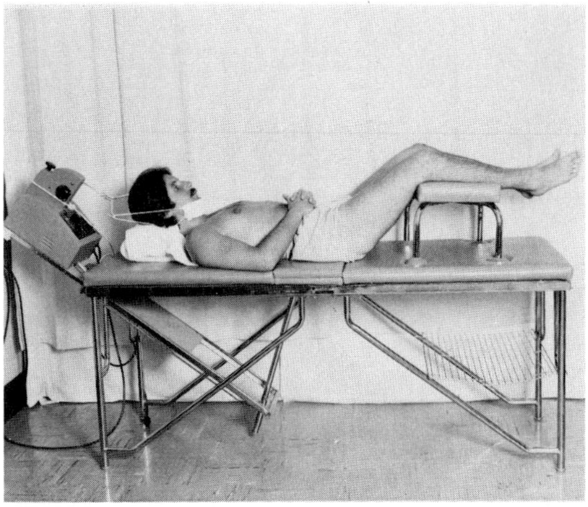

Figure 7-18. Cervical spine traction. For efficient traction the cervical spine must be flattened by dropping the chin into the neck. This helps to ensure that most of the pull of the head halter comes from the occiput. (Courtesy of Tru-Eze Manufacturing Co.)

MANUAL TRACTION

The logical way to open up the spinal segments is by manual traction. At the moment of pull, and depending upon its momentum, there is an undetermined poundage of pull and an optimal opening up of the interlaminar joints. Manual traction, then, simply produces the movement of long axis extension in the range of joint-play movement. When used as a single manipulative maneuver, it is an intervertebral interlaminar joint movement performed in a rather nonspecific way but it may be effective in relieving symptoms. Any muscle

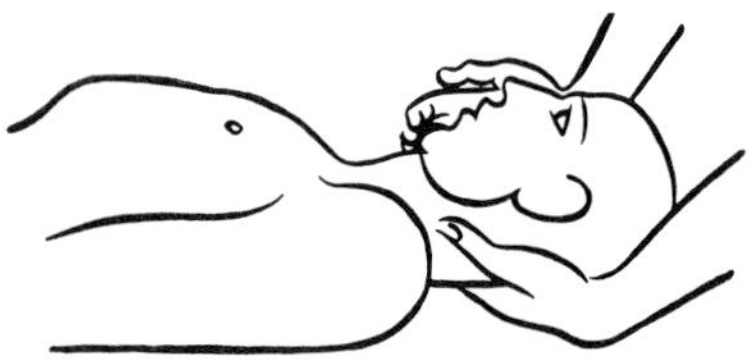

Figure 7-19. Manual cervical traction. Position adopted prior to exerting the traction. The physician's left hand pulls from beneath the occiput while his right hand steadies the chin, which is dropped down onto the neck to flatten the cervical lordosis.

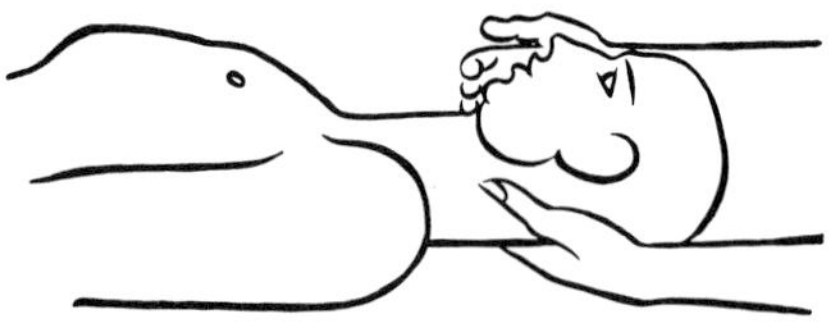

Figure 7-21. Completion of manual traction of the cervical spine. No flexion or extension of the head on the neck is permitted. However, in a patient with an exaggerated lordosis it may appear that the pull is made with the cervical spine somewhat flexed. Note that most of the pull is from the occiput.

stretch effect in manual traction is usually incidental. Figures 7-19 through 7-23 illustrate the technique for performing manual cervical traction. A prerequisite for successful use is the flattening of the normal cervical lordosis. Figures 7-24 and 7-25 illustrate the equivalents of manual cervical traction on the thoracic and lumbar spines respectively.

USUAL EFFECT OF TRACTION

The effect of traction is unpredictable if it is mechanically applied. Constant traction for the most part produces immobilization and rest from function—a reasonable prelude to more active treatment while acute inflammatory changes are resolving.

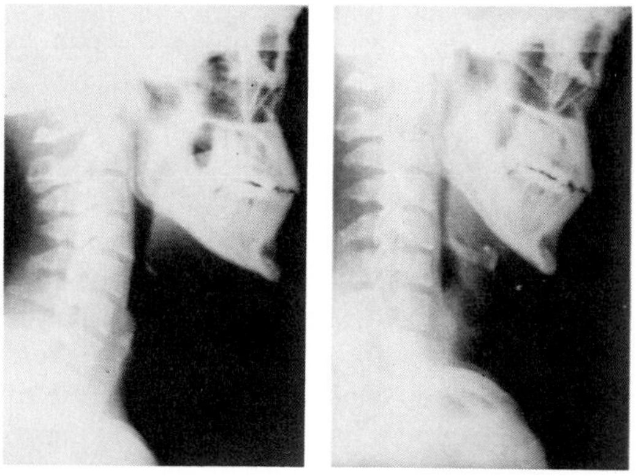

A *B*

Figure 7-20. *A*, X-ray picture of normal cervical lordosis. *B*, X-ray picture showing flattening of the cervical lordosis achieved simply by dropping the chin into the neck. This position is prerequisite to exerting manual traction as illustrated in Figure 7-21.

Traction, be it mechanical or manual, constant or intermittent, has a place in therapy if it is prescribed for the right reason. Unfortunately, it is usually prescribed empirically, and its beneficial effect may be fortuitous rather than a specific result of therapeutic planning. Often there are better and more predictable ways of producing the hoped-for effects of this modality of therapy. We do not believe that traction has any primary effect on the intervertebral disc.

If there is loss of joint play in a given cervical facet joint, the traction, in trying to open up all the joints, will either open them all up

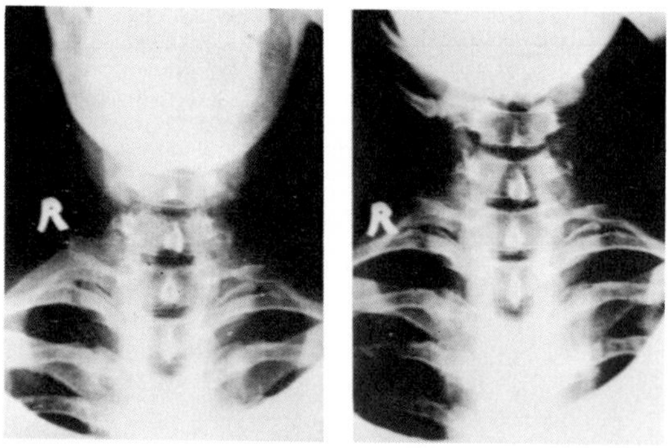

A *B*

Figure 7-22. *A*, X-ray picture of head on neck at rest before manual traction is applied. *B*, X-ray picture at completion of manual traction. The lower edge of the mandible is now one and a half vertebrae higher up the spine. This is an optimal movement. Poundage of pull using manual traction is immaterial. The effect is achieved more comfortably and efficiently than by any preconceived poundage of pull using mechanical traction.

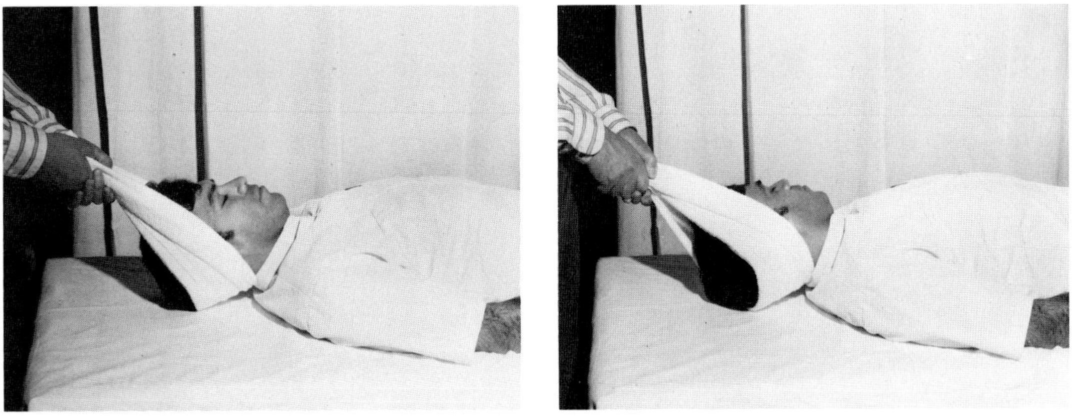

A *B*

Figure 7-23. *A,* Convenient and comfortable way of obtaining manual cervical traction using a Turkish towel under the occiput. *B,* Rotation of the head on the axis and atlas added to the straight pull of traction.

or, because of compensation in the joints above and below, fail to move the affected facet joint at all and may further strain it and make it worse. Hence, traction used in cervical pain to move or to open up a locked facet joint is not the best choice of treatment. Joint manipulation is preferable.

JOINT MANIPULATION

Joint manipulation remains anathema to many physicians in spite of its undoubted usefulness in physical treatment. The reason is largely that no two people mean the same thing when they talk about it.

Clearly joint manipulation must be confined to something mechanical that is done by another person to a synovial joint. Clearly that something is a movement and if that movement could be performed by muscle action there would be no need for another person to do it. So, manipulative movements are not the same as the voluntary movements produced by muscle action. Ideas of movement in orthodox medical thinking are clouded by the usual practice of demonstrating something in the postmortem room and/or under the microscope, or in the living patient, by a shadow change in an x-ray film. The science of mechanics tends to be forgotten when one thinks of normal moving human beings.

Everything that man makes to move has a built-in factor of play between the moving parts that is essential to efficient, noiseless (painless) movement. In the mechanical parts of the human being, every moving synovial joint—and all synovial joints move—has a range of play whose integrity is essential and prerequisite to painless functional joint movement. If the play in the joint is lost, the function of the joint is impaired and what movement remains is painful.

The etiological factors that precipitate the loss of play between the moving surfaces of the joints are intrinsic trauma and immobilization (including the effects of disuse and aging); it may also follow the resolution of some more serious joint pathology. The treatment for the loss of play between the moving joint surfaces, a mechanical problem, is to restore the lost movements by mechanical means.

The play movements that must be learned so that their loss can be diagnosed are called the movements of joint play. The etiological causes of their being lost have been mentioned above. Their loss, which produces the symptoms and signs of loss of function and pain (or painful function) is called joint dysfunction;

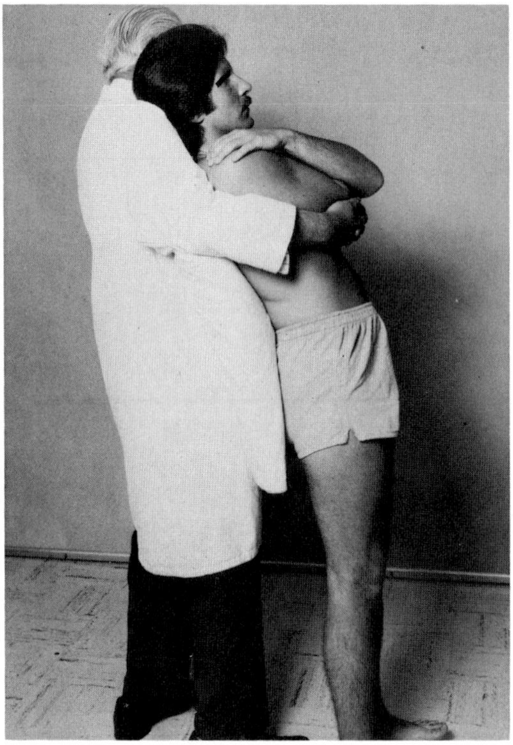

Figure 7-24. Thoracic traction.

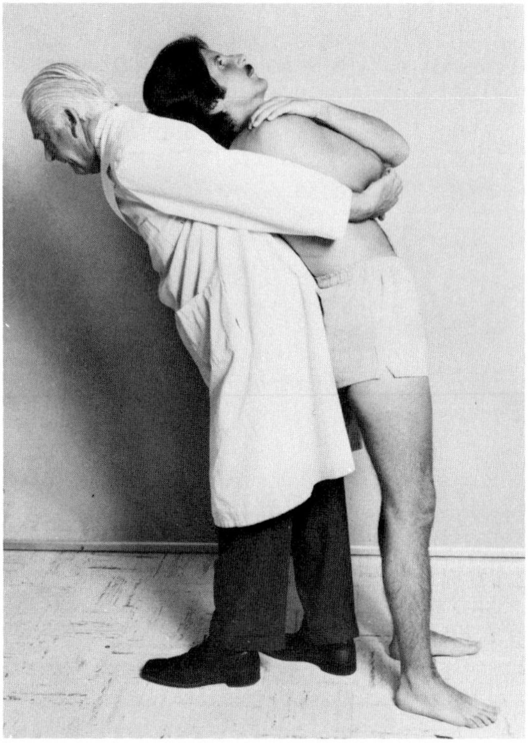

Figure 7-25. A means of exerting pelvic traction at the lumbosacral junction. Moving the buttocks up allows the model's lumbar spine traction to be specifically localized at any intervertebral junction in the lumbar spine. The model does not have to be lifted off the floor. This traction is, of course, long axis extension of the facet joints at each level, and a joint-play tilt can be added if the physician performs a gentle reverse pelvic tilt.

the pathology is mechanical, and the treatment is the mechanical treatment, joint manipulation.

Thus, no conflict arises with other medical hypotheses or scientific concepts in medicine. There is a normal anatomical fact (but in the mechanical field). There is a (mechanical) pathological condition that is predictably demonstrable. There are etiological factors that produce the (mechanical) pathological condition. There is a specific (mechanical) treatment to reverse the pathological condition and restore normalcy, and it is coincidentally associated with the restoration of function and the relief of pain.

In the joint play–joint dysfunction–joint manipulation concept, topographical anatomical areas are nonexistent. There is no such thing as a wrist, an ankle, an elbow, or a back, for instance. Each individual joint in each specific area has to be isolated on examination. Each joint has its specific range of joint-play movements. New examining techniques must be learned in searching for the presence or absence of joint play. Examination for normal joint play and its loss has been described in other texts.

In treatment, the manipulative techniques are confined solely to the performance of joint-play movements. The forcing of a lost functional movement is not manipulation and can only produce traumatic injury to either the joint, the bones that make up the joint, or the capsule or supporting ligaments of the joint.

Until this concept of joint manipulation is

fully appreciated, the use of this physical treatment modality should remain in question and probably be avoided. Once the concept is acceptable, it becomes the logical and only form of treatment for a specific cause of musculoskeletal pain producing a disabling problem.

There is nothing easy about joint manipulation. Joint-play movements are for the most part less than ⅛ inch in extent. This tiny range is difficult for the average physician or therapist to accept when he is accustomed to thinking of movements through an arc of up to 180 degrees. But recognition of the fact that the movements to be restored by manipulative therapy are very small is prerequisite to the safe and effective use of this treatment modality.

HYDROTHERAPY

Properties of Water

The two major properties of water are its cleansing action and its buoyancy. The former property needs little clarification, yet its use is often overlooked when changing adherent dressings whereas it is almost commonplace in the treatment of burns and ulcers.

The greater use of water in physical therapy stems from the fact that water reduces the effects of gravity from the whole body or any part of the body. The relative weightlessness allows muscles and joints to be used without function and thus facilitates putting into practice the principles of physical therapy while the primary pathological condition is healing. Hydrotherapy is often equally essential during the initial phases of restoration of function and reeducation in rehabilitation.

Recently (1967–1969) it has been reported by Young at the University of Pennsylvania that patients who have had hip surgery suffer virtually no postoperative complications either locally or systemically when they are allowed to walk in a walking tank with the water up above their shoulders within a day or two of surgery. Furthermore, healing of the fracture is solid within six weeks, a great improvement on the usual postoperative course and rehabilitation program.

The presence or absence of certain elements or salts in water is of little significance except that the higher the specific gravity of the water, the more buoyant it is. In other words, salt water is more buoyant than fresh water. The presence of sulfur in the water promotes the superficial circulation by surface irritation; it also promotes sweating, which is a useful excretory mechanism. If the water is warm (up to 102° F), these effects will be enhanced and produce relaxation as well. But one should beware of bath fatigue. Too much time in too hot water may cause serious collapse in older patients. If the water is agitated, unskilled massage is added to the treatment. In patients with acute back pain, treatment initiated under water may reduce the time of disability very considerably as they can perform corrective movements that they would find impossible under any other circumstances.

However, the thoughtless prescription of so-called whirlpool treatment for pain or following surgery or trauma to an extremity often does more harm than good. If the extremity is placed vertically in the whirlpool, the heat and massage produce increased congestion and swelling in the dependent part. There is a commercial arm holder to prevent dependency which should be used when necessary.

Contrast Baths

Contrast baths cause gymnastics in the vascular tree and are invaluable in treating musculoskeletal pain, especially in the patient who has a muscle problem with deficient blood supply from disuse. They are also useful following acute sprains and strains, particularly for foot pain from these causes. However, they are contraindicated in the presence of arterial disease. The time the part is immersed in the hot water should be about twice the time spent in the cold water. The heat and cold should be tolerable to the normal hand. The duration of treatment should be 10 to 15 minutes, and the last bath should be cold.

COMPRESSION

A wide variety of compression devices have been manufactured in an attempt to improve arterial and venous circulation, the flow of lymph, and the reduction of edema fluid. Some are simply static splints filled with air, but most are mechanical devices with alternate waves of pressure or gradients of pressure. Some are timed to the heartbeat.

We have not found any of these devices useful in improving peripheral arterial circulation. They do seem to be of benefit in the treatment of venous stasis, particularly when accompanied by edema. Lymphedema due to various causes has been definitely benefited by the use of a Jobst gradient pressure pump along with other measures such as an elastic stocking or arm sleeve, exercises, and care of the extremity. The compression may make manifest a subacute lymphangitis, in which case antibiotic coverage will be necessary. Edema due to trauma lessens when a course of pumping is undertaken, and pumping may be especially beneficial to postoperative hand surgery patients.

LIGHT THERAPY

There are three types of light therapy: heliotherapy, actinotherapy, and x-ray therapy.

Heliotherapy

The benefits of sunlight are lost in the twentieth century because fashion dictates a "healthy" suntan. In fact, a suntan is potentially unhealthy; there is no better way of preventing the absorption of ultraviolet light than having an overabundance of skin pigments. Also, a holdover of suntan into the winter months deprives the body of adequate ultraviolet light, thereby contributing to the susceptibility to winter infections.

Actinotherapy

Ultraviolet light is used in the treatment of osteomalacia and rickets. It is also used for its bactericidal effects in the treatment of bedsores and indolent ulcers but here its effect is nullified if the sores are not debrided mechanically first, for the light waves are absorbed by the superficial wound debris. Body ultraviolet light therapy without tanning should be used for its "tonic" effect in debilitated patients and in patients with anemia and osteoporosis.

A useful but little-known property of ultraviolet light is its counterirritant effect in the treatment of deep scars, particularly following periosteal bruising, especially at the elbow and ankle. Ultraviolet light also has a diagnostic and therapeutic place in the handling of some skin conditions, but this use is beyond the scope of this book.

X-ray Therapy

X-ray therapy merits mention in this section though it has been all but abandoned in the treatment of musculoskeletal problems. While it has a role to play in the treatment of calcific tendinitis, myositis ossificans, and ankylosing spondylitis, its use should be reserved for cases that are recalcitrant to other less potentially hazardous forms of treatment.

GENERAL OBSERVATIONS CONCERNING PHYSICAL THERAPY

The prescription of one modality of physical therapy, a common practice, is usually a futile exercise on the part of the prescriber and an abuse of both physical treatment and the patient.

Relaxation Therapy

In the prescription of physical therapy for the acute phase of pain in the musculoskeletal system, whether the pathological cause is acute or chronic, the constant goal is to produce relaxation. This is the aim whether the pain element is treated or whether the reactive spasm arising from the pain is attacked. Specifically designed to achieve relaxation is Jacobson's relaxation exercise regimen (p. 126). There is little doubt, moreover, that hypnosis has a place in the treatment of pain and in producing relaxation.

To achieve the maximal benefit from any form of physical therapy, the therapist must be allowed discretionary latitude not only in the choice of modality but also in the time and intensity of its use. Changes in treatment must be made in response to a patient's reaction, which may vary from day to day.

We have remarked that if a patient does not show a predictable response to physical treatment in a predictable time the diagnosis is wrong, the treatment is wrong, or both. Patients can become addicted to physical treatment.

It should be remembered that any benefit derived from a physical therapy session of, say, 30 minutes a day is readily lost unless the principles of the treatment are adhered to throughout the other 23½ hours of the day. Traveling any distance to and from therapy also may completely undo whatever good was done. Most programs of physical treatment have a summation of effects as day succeeds day.

Physical treatment is time-consuming but, properly applied, it is not only economical from a financial point of view but a material factor in lessening the duration of disability from musculoskeletal pathology, especially that arising from trauma. To delay the prescription of physical treatment is to invite morbidity. It is very satisfactory to know how to restore a lost bodily function, but timely physical treatment can often prevent loss of function or reduce the degree of loss—and that outcome is even more important and satisfying. This is a lesson for third-party providers of medical care to learn.

BIBLIOGRAPHY

Barbor, R. G. Rationale of manipulation of joints. *Arch. Phys. Med. Rehabil.* 43:615, 1962.

Barbor, R. G. Sclerosing injections for lax ligaments. Proceedings of the Fourth International Congress of International Federation of Manual Medicine, 1972.

Bennett, R. L., Hines, E. A., and Krusen, F. H. Effect of short-wave diathermy on the cutaneous temperatures of the feet. *Am. Heart J.* 21:490, 1941.

Bierman, W., and Licht, S. *Physical Medicine in General Practice* (3rd ed.). New York: Hoeber, 1960.

Blount, W. P. Don't throw away the cane. *J. Bone Joint Surg. (Am.)* 38A:695, 1956.

Bonica, J. J. Acupuncture anesthesia in the People's Republic of China: Implications for American medicine. *J.A.M.A.* 229:1317, 1974.

Brudny, J., Grynbaum, B. B., and Korein, G. Spasmodic torticollis: Treatment by feedback display of the EMG. *Arch. Phys. Med. Rehabil.* 55:403, 1974.

DeLateur, B. J., Lehmann, J. F., Warren, C. G., Stonebridge, J., Funita, G., Cokelet, K., and Egbert, H. Comparison of effectiveness of isokinetic and isotonic exercise in quadriceps strengthening. *Arch. Phys. Med. Rehabil.* 53:60, 1972.

Distefano, V., and Nixon, J. E. An improved method of taping. *J. Sports Med.* 11:209, 1974.

Downey, J. A., and Darling, R. C. *Physiological Basis of Rehabilitation Medicine.* Philadelphia: Saunders, 1971.

Fordyce, W. E., Fowler, R. S., Lehmann, J. F., DeLateur, B. J., Sand, P. L., and Treischmann, R. B. Operant conditioning in the treatment of chronic pain. *Arch. Phys. Med. Rehabil.* 54:399, 1973.

Fowler, R. S., and Kraft, G. H. Tension perception in patients having pain associated with chronic muscle tension. *Arch. Phys. Med. Rehabil.* 55:28, 1974.

Glick, E. N., and Lucas, M. Ice therapy. *Ann. Phys. Med.* 10:70, 1969.

Grant, A. E. Massage with ice (cryokinetics) and treatment of painful conditions of the musculoskeletal system. *Arch. Phys. Med. Rehabil.* 45:233, 1964.

Greenhoot, J. H. The management of intractable pain. *Va. Med. Monthly* 97:117, 1970.

Hardy, J. D. Physiological responses to heat and cold. *Ann. Rev. Physiol.* 12:119, 1950.

Hollander, J. *Arthritis.* Philadelphia: Lea & Febiger, 1972.

Jacobson, E. *Progressive Relaxation.* Chicago: Univ. Chicago Press, 1938.

Kane, R. L., Olsen, D., Leymaster, C., Woolley, F. R., and Fisher, F. D. Manipulating the patient: A comparison of the effectiveness of physician and chiropractor care. *Lancet* 1:1333, 1974.

Lehmann, J. F., McMillan, J. A., Brunner, G. D., and Blumberg, J. B. Comparative study of the efficiency of short-wave, microwave and ultrasonic diathermy in heating the hip joint. *Arch. Phys. Med. Rehabil.* 40:510, 1959.

Livingston, W. K. *Pain Mechanisms.* New York: Macmillan, 1943.

Melzack, R., and Wall, P. D. Pain mechanisms: A new theory. *Science* 150:971, 1965.
Meyer, G. A., and Fields, H. L. Causalgia treated by selective large fibre stimulation of peripheral nerve. *Brain* 95:163, 1972.
Shaffer, D. V., Branes, G. K., Wakim, K. G., Sayre, G. P., and Krusen, F. H. The influence of electric stimulation on the course of denervation atrophy. *Arch. Phys. Med. Rehabil.* 35:491, 1954.
Steinbrocker, O., and Neustadt, D. H. *Aspiration and Injection Therapy in Arthritis and Musculoskeletal Disorders.* New York: Harper & Row, 1972.
Stravino, V. D. The nature of pain. *Arch. Phys. Med. Rehabil.* 51:37, 1970.
Sweet, W. H., and Wepsic, J. G. Treatment of chronic pain by stimulation of fibers of primary afferent neuron. *Trans. Am. Neur. Assoc.* 93: 103, 1968.
Swezey, R. L., and Silverman, T. R. Radiographic demonstration of induced vertebral facet displacements. *Arch. Phys. Med. Rehabil.* 52:244, 1971.
Wale, J. O. *Tidy's Massage and Remedial Exercises* (11th ed.). Bristol, England: Wright, 1968.
Williams, P. C. The Conservative Management of Lesions of the Lumbosacral Spine. In *Instructional Course Lecture.* Ann Arbor, Mich.: American Academy of Orthopaedic Surgeons, 1953. Vol. 10, p. 90.
Winnie, A. O., Hartman, J. T., Meyers, H. L., Ramamurthy, S., and Barangan, V. Pain clinic II: Intradural and extradural corticosteroids for sciatica. *Anesth. Analg.* 51:990, 1972.
Young, D. The medical hydrology approach in post-operative hip fracture care. *Med. Hydrology* Q.2:1, 1970.
Zankel, H. T., Cress, R. H., and Kamin, H. Iontophoresis studies with a radioactive tracer. *Arch. Phys. Med. Rehabil.* 40:193, 1959.

8
Crossmatching Clinical Diagnoses with Treatment Principles

The musculoskeletal differential diagnosis system and principles of physical treatment are considered in the context of a specific joint to demonstrate their practicability. The shoulder is chosen because it is not only the joint that is a very common source of musculoskeletal symptoms but also the one to which pain from pathological conditions is referred more often than any other, except perhaps the spinal joints. Thus, the differential diagnosis of shoulder joint pain presents more problems to the physician than any other joint.

CAUSES OF PAIN LOCAL TO THE GLENOHUMERAL JOINT

Fracture

A fracture of the surgical neck of the humerus is not always immediately revealed by radiographic examination; the diagnosis may have to be made on clinical grounds. Impaired sound conduction of bone may be the clue (see Figure 4-2, p. 54); or Stimson's sign, which elicits pain at the site of fracture in a long bone when it is percussed in its long axis, may be present.

An avulsion fracture of the greater tuberosity of the humerus is considered under Tear of the Rotator Cuff, p. 157.

Treatment. The use of the hanging cast in the treatment of a fracture of the neck of the humerus makes one wonder whether strict immobilization of fractures should really be the prime object in promoting their healing. A fractured clavicle cannot be very well immobilized either, yet delayed union or nonunion of these two fractures seldom occurs. In any case, physical therapy modalities should be used while bone healing is taking place to prevent morbidity in the wrist, the hand, and if possible, the elbow, so that restoration of unnecessarily lost function is avoided. Joint manipulation of the glenohumeral joint may have to be undertaken in the restoration phase of therapy.

Dislocation

There is usually no problem in diagnosing the common anterior dislocation of the head of the humerus on the glenoid, but the posterior dislocation is often overlooked, and the diagnosis of this condition has been described in Chapter 5, p. 69.

Treatment. Treatment of the usual anteroinferior dislocation has the initial purpose, of course, of reducing the dislocation. Reduction can be achieved if the sedated patient lies prone, holding a weight over the edge of the table on the injured side. There are, in addition, two classic maneuvers: the Kocher method and the hippocratic method. Both are traumatic and subject the capsule and the muscles

—and no doubt the blood vessels and nerves in relation to the joint—to undue stretching and even tearing. The techniques employed in examining for joint-play movements of the glenohumeral joint are highly controlled, equally effective, and atraumatic.

Following reduction it is common practice to immobilize the joint in a Velpeau bandage or plaster for 14 to 21 days on the grounds that the capsule has been severely injured and probably torn. Whether this is the case or not (and the subscapular bursa part of the capsule may surely accommodate the displaced humeral head without the occurrence of a tear), immobilization following dislocation contravenes our principles of physical therapy and promotes morbidity. There may be some bleeding into the joint, and certainly traumatic synovitis supervenes. There is every reason not to immobilize the joint and every reason to treat the whole arm with physical therapy modalities from the start, including the use of the correct dosage of passive movement to the joint itself. To relax the shoulder girdle muscles and to allow proper therapy to be given, the patient must be treated in the supine position with the arm resting on a pillow or supported at all times by the therapist, thus avoiding the strain of shoulder extension. Between treatments the arm should be rested in a sling, but the sling should be put on while the patient is still recumbent to avoid having the weight of the arm fall on the shoulder by the force of gravity when he resumes the upright position. At night the arm should rest on a pillow as it does during therapy sessions. The reason is that anyone lying supine has both shoulders in extension to some degree; thus relaxation and rest are prevented because all the soft tissues in and around the joint are being stretched.

The choice of physical therapy modalities is predicated on the amount of pain to be relieved, the amount of excess synovial fluid within the joint, whether one suspects there is blood in the joint, and the amount of reflex muscle spasm that is present. Anodal galvanism relieves pain and encourages the absorption of synovial fluid. Ice massage of muscles in spasm reduces and relieves the spasm. Passive movement in the range of all the voluntary movements and within the limit of pain should be used, and each treatment session should be completed with active movement of all the other joints of the limb, assisted as necessary. Whole-limb stroking massage for relaxation completes the treatment and is best given with the patient lying on his uninjured side. The injured arm can then rest on the chest wall while the massage is being applied.

If the injury was a posterior dislocation, surgical reduction may have been necessary; but the physical therapy program follows the same outline, and there is no need to wait for the stitches to be removed before therapy is started. If closed reduction is achieved, there must be intracapsular bleeding because the posterior dislocation is usually associated with a fracture into the head of the humerus. In this case aspiration of the blood as early as possible is essential to prevent morbidity and to promote the early success of treatment. The patient should be taught to do isometric exercises with the muscles in relation to the injured joint between treatments and to maintain voluntary movement in the uninvolved joints and muscles.

Synovitis

Almost all joint trauma is associated with some degree of traumatic synovitis. If the injury is severe, bleeding into the joint will occur and hemarthrosis is the primary pathology.

Treatment. Treatment is described under Hemarthrosis, below.

Hemarthrosis

Blood within a joint should be considered an urgent matter, for blood has a lysin in it (as does pus) which readily destroys hyaline cartilage. The early removal of blood from a joint is mandatory.

It is therefore important to arrive early at a clinical decision as to whether or not there is blood in the joint. Table 8-1 indicates the dif-

Table 8-1. Differential Diagnosis, Synovitis, and Hemarthrosis

Symptoms and Signs	Synovitis	Hemarthrosis
Onset	Slow: up to 24 hours	Fast: within minutes
Pain	Aching	Pain a feature
Size	May be very large	Small initially
Heat	Warm	Hot

ferential points of diagnosis between traumatic synovitis and traumatic hemarthrosis.

Blood, being foreign to the joint, sets up acute inflammation of the capsule; hence the pain and heat. The swelling is relatively small because the clotting mechanism stops the bleeding and pressure is built up within the joint.

Treatment. In hemarthrosis, blood must be aspirated. Fat in the aspirated blood indicates an associated fracture. One of the best pain-relieving measures is to replace the aspirated blood with air. Why this relieves pain is not known but it is a clinical fact. Some hyaluronidase preparation may be instilled with advantage before the needle is removed. A triple-layer cotton wadding and Ace bandage pressure dressing may be applied, as well as ice packs, again with some advantage. After the blood has been removed, traumatic synovitis supervenes and the physical therapy program for both conditions becomes the same.

Excess synovial fluid should not be aspirated from a joint (except for diagnostic purposes) because synovial fluid is normal to a joint and its excess will be absorbed as healing occurs. Its disappearance is the best indication that the diagnosis is correct and that function of the joint may be resumed safely. Should the swelling of synovitis become very painful or so large as to risk the viability of the soft tissues or the efficiency of circulation, aspiration is indicated.

The proper program of physical therapy for traumatic synovitis is given in detail in our discussion of dislocations, and the reader is referred to that section on p. 153.

Tendinitis

SUPRASPINATUS TENDINITIS

Since no significant ligaments support the glenohumeral joint, the next cause of pain local to the joint is something arising in muscle. Stabilization of the joint while the arm is being used is achieved by the rotator cuff muscles, which by and large are well designed for this purpose, except for the supraspinatus. The supraspinatus is badly designed because of its poor leverage for stabilization and its three additional functions: (1) external rotation of the shoulder, (2) initiation of abduction of the shoulder, (3) strength of action when one is carrying heavy objects with the arm outstretched by the side. The muscle and its tendon are grossly overtaxed in everyday living, and the musculotendinous junction is a frequent site of early degeneration and inflammation: tendinitis. Inflammation of the tendon at its insertion causes a similar syndrome.

There is one place where stress can be laid on the musculotendinous junction by palpation. The examiner's index fingertip is placed in a triangle bounded laterally by the acromion, anteriorly by the clavicle, and posteriorly by the spine of the scapula (Figure 8-1). Tenderness on palpation here that is aggravated by the initiation of an isometric contraction of the muscle in abduction indicates the diagnosis of tendinitis.

There is a painful arc of movement which is helpful in the diagnosis of tendinitis. To test for this, the examiner asks the patient to abduct his arm slowly; he does so until he reaches a point where he complains of pain and cannot move it higher. The examiner then takes the arm and gently elevates it a few degrees. Then the patient is able to complete the range of motion. The same observation is made as the arm is lowered. The explanation of this sign is that the tendon is inflamed and swollen and is unable to glide under the acromion without

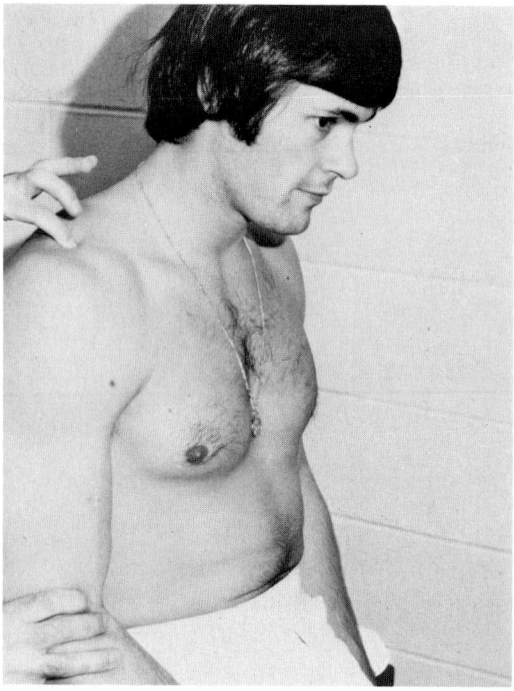

Figure 8-1. Palpating the musculotendinous junction of the supraspinatus muscle in a space bounded laterally by the acromion process, posteriorly by the spine of the scapula, and anteriorly by the outer end of the clavicle. Pain under the palpating index finger made worse by the initiation of shoulder abduction resisted by the examiner is diagnostic of supraspinatus tendinitis.

being pinched when it is contracting actively. When it is at rest, it can be eased under the acromion, after which there is room to accommodate it.

Radiographically there is nothing to help make this diagnosis, but ectopic calcium in relation to the shoulder joint is usually a calcareous deposit in the tendon. The presence of calcium in such a situation denotes an attempt by nature to heal and does not alter the treatment.

Treatment. The principles of physical therapy must be followed in deciding on a program of treatment; rest from function of the involved muscle is prerequisite to success, whatever other modalities are used, even the injection of hydrocortisone. The latter is useful providing the injection is made at the musculotendinous junction if that is the point of maximum tenderness, rather than solely at the insertion of the tendon. Prevention of morbidity—in this case stiffening of the joint—must be avoided by passive movement of the joint within the limits of pain. It can be achieved by the patient, between treatments, lying supine in a bath or pool and moving the shoulder in its painless range of movement with the assistance of the other arm. Rhythmic stabilization exercise is particularly helpful in restoring range of movement and increasing muscle strength. Codman's pendulum circumduction exercises are also useful, but the pendulum swing must be done with the muscles completely relaxed and in a clockwise and counterclockwise fashion (see Figure 7-2, p. 123). The treatment of choice is ice massage over the involved area of muscle followed by whole-limb massage for relaxation. A sling should be worn between treatments, put on while the patient is lying down. Maintenance of elbow, wrist, and finger joint activity is essential throughout the period of treatment. If trigger points in the surrounding muscles are painful, they must be treated (see pp. 158 and 190). When ectopic calcification is present, some physicians advocate the use of surgery, washing it out by needling, or x-ray therapy. However, in the majority of cases properly prescribed physical therapy is the treatment of choice. Further, it should be remembered that the ectopic calcium seen in a radiograph may not be the cause of the patient's symptoms at all.

TENDINITIS OF SHORT AND LONG HEAD OF BICEPS

Tendinitis of the short head of the biceps is another frequent cause of local shoulder pain. The diagnosis is made by palpation of the musculotendinous junction by the thumb, which presses in the deltopectoral groove just below the coracoid process of the scapula. If tenderness is elicited and it is increased by initiation of an isometric contraction of the biceps muscle, the diagnosis is plain. The tendon of the long head of the biceps is less commonly involved but it should also be checked by palpation in the bicipital groove; increased pain on resisted movement is diagnostic in a routine clarifying examination. Referred pain from a distant trigger point may be mistaken for tendinitis but location of pain on palpation of the trigger point in its predictable location should make the diagnosis clear.

Treatment. The physical treatment of tendinitis has been detailed in the foregoing paragraphs on supraspinatus tendinitis, to which the reader is referred. In treating tendinitis anywhere in the musculoskeletal system, anti-inflammatory medications are a helpful adjunct to physical therapy.

Tenosynovitis of Long Head of Biceps

Though the tendon of the long head of the biceps may be the seat of tendinitis, this tendon has a sheath which may be the seat of inflammation causing the symptoms. With a full-blown tenosynovitis, tenderness on palpation and an associated feeling of sticky crepitus on movement of the muscle make the diagnosis certain. However, it is sometimes very difficult to differentiate tendinitis from tenosynovitis. One may have to make diagnostic use of local anesthesia blocking the suprascapular nerve, which cuts out pain from the tendon sheath but not from the tendon, the latter being innervated by the musculospiral nerve. The importance of diagnosing tenosynovitis lies in the fact that this condition may herald the onset of systemic disease—a collagen vascular disease or tuberculosis, for instance.

The long head of the biceps not uncommonly ruptures and is the cause of acute shoulder pain.

Treatment. Traumatic tenosynovitis is treated by following the principles of physical therapy, namely, rest from function while healing is taking place and the avoidance of morbidity. To prevent adherence of the tendon to its sheath, movement within the limit of pain of the tendon within the sheath is important. It can be achieved by isometric contraction of the muscle, but often this is too gross. A more controlled and gentle movement is obtained by surging faradic stimulation. Medical galvanism over the tendon sheath is pain-relieving and promotes the absorption of the synovial exudate. To relieve tension on the inflamed tendon a sling is worn between treatments. The injection of hydrocortisone into the sheath is a good treatment providing it is not relied upon as the sole method of treatment. Hydrocortisone can be instilled locally by phonophoresis. A 10 percent hydrocortisone cream is used as a coupling agent, and ultrasound is applied using 0.75 to 1.5 watts per square centimeter for two minutes initially, gradually increasing the duration to four minutes.

If the tenosynovitis is part of a systemic disease process, the disease must be treated and local therapy avoided except to maintain the gliding of the tendon within the sheath. In differential diagnosis it should be remembered that tendon sheaths may be involved in neoplastic changes. Xanthomas, giant cell tumors, and malignant neoplasms are not unknown.

Tear of the Rotator Cuff

It has long been taught that a blow or fall onto the point of the shoulder is what causes a rotator cuff tear, but spontaneous tears are common in patients over 50 years of age. If the same trauma—a blow or fall onto the point of the shoulder—produces an avulsion fracture of the greater tuberosity of the humerus, it is not associated with a tear of the rotator cuff. Healing of such a fracture will occur without difficulty so long as the tuberosity separation is less than ⅛ inch as shown radiographically. With a separation of ⅛ inch or more, fixation of the fragment with a screw is recommended. Surgical intervention does not materially alter treatment using physical therapy modalities.

The diagnosis of tear of the rotator cuff is arrived at clinically with the loss of the stabilizing function of the rotator cuff. Thus the patient cannot maintain abduction passively attained. This loss of function is not specific for this condition; suprascapular or axillary nerve neuritis or injury produces the same clinical picture. A radiculitis involving the fifth cervical root presents a similar picture. Another sign of rotator cuff tear is available. With the patient supine, 40 degrees or more of abduction is relatively easily and painlessly achieved by passive means. If, at the extreme of painless abduction, the arm (well supported) is taken gently through external and then internal rotation by the examiner, the patient experiences a sharp pain, and there is a click at the same point in external rotation and return in internal rotation at the point of pain, the diagnosis of a tear of the rotator cuff is established. Arthrography confirms the diagnosis (Figure 5-23, p. 77).

Treatment. Surgeons claim that surgical repair of a torn rotator cuff is necessary. From the study referred to on p. 29, it must be obvious that this is not the case. A large tear probably should be repaired surgically, but for the most part surgical repair is not necessary if the patient does not require a full anatomical range of movement for his activities of daily living. A small tear appears to heal with a well-designed program of physical therapy that does not violate the principles of therapy. It is outlined under Dislocation, p. 153. The difference is that with a tear one can expect a healing time of about eight weeks, and the progress of the program must be adjusted to this.

Fibrositis of the Muscles of the Shoulder

Any of the muscles of the shoulder girdle may be infiltrated with fibrositic nodules, resulting in pain and loss of function of the glenohumeral joint. The diagnosis is arrived at by palpation, which reveals multiple tender nodules, each locally painful. Skin rolling is tight and tender over them. Referred pain from the nodules does not occur in any predictable pattern.

Treatment. When cold is the causative factor of fibrositis, treatment by wet heat followed by kneading massage with or without skin rolling and then by light stroking massage for relaxation usually brings ready relief for the patient. It may be necessary to treat an especially painful or recalcitrant nodule with ultrasound applied for perhaps as little as a minute using a low wattage; the application of medical galvanism over a nodule has the same effect. When a patient does not respond to therapy as expected, it may be that a distant focus of infection is keeping the process alive. In this case the focus of infection must be looked for and eradicated if it is found. Then physical therapy should be effective.

The use of the moist air bath is helpful in the treatment of this condition; see Chapter 7, p. 118. When the muscles lose their tenderness, a program of conditioning exercises helps to bring back the normal pumping action in them and restores their normal function. Between the time of near comfort in the muscles and starting an active exercise program, ultrasound with simultaneous electrical stimulation may be used to condition the muscles to their resumption of normal function.

The Trigger Point

The diagnostic differentiation between a trigger point and the nodule(s) of fibrositis is based on the fact that the trigger point is isolated and is in a location which is predictable by study of the pattern of pain described by the patient. Further, palpation of the trigger point sets off the predictable pattern of referred pain that the patient has described. In contrast, the location of fibrositic nodules is not predictable, nor is the pattern of pain referred from them.

Treatment. The vapocoolant spray (Fluori-methane*) and muscle stretch technique make up the most satisfactory treatment of pain and spasm arising from an irritable trigger point. It is not entirely clear why this treatment works, but the rapid alleviation of pain indicates that the effect of the spray is probably counterirritant. In fact, the Fluori-methane is

* Available from Gebauer Chemical Company, St. Catherine Road, Cleveland, Ohio, 44104.

delivered from a bottle through a calibrated nozzle as a jet stream, and we feel that the constant light touch is an important part of the effective mechanism. Using ice in the same manner is not as effective—partly, it seems, because of the inconstancy of the touch sensation. For this reason we employ the term *touch-cold sensation* when describing the use of Fluori-methane. Ethyl chloride, which is too cold, a local and general anesthetic, flammable, and explosive, is less effective than Fluori-methane and should not be used.

The working explanation of the therapeutic effect of Fluori-methane and muscle stretch technique is that the spray impulses from the skin (cold) are transmitted to some pain center, possibly in the thalamus, by larger afferents via the lateral spinothalamic tracts faster than are the noxious protopathic pain impulses from muscle spindle dysfunction, which initially are transmitted by the smaller gamma afferents. In this way, the nociceptive impulses from the skin arrive at the center quicker than the noxious impulses from the muscle, thereby setting up a refractory period during which the reception of the noxious impulses is blocked. This essentially is the gate theory of blocking pain. During the period of "block" the muscle can relax and be stretched to its normal resting length. The normal resting length of muscle being a pain-free state, the painful spasm is overcome, the spindle dysfunction is overcome, and normal tonic reflexes are restored, resulting in normal pain-free function. In this explanation we have to ignore the effect of the constant "touch" of the jet stream as touch sensations travel via the posterior columns. But clinically its effect does seem to be important.

Sometimes the use of the touch-cold method is not effective when theoretically and clinically it should be. In that event we inject the trigger point with ½ percent Novocain. This seems to "explode" the trigger point. The effect is probably due to the intense pain of the needling, which becomes the competing nociceptive impulse, rather than to the influence of the local anesthetic of the injected Novocain. Novocain is very rapidly metabolized, though it does have a curare-like effect which may play a part in the outcome. Stretching the involved muscle is still essential to the success of treatment.

With either of these methods of treatment it is essential to remember that, though pain is relieved, the primary cause of the pain is not necessarily determined. Moreover, a program of physical therapy is often required to maintain relief of symptoms. It must consist in the application of gentle moist heat, gentle stretching exercises, and a gradual return to functional activity.

The vapocoolant spray and stretch (or injection and stretch) treatment may fail if too intense cold is used (cold causes spasm), if the muscle is too vigorously stretched (overstretch of muscle causes spasm), if the spasm is primarily due to nerve pain (the stretch phase of treatment stretches the involved nerve elements and causes more pain), and if the spasm is residual to a healed fracture in which there is loss of bone length (the muscle cannot be stretched to its normal resting length but only to an adapted resting length). It also fails if a satellite trigger point rather than the primary trigger point is treated. It also fails if the area of referred pain is treated and not the trigger point itself. It also fails if the somatic component of a visceral pain is treated but the primary visceral causes are not eliminated.

To treat pain from a trigger point effectively by use of the vapocoolant spray and stretch method, the jet stream is aimed at the trigger point, and the therapist sweeps the jet stream from the trigger point to the insertion of the affected muscle. Then the area of referred pain is sprayed from the trigger point. If simple muscle spasm is being treated, the sweep of the jet stream is from origin to insertion of the muscle. Three or four unidirectional sweeps of the jet stream are usually sufficient. The bottle is held about 18 inches from the part of the body being sprayed, and the sweep rate should be about 4 inches per second. The jet stream should hit the skin at a 45-degree angle. The muscle stretch is superimposed as each sweep of the jet stream approaches its end point. Figure 7-4*A*, p. 127, illustrates the position adopted in spraying the right splenius capitis muscle; 7-4*B* illustrates the position adopted in spraying the shawl area of the right trapezius muscle. Figure 7-5, p. 129, illustrates the injection of a trigger point in the right infraspinatus muscle. Figures 9-12 to 9-15 on pp. 191–194 are Dr. Janet Travell's illustrations of common trigger points and their predictable patterns of pain.

Pyarthrosis

Whatever the cause of pyarthrosis, pus must be surgically drained from the affected joint. Antibiotics may be instilled into it following aspiration though there is evidence to suggest that this maneuver is no more efficacious than the giving of antibiotics systemically. Early removal of pus spares the articular hyaline cartilage, which is readily destroyed by it. The differential diagnosis of pus from blood and from excess synovial fluid within the joint should be possible by consideration of the history of onset of the symptoms and the signs obtained by clinical examination. Then a diagnostic aspiration may have to be undertaken to determine the organism and its sensitivity to antibiotics.

Treatment. Once drainage of the joint is achieved, antibiotic therapy is established, and all signs of infection have subsided, treatment by physical therapy is started following the principles of treatment during the healing phase. When healing is completed, treatment progresses to the restorative phase to restore the almost inevitable loss of function.

Capsulitis

Capsulitis is synonymous with frozen shoulder. However, it is well to define clearly our concept of capsulitis.

The normal capsule of the glenohumeral joint passes from the glenoid superiorly to the junction of the head of the humerus and the greater tuberosity. Inferiorly, it passes from the glenoid junction of the head of the humerus and the lesser tuberosity. In its inferior aspect it is a large synovial pouch erroneously termed the subscapular bursa. This pouch is obviously present so that the arm can be raised above the head either through forward flexion, full abduction, or any mixture of both without stretching the capsule. Figure 8-2 diagrammatically illustrates this concept. The large subscapular pouch contains a normal amount of normal circulating synovial fluid.

Now, if for any reason a patient has a feeling of pain in the shoulder joint, the structures surrounding and supporting the joint—namely, the muscles—go into reflex spasm. This is a

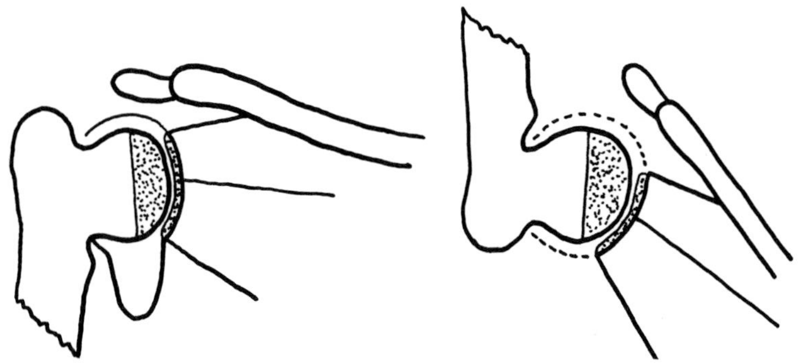

A *B*

Figure 8-2. *A*, A diagrammatic picture of the normal capsule of the glenohumeral joint. *B*, The arm extended above the head illustrating how the subscapular "bursa" is no longer pouched and how external rotation of the head of the humerus has occurred, allowing the use of the articular surface of the head of the humerus for a second time. The scapula has rotated through an arc of 40 degrees and the clavicle has rotated on its own axis.

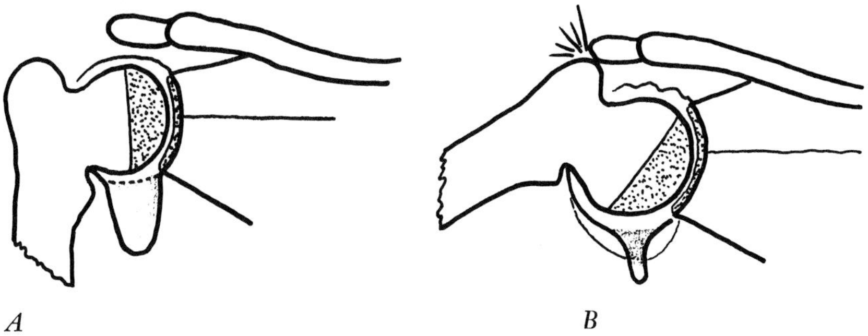

Figure 8-3. *A,* Stippling in the subscapular bursa indicating stagnant viscous synovial fluid which eventually effectively closes off the bursa as indicated in *B.* When the bursa is "closed off" the head of the humerus cannot drop downward or backward on the glenoid nor externally rotate on it. The diagram also indicates how the greater trochanter now impinges on the acromion, pinching the structures lying beneath, namely, the tendon of the supraspinatus muscle and the subacromial bursa.

normal reflex reaction to pain in any part of the musculoskeletal system. Movement of the splinted joint becomes impaired or lost, and the synovial fluid in the pouch ceases to circulate and becomes stagnant. Stagnant synovial fluid gets more and more viscous, and eventually the walls of the pouch become adherent, leaving an ineffective capsule without a pouch and mechanically much too small for freedom of movement of the head of the humerus within it to allow normal function (Figure 8-3).

Furthermore, the normal scapulohumeral rhythm of the 17 muscles attached to the scapula is disrupted. The ability of the head of the humerus to glide downward and backward on the glenoid is lost. Figure 8-4 illustrates the position of the humerus on the glenoid—*A,* at rest and *B,* in function. The joint-play rotation of the head of the humerus on the glenoid is also lost.

Because the head of the humerus does not drop downward and backward on the glenoid, and also because it loses its rotation, the greater tuberosity of the humerus cannot clear the acromion on abduction and forward flexion. Before 40 degrees of either of these movements is achieved, the tuberosity impinges on the acromion and pinches anything lying between them, namely, the subacromial bursa and the supraspinatus tendon (see Figure 8-3). If attempts are made to increase the range of anatomical glenohumeral joint movement under these circumstances, a traumatic bursitis and a traumatic tendinitis ensue. These conditions cause more pain, reactive muscle spasm becomes more intense, and now morbid changes take place in the muscles. Diffuse fibrositis may occur with the unrelieved muscle spasm because of loss of normal pumping action. Trigger points are activated, and referred pain patterns, typically in the infraspinatus, anterior deltoid, and pectoralis muscles, become established. Muscle atrophy and weakness from disuse occur. The blood and lymphatic drainage also becomes impaired.

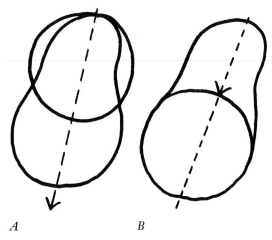

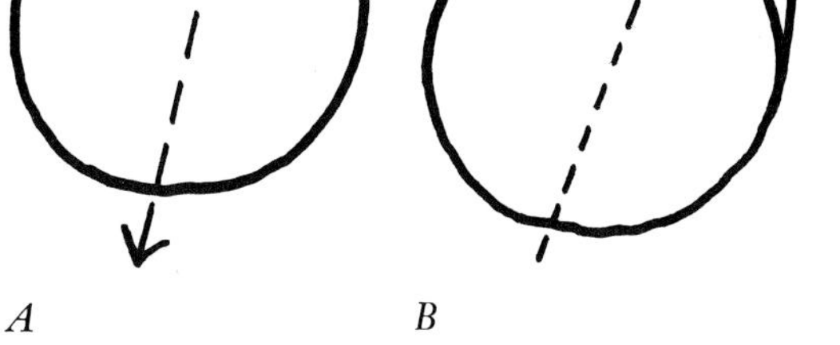

Figure 8-4. *A,* Position of the head of the humerus articulating on the small upper part of the pear-shaped glenoid at rest. *B,* The head of the humerus glides downward and backward onto the large dependent part of the glenoid in all functional movements of the glenohumeral joint.

This is the full-blown clinical picture of the frozen shoulder, with changes in virtually every anatomical structure local to the shoulder girdle and secondary changes in the physiological systems. If corrective measures are delayed, the neck structures, sometimes including the autonomic nervous system, the upper thoracic spine, and the chest wall structures become involved as well with pathophysiological morbid changes. One of the most difficult to reverse is adherence of the scapula to the chest wall. The complete shoulder-hand syndrome may result.

Treatment. The primary treatment of capsulitis should be its prevention by the early institution of proper treatment of the pathological condition originally causing pain to be felt in the shoulder. Besides those causes of pain in the shoulder arising from pathological conditions local to the shoulder, the list of which has still to be completed, the causes of pain referred to the shoulder must be sought for and treated and morbid changes that may occur in the shoulder must be prevented by good prophylactic treatment.

It should be obvious that the simple prescription of an exercise program or a simple injection of steroids is unlikely to overcome the problems of capsulitis once it is established. A therapy program for a patient with capsulitis may have to use any and every modality of physical therapy available according to the principles of physical therapy already detailed. The modalities may have to be changed from day to day, according to the discretion of the professional physical therapist, and a prewritten plan of specific treatment cannot be followed if success is to be achieved. At sometime in this program manipulative therapy must be instituted to restore the lost joint-play movements of the head of the humerus, especially those involving the downward, backward, and rotation movements on the glenoid, without which voluntary functional movements are impossible to perform. The time when manipulation becomes appropriate is determined by two factors: First, the patient's pain is relieved by rest; second, a rule of thumb illustrated by a simple X pattern relating pain to movement is applied (Figure 8-5). For specific usages of therapy modalities the reader is referred to the appropriate section on treatment of the various structures involved in primary pathological or secondary morbid changes. In treatment of a frozen shoulder, a stubborn tender point on palpation is found just anterior to the greater tuberosity, which responds only to infiltration with a local anesthetic agent. There may be equally recalcitrant points of tenderness in the infraspinatus; they must be treated in the same way.

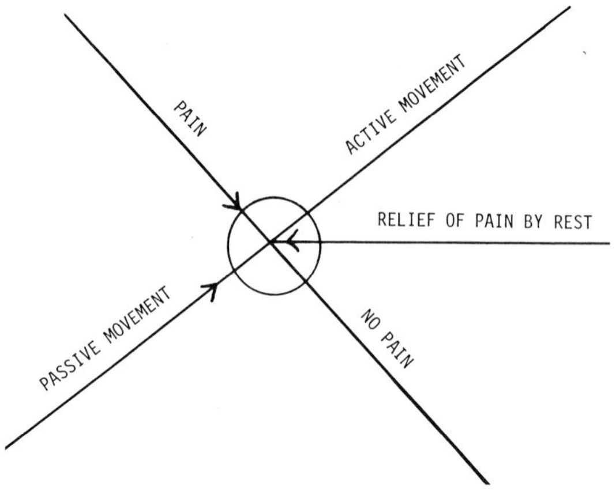

Figure 8-5. The X pattern used to decide at what point to change from a sedative physical therapy program to an active program including manipulation of a frozen shoulder. It is time to change when the symptom of pain is relieved by rest.

Osteoarthritis

The diagnosis of osteoarthritis as a cause of pain merits little consideration for reasons stated on p. 22. In any event, x-ray examination seldom shows gross changes in the glenohumeral joint characteristic of osteoarthritis even after severe preexisting injury.

Treatment. Treatment of pain in an osteoarthritic joint should be directed at the joint structure from which the pain is shown to be arising and follow the principles of therapy outlined in the appropriate section.

Bursitis

Primary bursitis at the shoulder is rare. When present, the symptoms and signs are clearly of that condition and are given in detail in Chapter 2, p. 32. That many patients who have shoulder pain develop secondary bursitis because of poorly conceived physical treatment is undoubted, although this is not the only cause of secondary bursitis. One can therefore perhaps accept the lay rather than professional term *bursitis* loosely used to describe local shoulder pain problems. Usually, however, the term is as meaningless diagnostically as is *whiplash injury* for problems giving rise to pain in the neck.

Treatment. Bursitis may herald the onset of systemic disease, particularly rheumatoid arthritis and gout. If this is the case, treatment of the cause must be instituted as well as local therapy for the bursitis, which may include ice packs, hot packs, aspiration, the instillation of steroids, or even surgical excision. Prevention of morbidity in the structures primarily unaffected by the pathology by proper use of physical treatment is necessary, as outlined previously.

Ligament Injury

No ligaments support the glenohumeral joint itself but we must mention this structure in connection with maintaining the crossmatching exercise. Indeed, there is a ligament adjacent to the joint which, if injured by strain, tear, or rupture produces glenohumeral loss of function and pain. This is the acromioclavicular ligament. Complete rupture produces a classic clinical picture, the step-down between the outer end of the clavicle and the acromion lateral to it. A tear or sprain of that ligament is detected by palpation and painful movement of the acromioclavicular joint in its joint-play range.

Treatment. The acromion and the outer end of the clavicle are brought together by strapping over the separated joint. Passing the strapping under the flexed elbow and over the acromial end of the clavicle lifts the arm up and pushes the clavicle down. With this treatment healing of the ligament may occur in about six weeks. Felt pads are placed under the strapping over the acromioclavicular junction and over the proximal end of the ulna to prevent skin changes from pressure and adhesive tape burns. While the acromioclavicular joint is thus immobilized (Figure 8-6), morbidity must be prevented in the joints of the elbow, hand, and wrist and in the musculature of the shoulder girdle, arm, and hand. With the patient supine, movement of the glenohumeral joint within the limit of pain must be maintained passively, while the acromion and clavicle are held in apposition. The choice of modalities to be used is determined by the principles of physical therapy already outlined. If healing does not occur, function of the shoulder is still possible with the arm up to about shoulder level. If more is required or the injury continues to be sufficiently painful, or if it becomes cosmetically distressing to the patient, surgical repair may have to be undertaken.

We subscribe to the proposition that, if surgical repair of a ligament is contemplated, the best results follow early surgery.

Meniscal Injury

There is no meniscus within the glenohumeral joint. Nevertheless, in maintaining the crossmatching exercise one must be aware that a meniscus close by may affect function of the glenohumeral joint, and indeed there is a meniscus in the sternoclavicular joint. Injury to this meniscus locks that joint, producing pain and dysfunction, and also impairs function in the glenohumeral joint, with pain on attempted movement.

Treatment. The restoration of joint play and normal function of the sternoclavicular joint is all that is required. Besides using the local manipulative movements, one may have to add the movement of long axis extension, which is achieved by exerting a therapeutic pull on the

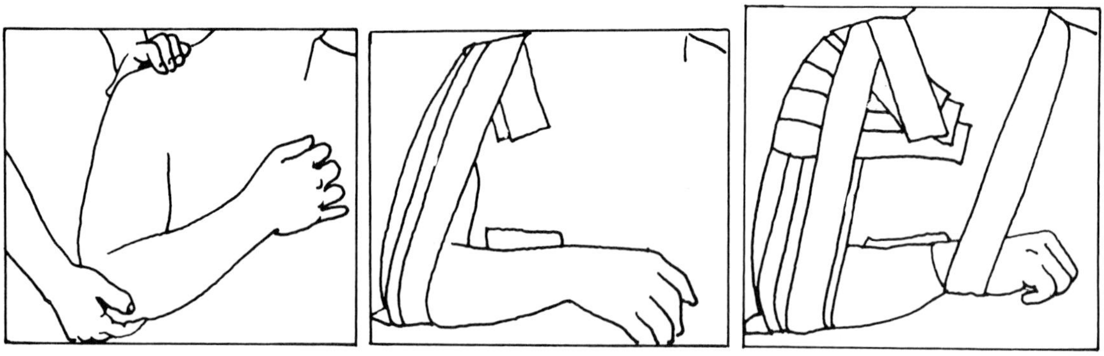

A B C

Figure 8-6. *A,* Patient sits with injured arm in a flexed position. The acromion and clavicle must be pushed together to approximate normal position of the joint. A felt pad (2 × 4 inch) or a low-density foam pad (of the same size) is placed over the acromioclavicular joint and a piece of cotton or a 4 × 4 inch gauze pad is placed under the elbow. *B,* The first strip of tape (1½ inches or 2 inches wide) is started just below the clavicle in front of the pad, carried over the pad, down the back of the arm, under the elbow, and up the front of the arm to the starting point. Maximal pressure is employed in applying the tape over the pad, to ensure stability at the acromioclavicular joint. Three more strips of tape are applied in the same manner over the first strip for added strength and support. *C,* Further stabilization can be achieved by applying three or four strips of 1½ inch wide tape around the upper arm, chest, and back, but not encircling the body. A final strip of 1½ inch wide tape can be applied to act as a sling and hold the wrist and forearm in place. This final strip starts with attachment to the wrist, is pressed against the chest so that adhesion can add stability, and ends at the bottom of the scapula. (Reprinted by permission of the copyright holder, © Johnson & Johnson, 1973.)

arm abducted to 90 degrees. Local pain in the sternoclavicular joint and any synovial reaction must be treated with heat or an ice pack, medical galvanism, and intraarticular injection with steroids. Rest from function (i.e., of the arm) should be part of therapy until the local joint reaction has subsided.

Metabolic Disease Affecting Menisci

Chondrocalcinosis and ochronosis may affect the sternoclavicular meniscus and prevent movement of the glenohumeral joint. There is no specific physical treatment for these conditions.

Freedom of Movement of Other Joints and the Scapula

For the glenohumeral joint to be free to move, there must be freedom of movement in the sternoclavicular joint and the acromioclavicular joint. Figure 8-7 illustrates performance of joint-play movement at the sternoclavicular joint. Figure 8-8 illustrates performance of joint-play movement at the acromioclavicular joint. Further, the scapula must be free to move on the chest wall through an arc of 40 degrees independent of arm movement. Movements in the two joints are pure joint-play

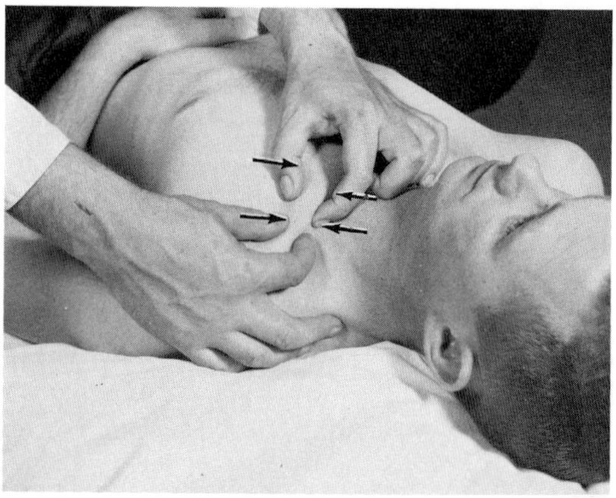

Figure 8-7. Performing joint-play movement of anteroposterior glide at the sternoclavicular joint. This movement must be free to allow painless full movement of the glenohumeral joint.

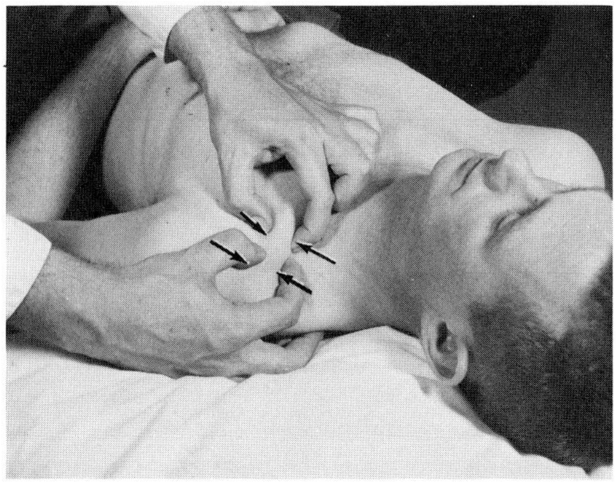

Figure 8-8. Performing joint-play movement of anteroposterior glide at the acromioclavicular joint. This movement must be free to allow painless full movement of the glenohumeral joint.

movements but allow rotation of the clavicle on its own axis of about 60 degrees. The way to examine for scapula movement on the chest wall is illustrated in Figure 8-9.

Treatment. If movement of the scapula on the chest wall is lost or if joint dysfunction is present, the lost movement must be restored by manipulative techniques before attempts are made to restore lost glenohumeral joint movement.

Eighteen diagnosable causes of shoulder pain that are local to the shoulder, or at least to the shoulder girdle, have now been discussed. The diagnoses of all these conditions must be on clinical grounds because, even in the cases of fracture of the neck of the humerus and of posterior dislocation of the glenohumeral joint, x-ray photographs may show no diagnostic changes early. Even in pyarthrosis the usual laboratory studies undertaken in patients with pain may, and probably will, initially show no diagnostic changes. Yet each cause of pain is diagnosable, and proper treatment brings freedom from pain and disability.

The more specifically orthopedic causes of shoulder pain local to the shoulder in which the primary approach to treatment is orthopedic surgery are not discussed. However, following surgical procedures, better and quicker recovery is achieved if the patient is afforded restorative physical treatment, the principles of which have already been outlined and are not repeated here. The earlier proper physical treatment is instituted, the fewer complications arise. It is much easier to prevent loss of function by physical therapy than to restore function when it is lost. Much disabling morbidity can thus be avoided and the patient re-

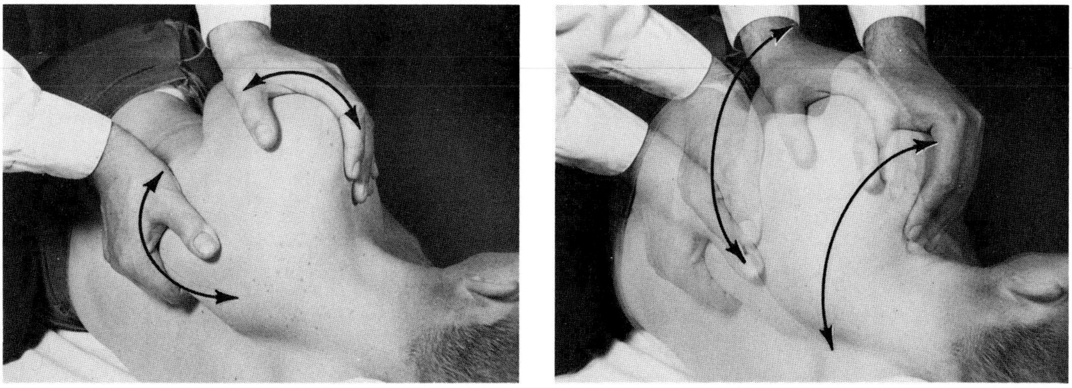

A *B*

Figure 8-9. *A,* How to grasp the blade of the scapula in order to rotate it on the chest wall without moving the arm. The scapula must freely move through an arc of 40 degrees on the chest wall to allow full pain-free movement of the glenohumeral joint. *B,* A double exposure illustrating the extent of the movement.

turned to normal activity, often in a fraction of the time to which surgeons are accustomed, and without so-called residuals. There are two exceptions to this plea for early physical therapy. In osteomyelitis and pyarthrosis absolute immobilization of the part involved is required, in addition to necessary drainage and antibiotic treatment. Care must be taken that the unaffected parts of the body are not neglected, even in these conditions.

Osteomyelitis

Osteomyelitis may affect any bone in the shoulder complex, either by the hematogenous route or by direct infection through a wound, causing pain and loss of function of the shoulder and arm. The differential diagnosis of this condition is discussed in Chapter 2, p. 17.

Osteochondritis

Osteochondritis dissecans or juvenilis may affect any joint, not excepting the shoulder. There is no specific physical treatment for this condition.

Neoplasms

Tumors of soft tissue or bone, primary or secondary, benign or malignant, may be the cause of pain in the shoulder from a structure local to the shoulder. There is no specific physical treatment for these conditions.

Metabolic Disease of Bone

The metabolic diseases of bone which may produce local pain to the shoulder are those mentioned in Chapter 2, pp. 18–20. Part of the treatment of osteoporosis senilis is properly applied physical therapy, and this is discussed on p. 188.

Metabolic Disease of Muscle

The differential diagnosis of the very rare metabolic diseases of muscle has been discussed in Chapter 2, p. 29. There is no specific physical treatment for these conditions.

Joint Dysfunction

To many physicians the mechanical condition of joint dysfunction will be the only new concept of a cause of pain in the shoulder local to the shoulder. The simple loss of one or more movements of joint play is a common cause of joint pain local to the shoulder and follows intrinsic joint injury. It is an even more common cause of pain in the shoulder residual to the healing of more serious pathological conditions. X-ray films do not help in its diagnosis, which is discussed at length in Chapter 1, pp. 8–10.

Treatment. The treatment of joint dysfunction is joint manipulation.

Though the shoulder has been taken as an example to explain the hypothetical cross-matching of anatomical structures with possible pathological conditions in an exercise of differential diagnosis, it is now stressed that, though specific attention is given to the glenohumeral joint, the differential causes of pain anywhere in the musculoskeletal system, whether in the foot, the ankle, the knee, the hip, the elbow, the wrist, the hand, or the back, have in fact been discussed. The back is no different from any other part of the musculoskeletal system. It is composed of the same anatomical structures which, in turn, respond to the same pathological processes in the same way.

Were the differential diagnosis of musculoskeletal pain confined to causes local to that part of the musculoskeletal system in which the pain is felt, successful patient care in this field would not be too difficult. Unfortunately, visceral pathology may refer pain to any part of the musculoskeletal system and may therefore be mistaken for a musculoskeletal condition.

Again using the shoulder joint as an example, we point out that pain from a diseased viscus may be referred to a structure of the musculoskeletal system so that the visceral problem masquerades as a somatic problem and remains undiagnosed, with perhaps dire consequences to the sick patient. Besides pain referred by visceral disease, problems involving one part of the musculoskeletal system may refer the symptom of pain to another part of it, and in this context pain is very frequently referred to the shoulder.

We do not intend to discourse on the differential diagnosis of the medical conditions we have seen over the years that present with a major symptom of shoulder joint pain. The conditions are simply listed according to the topographical area from which pain is referred. The objective is to provoke the physician to think more about differential diagnosis when facing a patient who complains of musculoskeletal pain. When it becomes apparent on examination that the pain is not arising from the place where the patient says it is, the physician must look elsewhere for the cause.

CAUSES OF PAIN FROM THE NECK

Shoulder pains may be referred from the neck. The following is a list of conditions in the neck which we have seen with a major complaint of shoulder pain:

1. Joint dysfunction in a joint or joints of the cervical spine
2. Bone tumors and disease in a cervical vertebral body including metastases, a hemangioma, tuberculosis
3. An undiagnosed healed laminar fracture, and cervical vertebral subluxation
4. Cervical ribs
5. The anterior scalene syndrome
6. Spinal cord tumor
7. Nodes in the neck from metastases, leukemia, and Hodgkin's disease
8. A retropharyngeal abscess
9. Irritation of the stellate ganglion from bleeding, an abscess, and a tumor
10. Cervical disc prolapse; neuritis and radiculitis
11. Fracture-dislocation in the cervical spine long after healing with residual dysfunction
12. Irritable trigger points in the neck muscles

CAUSES OF PAIN FROM THE ARM

Shoulder pain may be referred from conditions in the arm distal to the shoulder:

1. Carpal tunnel syndrome
2. Conditions about the elbow
3. Irritable trigger points in the upper arm muscles

CAUSES OF PAIN FROM THE CHEST

Shoulder pain may be referred from within the thorax or the thoracic spine. The following is a list of conditions in these areas which have been seen with a major complaint of shoulder pain:

1. Coronary artery disease
2. Pericarditis
3. Empyema and lung abscess
4. Aortic aneurysm
5. Pulmonary tuberculosis
6. Pancoast's tumor (superior sulcus tumor)
7. Mediastinitis–nodes in the mediastinum
8. Nodes in the axilla
9. Hiatal hernia
10. Breast cancer in the outer upper quadrant
11. Kümmell's osteochondritis
12. Pott's disease in the thoracic spine
13. Metastases in the thoracic vertebrae
14. Spinal cord tumor
15. Neurofibroma
16. Epidural abscess
17. Irritable trigger points in the thoracic wall musculature
18. Joint dysfunction in the upper thoracic spine and costovertebral joints

CAUSES OF PAIN FROM THE ABDOMEN

The following conditions have been seen with a major complaint of shoulder pain:

1. Gallbladder disease
2. Peptic ulcer
3. Subphrenic abscess
4. Cancer of the pancreas
5. Liver disease including an amebic abscess

6. Perisplenitis and ruptured spleen
7. Spinal metastases
8. Retroperitoneal sarcoma
9. A dissecting aortic aneurysm

CAUSES OF PAIN FROM SYSTEMIC DISEASES

Systemic diseases may select a shoulder joint in which to present symptoms of pain. The following diseases have been diagnosed that have at their onset the presenting symptoms of shoulder pain:

1. Collagen vascular diseases
2. Gout
3. Syphilis
4. Gonorrhea
5. Sickle cell anemia

Considering the possible implications of the symptoms of pain in the shoulder, it will be a brave physician indeed who states that the symptom is psychosomatic in origin. The more one knows about pain in the musculoskeletal system, the more often diligent search reveals a physical cause of it. Though we specify the shoulder, we are referring to any part of the musculoskeletal system.

BIBLIOGRAPHY

Awad, E. A. Interstitial myofibrositis: Hypothesis of the mechanism. *Arch. Phys. Med. Rehabil.* 54:449, 1973.

Bonica, J. J. Management of myofascial pain syndromes in general practice. *J.A.M.A.* 164:732, 1957.

Drutz, D. J., Schaffner, W., Hillman, J. W., and Koenig, M. G. The penetration of penicillin and other antimicrobials into joint fluid. *J. Bone Joint Surg.* (*Am.*) 49A:1415, 1967.

Ibrahim, G. A., Awad, E. A., and Kottke, F. J. Interstitial fibromyositis: Serum and muscle enzymes and lactic dehydrogenase isoenzymes. *Arch. Phys. Med. Rehabil.* 55:23, 1974.

Kraus, H. The use of surface anesthesia in the treatment of painful motion. *J.A.M.A.* 116:2582, 1941.

Spengler, D. N., Kirsh, M. M., and Kaufer, H. Orthopaedic aspects and early diagnosis of superior sulcus tumor of lung (Pancoast). *J. Bone Joint Surg.* (*Am.*) 55A:1645, 1973.

Steinbrocker, O., and Argyros, T. G. The shoulder hand syndrome: Present status as a diagnostic and therapeutic entity. *Med. Clin. North Am.* 42:1533, 1958.

Steinbrocker, O., and Argyros, T. G. Frozen shoulder: Treatment by local injections of depot corticosteroids. *Arch. Phys. Med. Rehabil.* 55:209, 1974.

Travell, J. G. Ethyl chloride spray for painful muscle spasm. *Arch. Phys. Med. Rehabil.* 33:291, 1952.

Travell, J. G. Referred pain from skeletal muscle: The pectoralis major syndrome of breast pain and soreness and the sternomastoid syndrome of headache and dizziness. *N.Y. J. Med.* 55:331, 1955.

Travell, J. G., and Rinzler, S. H. The myofascial genesis of pain. *Postgrad. Med.* 11:425, 1952.

Travell, J. G., Rinzler, S. H., and Herman, M. Pain and disability of the shoulder and arm: Treatment by intra-muscular infiltration with procaine hydrochloride. *J.A.M.A.* 120:417, 1942.

III
Miscellaneous Pain Conditions

9
Musculoskeletal Pain Conditions

Chapters 9 and 10 present selected painful conditions to illustrate the importance of differential diagnosis and the orderly prescription of physical therapy in their management. In addition, we have selected a few less common conditions because they are of special interest.

WHIPLASH INJURY

Most physicians decry the term *whiplash injury*, but its use is perpetuated by lawyers, the public, and, for convenience, many doctors. There is no reason why it should not be used to describe a mechanism of injury so long as it does not constitute a diagnosis.

We lend our voice to what should be a clamor against the use of this term in medicine. We advocate the following as an analysis of the pathological diagnoses representing the changes that may occur as the result of the described mechanism of injury.

First, we stress that the pathological changes resulting from this injury are not confined to the neck. Because the function of the spine is all or none, injury to any part of it may cause injury to any other part. Examination of a patient who has been subjected to this injury, therefore, should not be confined to the cervical spine and soft tissue structures of the neck but should embrace the whole back. Further, we remind the reader that x-ray pictures are helpful only when the injury results in frank fracture (and even this may remain hidden) and/or dislocation, or to determine whether there has been retropharyngeal hemorrhage (see Chapter 5, p. 90). The laboratory cannot be of any assistance.

The method of examination does not vary from that advocated throughout this book. For those who have difficulty in visualizing what may happen to the musculoskeletal structures following a "whiplash" of the spine, we recommend that consideration be given to what happens to a patient when he twists his ankle. This is a typical whiplash injury; the sequence of pathological events stemming from it is usually quite clearly assessed, and the outcome of treatment is usually predictable. The only differences in considering what may happen in the back result from the presence of the intervertebral discs, the proximity of the spinal cord, the elements of the autonomic nervous system, particularly the stellate ganglion, and the close proximity of the pharynx, the esophagus, the larynx, and the trachea.

Muscle Strain

Probably the simplest but most common cause of symptoms following whiplash injury to the neck is muscle strain with microscopic bleeding into the muscle(s). This results in muscle spasm that, within 24 hours of the injury, responds therapeutically to treatment by the vapocoolant spray and stretch techniques. In the initial hours after injury the application of ice packs and the Turkish towel collar should provide comfort. Unfortunately, the patient is not usually seen in this period of time. Easy exercises with a beach ball for assistance (the ball is not tightly inflated) or resistance (the ball is tightly inflated) and the use of a soft collar between treatments should resolve problems from muscle strain within a week.

Simple muscle strain may be complicated by joint dysfunction, particularly at the atlanto-occipital or the thoracocervical synovial joints but also at any other intervertebral junction. Pain referred from joint dysfunction may prove confusing in diagnosis, but the reader is referred to Travell's patterns of pain from trigger points (Figures 9-12 to 9-15, pp. 191–194); referred pain from joints is similar but it follows neural pathways. There should be no problem unless referred pain is confused with radiating pain.

The usual pattern of restricted movement with pain from joint dysfunction in the cervical spine is this: Rotation is limited in one direction and side bending is limited in the other direction. The pain is relieved by rest and aggravated by activity. Dysfunction at the cervicothoracic junction is often associated with a history indicating that the patient has difficulty in raising his head when arising from the supine position. Patients also particularly complain of pain on neck extension. Muscle spasm is likely to be added to the joint dysfunction, and then the neck may stiffen with rest. This development somewhat confuses the clinical picture. The treatment of simple joint dysfunction is the restoration of the lost joint-play movement in the involved joint by manipulation.

Synovitis

A slightly more severe injury produces synovitis in the interlaminar joint or joints that are most affected by the injury. The symptoms are some immediate discomfort in the neck at the affected level, then a period of perhaps hours of improvement or even freedom from pain, followed by increased stiffness and pain in the neck, often first noticed on waking next morning. If the synovial swelling of a facet joint becomes at all marked, irritation of the nerve root in relation to it may occur, giving the false impression of a disc prolapse because of the presence of pain radiation, paresthesias, possible muscle weakness, and possible sensory and reflex changes. The differentiation may not be easy, and a lumbar puncture may have to be done for determination of the spinal fluid protein content (which is not raised with synovitis) before a diagnosis may be made. However, with proper treatment, following the principles we have outlined, the signs from synovitis rapidly disappear within a few days, whereas the signs from a prolapsed disc persist and may worsen.

The proper treatment of synovitis is outlined in Chapter 8, p. 154. In the neck, treatment entails use of a cervical collar temporarily, intermittent traction, preferably with the patient lying down, and the application of heat two or three times a day. Anodal galvanism has a pain-relieving effect either on its own or following Novocain infiltration. Anodal galvanism also tends to reduce swelling. Some degree of joint dysfunction may remain following the resolution of the synovitis. Residual dysfunction becomes apparent when there is a change of symptoms from stiffening after rest to relief from discomfort after rest but recurrence of pain after renewed activity. Manipulative restoration of lost joint play is then indicated.

Hemarthrosis

The symptom of hemarthrosis is acute local pain, with onset almost immediately following the injury, that tends to worsen in the first few hours after the injury. Differentiation of this condition from a fracture or dislocation can be made only by the absence of x-ray evidence of bone or joint injury; and it should be remembered that the initial x-ray photograph, if it is negative, does not necessarily rule out the presence of a fracture.

The correct treatment of hemarthrosis is aspiration of blood from the involved joint. This is technically impossible when one is dealing with the interlaminar joints of the spine, so immobilization by continuous traction and the application of ice constitute the best treatment in the acute stage, together with adequate sedation. Once the acute stage has passed, treatment such as that used for synovitis is instituted, and any residual dysfunction must be overcome by manipulation. Synovial joint reaction is common if there is capsular adherence, and this in turn requires a period of appropriate physical therapy, in addition to maintenance of a range of motion restored each day until the therapeutic reaction is quiescent. Ultrasound may assist the absorption of blood from within a joint capsule.

Any of the above three pathological conditions may occur singly. With a severe injury, more than one pathological condition will occur, and each must be treated before the patient can be expected to be relieved of his symptoms.

Soft Tissue Injury

The next thing that may happen in the order of severity of the injury is tearing of the supporting soft tissues of the neck—first the muscles and then the ligaments, particularly the interspinous ligaments. Severe soft tissue injury may be associated with hemorrhage into the retropharyngeal space, and when this occurs, it takes precedence in treatment. Embarrassment of respiration may make this an acute surgical emergency. Embarrassment of deglutition may require intravenous feeding.

Having dealt with these complications as indicated, the physician turns to treatment of the muscle tear. Whereas treatment of hemorrhage includes the application of ice, after 24 hours this should be changed to the application of heat. Whereas immobilization is part of the treatment of hemorrhage, when one can assume it has stopped, movement of the damaged muscle by exercise without use is indicated. This is best achieved by faradic stimulation. The intensity and the frequency of the current should be graduated by the therapist, the indication for increasing the treatment being the absence of pain. The heat and faradism are followed by light stroking massage to attain muscle relaxation and to disperse any congestion remaining from the application of heat. In the restoration phase of treatment, exercises with a beach ball are most efficacious. When the ball is not fully inflated, the exercises are assisted. When it is fully blown up, they are resistive (Figure 9-1). The symptoms and signs of a muscle fiber tear are muscle spasm, inaccurately localized pain, and stiffening of the neck with rest. There is local acute tenderness in the muscle on palpation, as well as a feeling of a gathering of the torn muscle fibers. Infiltration of the torn area with Novocain, followed by anodal galvanism for ionization, may be necessary before the more active physical therapy is started.

A torn joint ligament can scarcely be present without a torn muscle or at least some muscle injury. A ligament tear requires a longer period of immobilization for healing to take place. The diagnosis is not easy, but tension on the ligament by passive movement without muscle stretch produces pain. A tear in an interspinous ligament can be detected on palpation by eliciting acute tenderness localized between two spinous processes.

The diagnosis of ruptured interlaminar joint ligament cannot be demonstrated clinically, but the condition does occur and can be recognized at surgery. Persistent or frequently recurring local pain from continuing joint instability always in the same place should suggest this possibility.

A ruptured interspinous ligament is detected by the presence of a dip on palpation between two spinous processes and clinical signs of intervertebral junction instability. We have seen one case each of these ruptured ligaments, and the only effective treatment was spine fusion at the affected junction.

An injury of similar severity may result in avulsion of a part of a spinous process without rupture of the attached ligament. We have

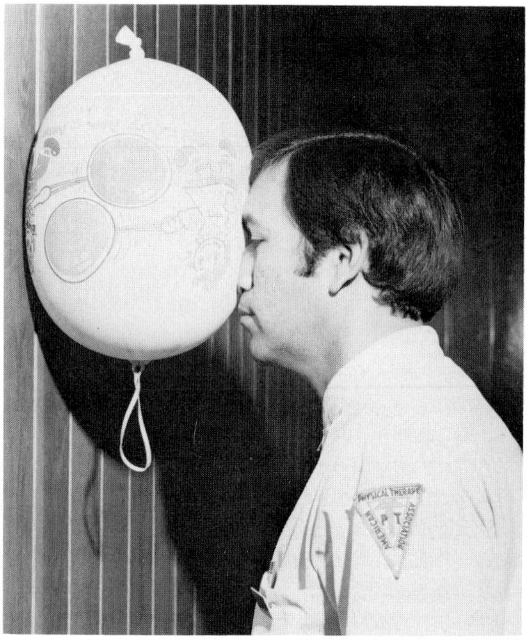

A

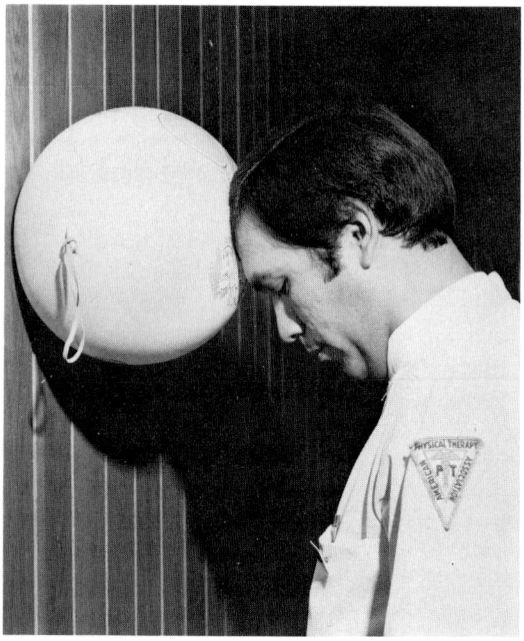

B

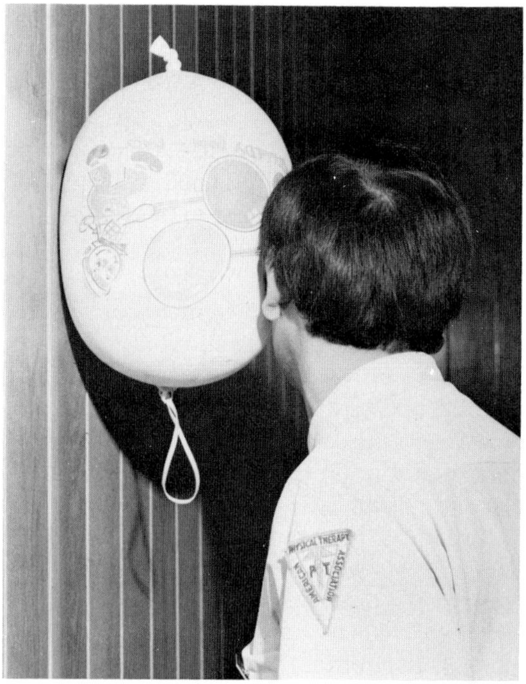

C

Figure 9-1. Use of a beach ball in neck reeducation exercise. *A*, Starting position for flexion, extension, and rotation exercise. *B*, The head and neck in flexion. *C*, The head and neck in rotation. The incomplete inflation of the ball makes the exercises assistive. Complete inflation of the ball makes the exercises resistive.

seen this only at the sixth and seventh cervical vertebrae following whiplash injury. Treatment by immobilization until the patient is symptom-free, followed by treatment of residual muscle pathology or joint dysfunction, is sufficient in such a case.

Dislocation and Fracture

An injury of about the same severity may produce dislocation of the facet joints. This can be detected radiologically and should promptly be reduced, the aftertreatment being the same as for any other dislocated weight-bearing joint.

An injury of any greater severity produces a vertebral body fracture, with or without dislocation; complications may be serious and treatment prolonged. Possible complications include acute prolapse of a disc, epidural hematoma, compromise of a vertebral artery, trauma to the anterior spinal artery resulting in myelomalacia, transient cord concussion, and cord severance. Treatment of these conditions lies in the specialist fields and will not be discussed here. But there should never be any reluctance to perform a lumbar puncture with manometric studies and determination of the protein and cellular content in the diagnostic work-up of patients following a whiplash injury. The examination of the patient must always include study of the nervous system.

Residual Symptoms—Treatment

The important thing is how to assess residual symptoms once recovery from the severe pathological conditions has occurred. We shall ignore residual quadriplegia, which requires prolonged rehabilitation techniques, and confine ourselves to local pain in the neck and pain referred from the neck.

The only basis on which residual symptoms can be assessed is adequate clinical examination. The common causes of residual symptoms are simple untreated joint dysfunction with or without pain referred into the head, adherent scar formation in the supporting muscles of the neck, irritated trigger points, unhealed ligament pathology resulting in joint or junction instability, irreversible disc pathology, muscle imbalance or weakness due to atrophy from prolonged spasm or transient paresis, and any mixture of any of these. The conditions can be differentiated clinically and treated specifically and successfully. Seldom, in our experience, are residual symptoms purely psychogenic; usually there is a physical basis for them. Any psychological overlay is likely to be a direct result of failure to diagnose and to relieve the patient of his symptoms, which may cause the development of a fear of permanent disability. The psychological aspect can be treated successfully only by removal of the physical cause. The physical disability can never be removed by psychological treatment, tranquilizers, or pain-killing drugs.

There is no doubt that residual symptoms often result from autonomic dysfunction. We believe that the most likely mechanism in this event is an irritation of the cervicothoracic or stellate sympathetic ganglia, either by pressure from local swelling due to blood or edema or reflexly by irritation of the nerve roots (particularly the first and second thoracic nerves) with which the ganglion is connected by its rami communicantes.

Symptoms and signs that direct one's attention to stellate ganglion irritation are causalgic shoulder and arm pain, trophic skin changes, especially of the hands and fingers, swelling and stiffness of the hands, excess palmar perspiration, dilatation of the ipsilateral pupil and the palpebral fissure, proptosis with oculopalpebral asynergia, and deficient convergence of the eyeballs. The arm symptoms and signs are common; the gross eye signs are not. Treatment by stellate ganglion block often brings dramatic relief, especially of the arm pains and the changes in the hands.

Headaches as residuals of whiplash are common. They usually arise from irritable muscle trigger points in the neck, muscle spasm (sometimes with entrapment of the great occipital nerves), and cervical joint dysfunction.

Figure 9-12 shows typical patterns of head pain from trigger points in the neck.

Vertigo is another residual symptom of which sufferers from the whiplash injury may complain. Body equilibrium is coordinated in the cerebellum by impulses from the labyrinth, the inner ears, the eyes, the skin, the muscles, and the joints. Maintenance of equilibrium is regulated by the cerebellum and is largely a muscular effect, influenced by the activity of the tonic neck reflexes among other things. Equilibrium can be affected by dysfunction in the joints and spasm in the muscles of the neck. Treatment of posttraumatic vertigo may well include mobilization of the joints of the cervical spine.

Residual pain in the neck sometimes arises from nonunion or fibrous union in fractures that have been overlooked and therefore not treated. These may occur in the laminae, at the base of the odontoid process of the axis or, we suspect, in the outer facets of the joints of Luschka. Treatment will probably have to be surgical—either spine fusion, neurectomy, or rhizotomy.

Radicular symptoms residual from a whiplash injury may be due equally to capsular swelling from either synovitis or hemarthrosis in an interlaminar joint or to pressure from a prolapsed nucleus pulposus. In our experience, cervical disc prolapses are uncommon. Similar symptoms may be due to pressure from callus formation (from an unrecognized fracture of a lamina, for instance) or to a stretch injury involving the brachial plexus which was unrecognized at the time of the acute injury.

There are other rare complications of spinal injuries that may produce residual or late symptoms. Among these are Kümmell's aseptic necrosis of a vertebral body, rotatory subluxation of the axis, and vertebral artery insufficiency giving rise to the "flop syndrome," when the patient looks up. Intracranial injuries may occur at the time of a whiplash injury and remain unrecognized until residual symptoms draw attention to them. Residual symptoms of late origin may occur from cervical spondylosis, and marked osteoarthritic changes in the interlaminar joints may give rise to symptoms later in life.

A whiplash injury after the acute symptoms have subsided may leave a patient with symptoms that he considers to be residual from the injury but that are, in fact, manifestations of some other disease whose symptoms have been masked heretofore. Such a situation poses many difficult problems between doctor and patient.

Finally, residual symptoms following a whiplash injury may be neurasthenic or psychoneurotic, but such a diagnosis should be considered only after all the possible physical causes outlined above have been ruled out. We are sure that residual symptoms from the whiplash injury of the neck would be less severe and less common if treatment of the pathology resulting from it were instituted from the time of injury.

MECHANICAL HEADACHE

Mechanical headache is a well-recognized but often inappropriately treated condition and many of its causes are overlooked. Neurological causes of headache must be carefully looked for before the diagnosis of mechanical headache is made.

Four etiological factors can be identified in mechanical headache.

1. Abnormalities of the muscles of the neck and head due to posture, tension, trauma, poor bite, and a host of other causes may refer pain to the head. Since the great occipital nerve lies free in the posterior muscles of the upper neck, it may be entrapped by muscle spasm and produce pain (see p. 202).
2. Abnormalities of the joints of the cervical spine due to osteoarthritis or joint dysfunction produce pain, associated muscle spasm, and headache.
3. Degenerative changes in the discs of the cervical spine alter the mechanics of the

neck and put abnormal stress on ligaments and muscle; headache often results.

4. Tension and/or depression.

All the above may be interrelated. For instance, the posture assumed by the depressed patient places an abnormal stress on the posterior neck muscles, producing headache. As mechanical headaches arising from trigger points are common, reference should be made to patterns of pain arising from them (Figure 9-12).

History. The diagnosis of muscle tension headache is usually one of exclusion, so other causes should be searched for and eliminated before this diagnosis is made. Mechanical headaches are often residual to whiplash injuries. Repetitive microtrauma—for example, the wearing of bifocals—may be sufficient to produce abnormalities in the muscles of the neck with resultant headache. The headache is often of long standing, often present on awakening, and usually worse with the passage of the day. It may awaken the patient at night. As a rule it is relieved or ameliorated by salicylates and other nonnarcotic analgesics and various muscle relaxants, sedatives, and mood elevators. The patient frequently appears depressed, but chronic pain in itself is sufficient cause for depression.

Most often the patient will describe the pain as starting in the posterior aspect of the neck and radiating to the back of the head, frequently with extension over the top of the head to the frontal area. Sometimes it is bitemporal in location. The classic description of tension headache is that of a band around the head. If an unusual description of pain circling around the ear or eye is given, a referred pain pattern from the neck should be suspected as the cause.

Physical Examination. The posture of the patient is noted. Range of motion of the cervical spine should be tested in all planes. Palpation for tenderness in the muscles of the face (masseter and temporalis) and the anterior and posterior cervical muscles may disclose points of significant localized tenderness. If so, these should be firmly pressed in an attempt to produce a pattern of referred pain. Particular attention should be paid to the posterior occipital muscles, which should be palpated in the supine position. Examination for joint dysfunction in the cervical and upper thoracic spine should also be included.

X-ray Pictures. X-ray examination of the cervical spine presents particular difficulties. Since mechanical headache is most common in middle-aged people, most will have some degree of degenerative changes in the neck revealed by x-ray pictures. Therefore, it is always tempting to describe these changes as the cause whether or not they really have any relationship to the problem. Significant disc space narrowing does not necessarily produce alterations in neck mechanics or chronic irritation of the soft tissues.

Treatment. Emphasis in treatment should be on the correction of the suspected etiological factors. If the problem appears to be primarily in the muscles, these should be treated by a combination of icing, vapocoolant spray and stretching (manually or with traction), massage, and local injections. This should be followed by the application of gentle heat and gentle exercises with the use of a beach ball. To prevent repetition of the problem, instruction in proper neck mechanics should be included toward the end of the treatment program.

If the symptoms are due primarily to joint dysfunction, manipulation should be performed. For headaches that appear secondary to degenerative changes in the neck, all the above measures would be appropriate. The use of a home traction unit is often helpful in maintaining the relief from pain. When the problem originates from psychological tension the patient should be taught relaxation. Biofeedback techniques are helpful in training muscles to relax. This can be continued at home using Jacobson's techniques (refer to Chapter 7, p. 126).

TENNIS ELBOW

As we have repeatedly emphasized, there is no place for such diagnostic terms such as *tennis elbow.* Because pain at the elbow is common and especially recalcitrant to empirical treatment, we feel that this is a clinical state meriting discussion.

Soft-Tissue Causes of "Tennis Elbow"

TEAR OR PAINFUL ADHERENT SCAR IN THE COMMON EXTENSOR TENDON

The most common cause of pain about the elbow is inflammation of the common extensor tendon at its insertion on the lateral epicondyle, somewhere in the common tendon, or at the musculotendinous junction. This is usually the result of sudden uncontrolled stress or repetitive stresses that exceed the tolerance of the tissues. It is a common occurrence in tennis players, golfers, and other athletes, but of course appears in a wide variety of other activities involving the same mechanism. If the process is chronic, it is due to painful scars from the healed tears (Figure 9-2).

Figure 9-2 is an x-ray picture of an elbow showing linear calcification adjacent to the lateral humeral condyle. The calcification represents an attempt by nature to reinforce an attenuated tendon weakened by stress. It is analogous to calcific tendinitis in the supraspinatus tendon in the subacromial area of a shoulder.

Frequently the patient will point directly to the site of the inflammation, tear, or scar. Sometimes he just complains of a generalized aching over the dorsum of the forearm or of a sensation of paresthesia over the dorsum of the hand. With the wrist in full flexion and the forearm in pronation a restriction of extension of the elbow is often observed. Stressing the extensor tendon by resisted wrist or finger extension with the arm extended or by forced grasp refers pain to the lateral epicondyle or

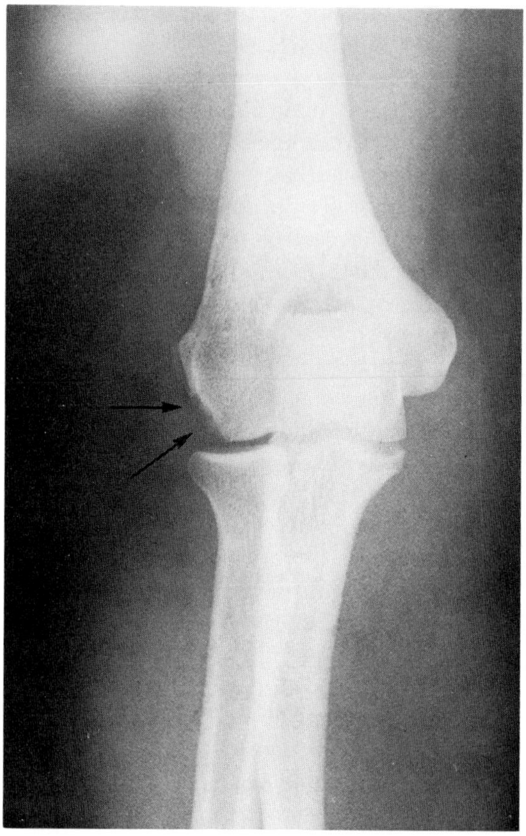

Figure 9-2. X-ray picture of the elbow. Arrows show calcification adjacent to the lateral epicondyle in the common extensor tendon as a result of chronic stress (tennis playing). There is also roughening of the adjacent lateral condyle of the humerus.

to the musculotendinous junction. Careful palpation reveals a very localized area of tenderness over the lateral epicondyle or below it somewhere in the tendon or at the musculotendinous junction. If necessary, this can be confirmed by the use of subtonal faradism in which the button electrode is passed over the area of pain in different directions. The intensity of the faradic current is that which just falls short of inducing muscle contraction in neighboring unaffected muscles, and the current is constant. Each time the electrode passes over the tear or scar, pain is felt in that place and only that place.

During the examination the forearm extensor muscles and the supinator brevis muscle should be carefully palpated for the presence of associated trigger points. So too should the scalene muscles be palpated on the affected side (refer to Trigger Points, below).

In the acute phase, treatment involves rest of the part (the inflamed extensor tendon), accomplished by the use of a cock-up splint for the wrist which permits pinch prehension but prevents grasp (see Figure 7-8, p. 139). Antiinflammatory medication may be used orally or by local injection into the part, and physical therapeutic measures to decrease the inflammation, as outlined in Chapter 7, may also be employed. Joint dysfunction of the elbow, if present, should be treated.

Treatment of the chronic phase should include all the above. If it is unsuccessful, tearing of the painful scar by manipulation should be done.* Having created a fresh injury, one now treats this as such, following the principles of physical treatment. When healing occurs, though there must be a new scar, it is neither adherent nor painfully contracted, and it does not interfere with normal function. Infiltration of the scar with local anesthetic may be required to allow the manipulative part of the treatment to be undertaken.

In addition, corrective measures should be employed, particularly if the cause, occupational or recreational, is known to be one that will continue. Forearm strapping (see Figure 7-14, p. 143) absorbs some of the force of a maximal contraction of the extensor muscles, thereby relieving some of the stress on the common extensor tendon. In a similar fashion, exercise designed to increase the flexibility and the strength of the entire extremity lessens the stress on the common extensor tendon of the wrist.

* To tear the offending scar the elbow is snapped into its last few degrees of extension while the forearm is fully pronated, the wrist fully flexed, and while tension on the scar is increased by pressure exerted over it with the thumb of the mobilizing hand. The forearm is not pulled back; rather, the lower end of the humerus is pushed forward.

TRIGGER POINTS

Irritable trigger points are a common cause of the pain of tennis elbow and may be associated with painful muscle extensor tendon scarring. Therefore, both causes should be examined for at the same time and treated at the same time. Epicondylar or condylar pain is often a manifestation of referred pain from a trigger point. In this case it is not relieved by local therapy.

These trigger points may be found locally in the muscles around the elbow but there is one distant trigger point that gives rise to referred pain in the area of the lateral humeral condyle at the elbow. It is to be found in anterior scalene muscle. We suspect that this is the reason it was once popular to designate tennis elbow as a manifestation of a cervical disc injury. When local trigger point irritation is the cause of elbow pain, the muscles usually involved are the supinator brevis, the flexor carpi radialis, the extensor of the middle finger, or one of the other extensor muscles. Epicondylar pain to percussion is a feature of this condition, and it disappears when the correct trigger point is treated, either by vapocoolant spray and stretching or by local injection and stretching.

CORONOID BURSITIS

There is a coronoid bursa that, when it is traumatized and swollen, can give rise to pain that may be diagnosed as tennis elbow. The swelling, not greater than the size of a pea, is located in the antecubital fossa in relation to the proximal border of the lacertus fibrosus. It is readily detectable with the elbow in extension. As a painful elbow is usually held in flexion, this swelling may be overlooked. Treatment is as for any traumatic bursitis.

Joint and Bone Conditions Causing Tennis Elbow

First it must be recognized that the elbow is made up of three different joints: the proximal radioulnar joint, the ulnohumeral joint, and the radiohumeral joint (Figure 9-3).

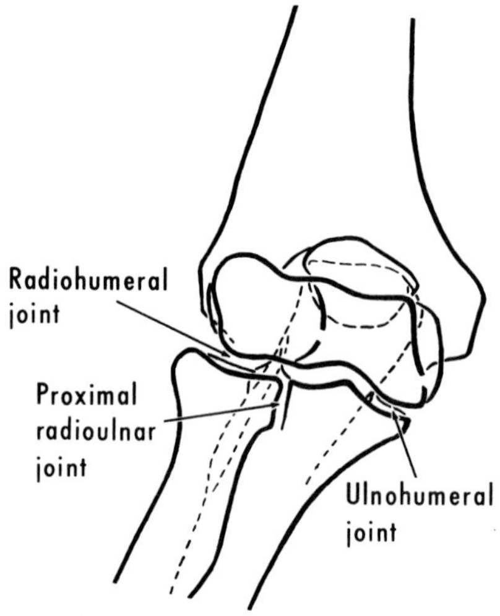

Figure 9-3. The three joints at the elbow.

THE PROXIMAL RADIOULNAR JOINT

The "Pulled Elbow." The pulled elbow in children, resulting in pseudoparesis of the arm, is a classic problem that many physicians and surgeons rightly claim they can diagnose over the telephone. The only treatment for it is manipulation. The explanation that this condition is painful because there is a subluxation of the head of the radius downward is almost certainly incorrect. We believe the cause of pain is simple intrinsic traumatic joint dysfunction and provides a classic example of the joint play, joint dysfunction, joint manipulation hypothesis stressed in our work.

The pulled elbow is not peculiar to children. In adults it is a cause of the pain of tennis elbow. Figure 9-4 illustrates the position adopted for manipulation to restore the lost joint play causing the pain associated with the pulled elbow. The thrust is made by the therapist's left hand—radius to radius with the patient's arm. As the thrust is made the patient's forearm is supinated as well.

The "Pushed Elbow." The pushed elbow is the reverse of the pulled elbow and may result from a fall on the outstretched arm. Whereas in the pulled elbow the movement of joint play lost is the movement of the head of the radius upward on the ulna at the superior radioulnar joint, now the movement of joint play lost is the movement of the head of the radius downward on the ulna at the superior radioulnar joint. The treatment of this

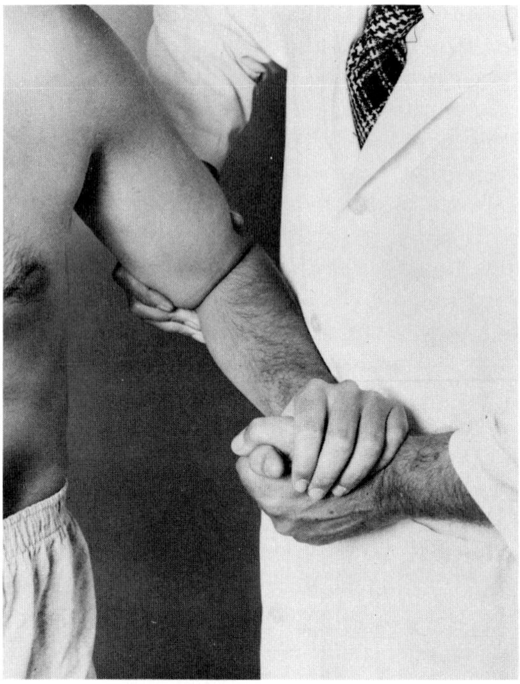

Figure 9-4. Position adopted to overcome the joint-play loss at the radioulnar joint to correct the pulled elbow.

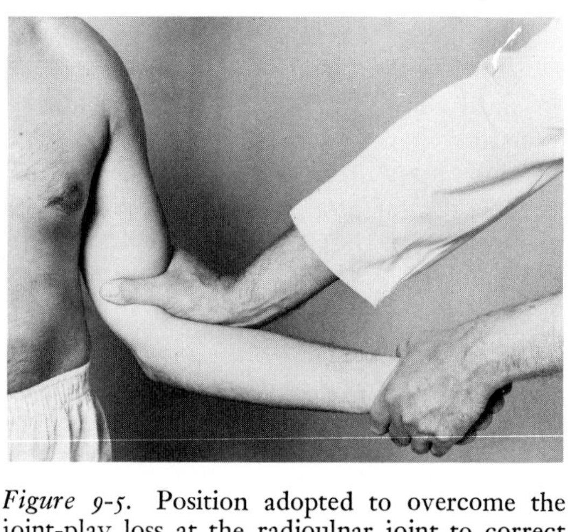

Figure 9-5. Position adopted to overcome the joint-play loss at the radioulnar joint to correct the pushed elbow.

condition is designed to restore the lost joint-play movement by manipulation (Figure 9-5). This is another cause of pain again resulting in the diagnosis of tennis elbow.

The "Rotated Elbow." In supination and pronation of the forearm there is a functional rotatory movement of the radius on the ulnar that is supposed to be limited to the proximal radioulnar joint. There is an analogous joint-play movement in rotation in the superior radioulnar joint of the head of the radius on the ulna. If this is lost, pain results. Treatment is directed to restoring the lost movement of joint play by manipulation. The ulna is mobilized on the head of the radius, which is stabilized. Figure 9-6 shows how this is accomplished.

Fracture of the Head of the Radius. Fracture of the head of the radius is one of the fractures which may not be revealed by x-ray pictures

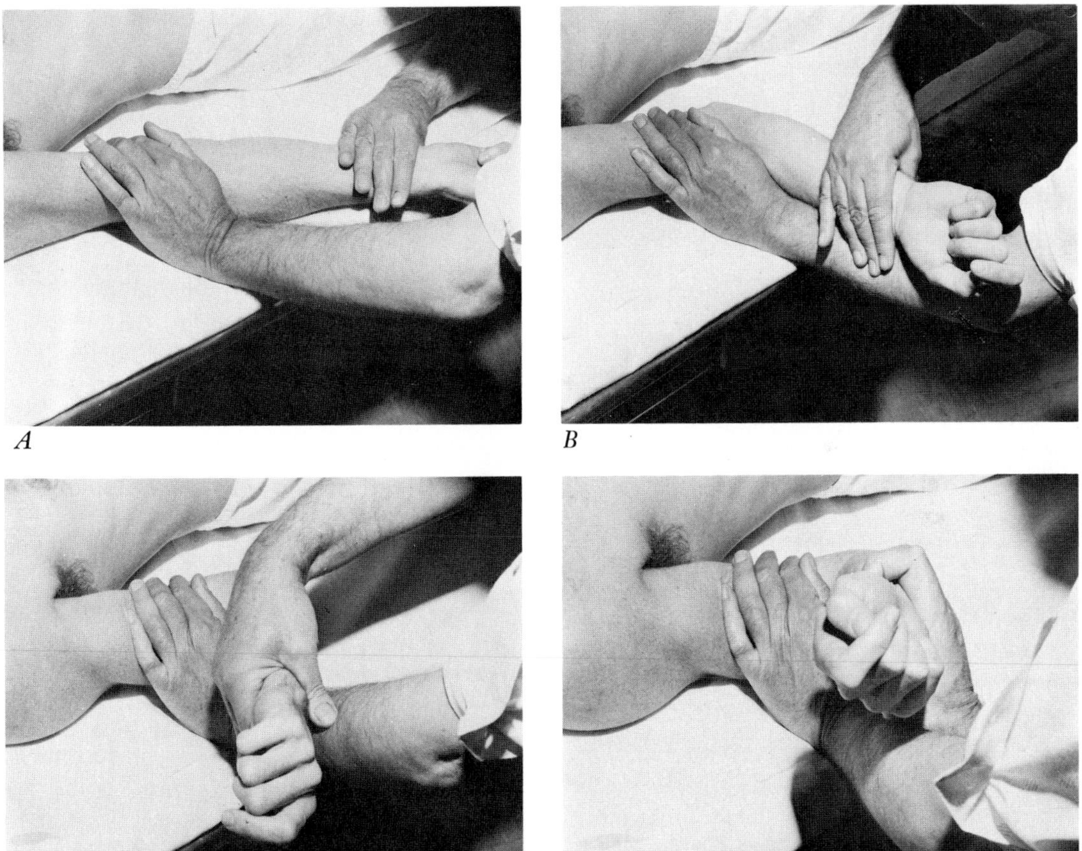

A *B* *C* *D*

Figure 9-6. Steps taken in adopting the correct position to overcome the pain of a rotated elbow. *A,* The physician's left thenar eminence is laid over the head of the radius with the arm in full supination. His right hand is the mobilizing hand throughout. *B,* The physician moves the forearm by increasing the carrying angle while maintaining the forearm in full supination until it pinches against the thumb of his left hand. *C,* At this angle the physician whips the forearm from supination into full pronation by using his left hand as indicated in *D* while holding the head of the radius back with his left thumb, which has not moved.

at the initial examination (see p. 66). Pain diagnosed as due to tennis elbow, therefore, may be arising from an undiagnosed fracture. Aspiration of blood from the joint should reveal fat in it and preclude such misdiagnosis. After healing of the fracture, joint dysfunction may be a cause of residual pain, and treatment follows the principles of physical therapy outlined in Chapter 8, p. 153. Treatment of residual dysfunction after healing must include manipulation.

THE ULNOHUMERAL JOINT

There are two anatomical parts to this joint. The larger and more important part is that between the olecranon beak of the ulna and the ulnar fossa of the humerus posteriorly. The other part of the joint is made up of the coronoid beak of the ulna and the coronoid fossa of the humerus anteriorly.

Pinched Synovial Fringe. A cause of pain which may be diagnosed as tennis elbow is the pinching of the synovial fringe by the beak of the olecranon as it enters the olecranon fossa on unguarded extension of the elbow. This results in a rather special kind of joint dysfunction which is relieved by sweeping the synovial fringe out of the way of the olecranon beak by means of a side-to-side joint-play movement of the beak of the olecranon while bringing the elbow from its angle of limitation in flexion into extension (Figure 9-7). The movement is elicited by moving the humeral condyles on the stabilized ulna.

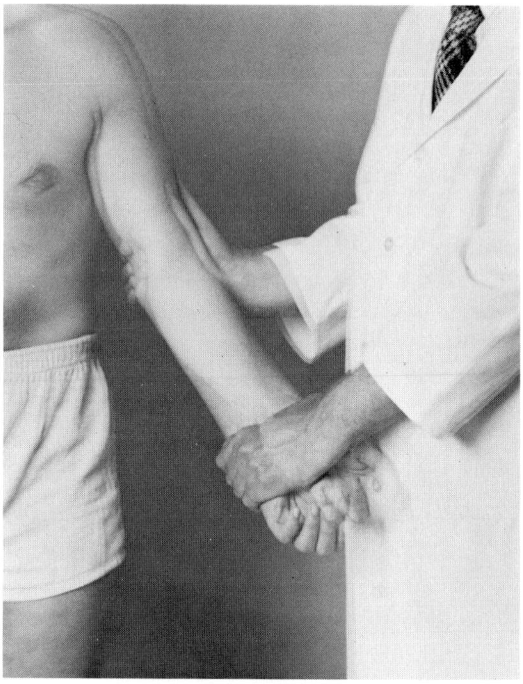

Figure 9-7. Position adopted by the physician (with the patient in the upright position) for performing the medial and lateral sweeping movements of the olecranon across the olecranon fossa of the humerus to push a pinched synovial fringe away from the fossa and allow the ulna olecranon to engage in it. The physician's right hand is the mobilizing hand. Therapeutically this manipulation should be performed with the patient in the recumbent position.

THE RADIOHUMERAL JOINT

Meniscal Lock. The radiohumeral joint is one of the five synovial joints in the body having an intraarticular meniscus. Some authorities say that there is no meniscus in this joint and that the condition simulating a meniscal injury here is due to an infolding of the synovium, which then acts like an injured meniscus, blocking joint function. Figure 9-8 illustrates how the therapist holds the head of the radius stable on the ulna with his right thumb as the patient's arm is swept from flexion into extension while maintaining the forearm in full pronation on the wrist fully flexed. The final manipulative thrust is made by the therapist's right hand. Clinically, it is irrelevant which explanation of pain and loss of movement is accepted. Joint dysfunction occurs at the radiohumeral joint and is diagnosed as tennis elbow. Figure 9-9 demonstrates the closeness of the location of palpable pain when one is differentiating a radiohumeral block and a painful scar or tear of the common extensor tendon. The thumb is on the radiohumeral joint; a cross indicates the area where pain would be palpated were it due to a painful extensor tendon scar.

Periosteal Bruise. Direct trauma to one of the subcutaneous areas of bone found at the elbow

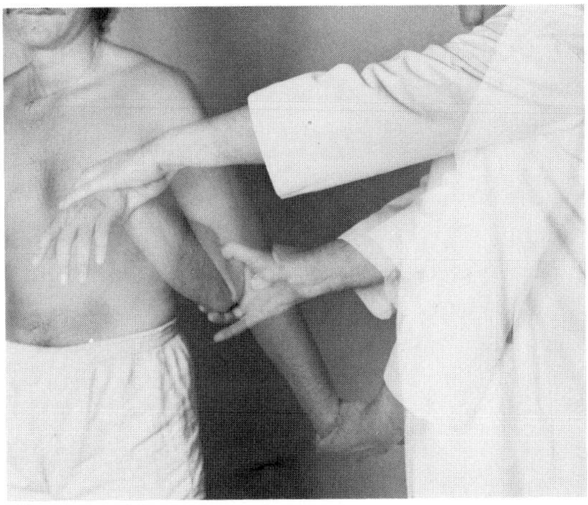

Figure 9-8. Double exposure illustrating therapeutic manipulation to overcome the dysfunction of a meniscal lock at the radiohumeral joint. The same procedure is used to tear a painful extensor tendon scar. In the first case the physician's thumb holds the head of the radius back; in the second case the physician increases tension over the scar by pressing his thumb on it.

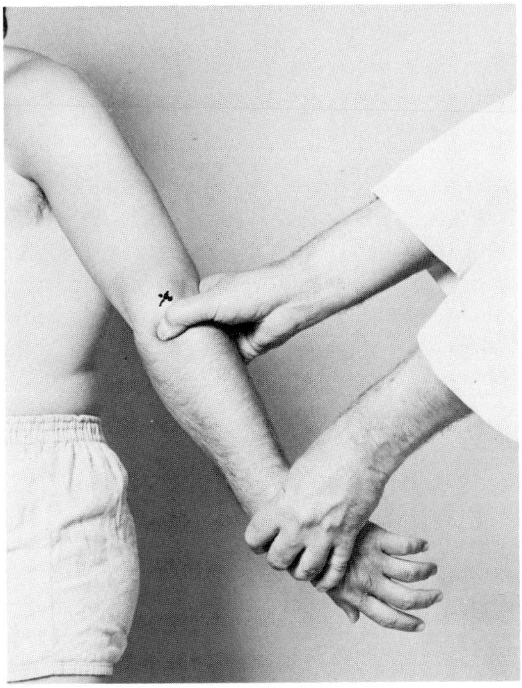

Figure 9-9. Location of the physician's right thumb over the head of the radius, where pain is elicited when a meniscal lock prevents full extension of the elbow. The cross indicates where his thumb would be if the pain were arising from a painful scar in the extensor tendon.

may result in a periosteal bruise. The history and localization of pain give the clue to the correct diagnosis. Treatment of periosteal bruises is described on p. 150.

We have drawn attention to a number of diagnosable conditions that give rise to pain at the elbow and are usually lumped together under the term *tennis elbow*. Each condition responds readily and predictably to proper treatment, chosen according to the proper principles of therapy, when the proper diagnosis is made. There are obviously many other causes of pain at the elbow; any bone, joint, capsular, ligamentous, muscular, cartilage, or bursal pathological condition or systemic disease that is characterized by joint manifestations may be culpable. We stress that this section refers to the nebulous but common or garden variety of pain arising from specific and diagnosable causes.

MUSCLE CRAMPS

Muscle cramps represent dysfunction in muscle physiology. Definitions of muscle cramps vary, but our discussion will be limited to muscle contractions that are both sustained and painful. There is an inevitable overlap between muscle cramp and muscle spasm. Causes of cramps may be divided into mechanical, metabolic, vascular, neuritic, and idiopathic. It is likely that the end result of all causes is interference with adequate muscle nutrition and oxygen supply, leading to cramp.

Mechanical Causes

Mechanical factors causing cramps include fatigue, strain, and overexertion. In such situations temporary anoxia of the muscle results in cramps. Swimmer's cramp is a dramatic example.

Treatment by stretching of the muscle to restore normal length, with or without use of ice massage or the vapocoolant spray, followed by adequate rest, is all that is needed for the acute condition. In chronic cases, search for and eradication of underlying causes such as poor body mechanics, muscle weakness, and body asymmetry should be undertaken.

Metabolic Causes

Alteration of electrolytes may lead to cramping. The ability of changes in the serum calcium concentration to produce tetany is well known. Less familiar changes resulting from alteration of the serum sodium may be at fault. Laborers who perspire freely during the summer months are liable to cramping, as are athletes under similar circumstances. Cardiac patients on a diuretic or salt-free diet may develop muscle cramps. Cramping that sometimes occurs during dialysis for chronic renal conditions is also believed to be due to hyponatremia. Prophylactic treatment consists of salt supplementation and adequate fluid intake. Hypomagnesemia during dialysis has been implicated as another cause of cramps and the injection of magnesium sulfate has relieved the cramping.

Persons who engage in heavy work or play during the summer months should also take the precaution of salt supplementation and adequate fluid intake. So far, there is inadequate knowledge of the role played by water and electrolyte imbalance in the production of pain of muscle origin. Yet close questioning of people with chronic musculoskeletal pain indicates that their fluid intake often is grossly inadequate.

Painful muscle cramps is one of the muscular manifestations of hypothyroidism. The cramping is usually nocturnal, affecting the calf muscles. The exact etiology of the cramping is uncertain.

The rare disorder of carbohydrate metabolism characterized by muscle cramping, pain, weakness, and stiffness after exercising, with subsidence after rest, is known as *McArdle's syndrome*. Laboratory evaluation shows a fall in the blood lactate and pyruvate after exercise, rather than a rise. This can be considered a diagnostic test. The basic pathology is felt to be an absence of phosphorylase (the enzyme which converts glycogen into glucose-1-phosphate).

Vascular Causes

Cramping from vascular causes may be severe, occurs with exertion, and is relieved by rest, in a combination of events known as *intermittent claudication*. The history is typical and generally points directly to the diagnosis. When possible, treatment is directed toward eradication of the underlying cause. Elimination of aggravating factors such as nicotine is desirable, and vasodilators may be tried although success here is far from universal. Physical therapeutic measures as described in the section on peripheral vascular disorders (pp. 205–207) are applicable here.

Neuritic Causes

Neuritic problems such as herniated nucleus pulposus, peripheral neuritis, and thoracic outlet syndrome may present with muscle cramping as one of the symptoms. In many of these disorders vascular dysfunction with impairment of the vasa nervorum probably plays a significant role. Upper motoneuron diseases may be manifested at their onset by what a patient describes as cramps. This complaint may delay the reaching of a diagnosis.

Idiopathic Causes

The conditions mentioned here have no definite relationship to one another. It is likely that with the passage of time the etiology for many will be uncovered.

OCCUPATIONAL CRAMPS

Occupational neuroses, as they are sometimes called, occur in writers, musicians, salesmen, etc., who use their hands extensively in their work. Rarely is an organic cause identifiable, and neurological deficit is absent. Occupational cramp is believed, therefore, to be a functional disorder. No specific treatment has been effective except the elimination of the precipitating repetitive act, a prohibition that often cannot be heeded since the act is usually vital to the patient's livelihood. The vapocoolant spray may abort episodes of such occupational cramps.

NOCTURNAL CRAMPS

Investigation has failed to reveal any specific relationship between nocturnal cramps and inadequate circulation, inadequate exercise, or loss of calcium, sodium, or any other mineral or nutrient. It is particularly common during pregnancy and in the elderly. It may have been preceded by unusually heavy exertion during the day. Treatment using quinine sulfate or aminophylline, sometimes given in combination, is often effective. Ice massage or vapocoolant spray, massage, and stretching may also be helpful. Calcium salts, antihistamines, muscle relaxants, and tranquilizers may be useful as adjunctive therapy. Vitamin E has been proposed as a method of therapy. Patients who have venous stasis may find elevation of the foot of the bed beneficial.

STIFF-MAN SYNDROME

This rare syndrome is characterized by progressive stiffness, rigidity, and painful muscle spasms, leading eventually to extreme disability including difficulty in moving the limbs and in swallowing. The neurological examination is normal but electromyographic studies show persistent firing of motor units even though the patient is at rest. Treatment is directed toward abolishing the persistent motor unit activity and the accompanying stiffness by the use of myoneural blocking agents such as succinylcholine, peripheral nerve blocks, and general anesthesias. Many patients respond favorably to diazepam.

SHIN SPLINTS

The term *shin splints* is a catchall phrase applied to a number of different conditions limited to the legs. Shin splints usually occur in athletes after strenuous workout on a hard surface. Tearing of the muscle fibers where they attach to the bone is the usual form of injury, and periostitis and irritation of the interosseous membrane may also occur. The tibialis anterior and tibialis posterior muscles are those ordinarily involved, the lesion occurring where they attach to the tibia. Some physicians feel that the involvement of the tibialis anterior may represent a forme fruste of the anterior tibial compartment syndrome. Recently, it has been suggested that the term *shin splints* be limited to strain of the flexor digitorum longus muscle, but we shall continue to use it in its broader sense.

Clinical Picture. Pain in the shin is the predominant symptom. It is exaggerated on active movement of the ankle. Tenderness is present on palpation along the attachment of the muscle to the bone, rather than over the muscle belly. Slight swelling and warmth may also be present. Laboratory and x-ray studies are essentially negative although chronic cases may show some calcification of the bone cortex at the site of irritation.

Treatment. Rest of the leg is the primary treatment. Ice packs may be applied in the acute stage, followed by the use of ultrasound to the area of tenderness. In a patient frequently subject to this problem, strapping of the longitudinal arch of the foot to protect the affected muscle or tendon may be beneficial (see Figure 7-15, p. 143).

ANTERIOR TIBIAL COMPARTMENT SYNDROME

Pathogenesis. The anterior muscular compartment of the leg contains the muscles that dorsiflex the ankle and the toes. Its boundaries are the tibia, interosseous membrane, lateral muscular compartment, and anterior crural fascia. These relatively unyielding walls predispose to the compression of structures within the compartment following severe or unaccustomed exertion. The condition is found most often in young males. Swelling of the musculature occurs. Increased pressure within the compartment produces circulatory impairment of the smaller vessels of the arterial and capillary tree. This leads to ischemia of the muscle followed by reactive hyperemia producing swelling, which impairs venous outflow, producing further swelling and initiating a vicious circle of cause and effect.

Clinical Picture. Severe pain occurring after exertion and located over the anterior compartment is the hallmark of the disorder. The skin overlying the compartment is glossy, and the compartment is tense and bulging. Severe tenderness on palpation over the muscle bellies is noted, and weakness of the dorsiflexor muscles of the ankle and toes frequently occurs. A sensory deficit in the distribution of the peroneal nerve may be found. The dorsalis pedis pulse is usually present. Low-grade fever, mild leukocytosis, and normal sedimentation rate are commonly seen while x-ray examination of the area reveals no abnormalities.

Most commonly confused with anterior tibial compartment syndrome is shin splints. Absence of muscle weakness and inflammatory reaction, and limitation of tenderness to the bone, rather than to the muscle bellies, serves to differentiate between the two conditions. Cellulitis is also frequently confused with this syndrome. A history of recent trauma, higher fever and greater toxicity, a visible portal of entry, inguinal adenopathy, and absence of neurological deficit help to establish the diagnosis of cellulitis. Other conditions to differentiate between are thrombophlebitis, osteomyelitis, stress fracture of the tibia, and arterial embolus.

Treatment. Diagnosis is clinical and should not await extensive laboratory and radiological investigation since recovery appears to be related to the promptness with which an extensive fasciotomy of the anterior crural fascia is performed. If treatment fails, loss of function of the muscles with a permanent footdrop results. Eventually tenodesis of the anterior compartment muscles may act to some degree as a natural brace. Electromyography, performed after surgery, serves as a prognostic guide to recovery since the process of reinnervation is observed electrically before it is observed clinically. The presence of complete electrical silence, both at rest and on exertion, coupled with a "woody" feeling on insertion of the needle electrode, signifies the replacement of contractile muscle tissue with fibrous tissue. Physical treatment such as electrical stimulation and muscle reeducation, which should be used if recovery is taking place, is futile if these irreversible signs are present.

Chronic Form. Chronic anterior compartment compression presents with long-standing anterior leg pain made worse by exertion. Following exertion, weakness of the dorsiflexor muscles, numbness, and swelling of the compartment with glossiness of the overlying skin occur. Electromyography shows the presence of denervation potentials, indicating neuromuscular damage.

Treatment consists in elevation, ice packs, and rest. These produce relief from the acute-phase symptoms. For the residual chronic symptoms appropriate treatment is the surgical excision of a portion of the crural fascia (partial fasciectomy).

PERONEAL (LATERAL COMPARTMENT) SYNDROME

A rare variant of the anterior tibial compartment syndrome involves the muscles of the lateral rather than the anterior compartment of the leg. The clinical picture is similar to that of the anterior tibial compartment syndrome, with pain, redness, swelling, tenderness, numbness, and induration present. The etiology is similar: increased pressure producing ischemia and, subsequently, edema.

Treatment. Once the diagnosis has been established, treatment consists in early fasciotomy of the lateral crural fascia.

MYOSITIS OSSIFICANS

Trauma, either solitary or repeated, with resultant muscle contusion and hematoma formation, is the antecedent cause for myositis ossificans. Such trauma need not be gross. In its most common form, ossification of the infiltrated blood takes place. This mass becomes attached directly to the bone, usually on a broad base. Less commonly, a plaque of bone forms in the muscle completely separated from the bone. Although all muscles may be affected, the quadriceps and brachialis muscles are most often involved in this condition.

Symptoms. Initially the symptoms are local swelling and pain, and, subsequently, limitation of motion of the adjacent joints. After several days a deep-seated mass may be palpated.

X-ray Pictures. X-ray films taken within the first two or three weeks will show nothing except perhaps some slight cloudiness in the muscle. Then a poorly defined, hazy calcification is noted (Figure 9-10). This is known as an immature callus. At this stage, differentiation from a sarcomatous tumor may become a problem. Myositis ossificans is diagnosed by a history of trauma, absence of improvement, and changes seen radiographically. These changes consist in a gradual maturing of the callus with ossification replacing the calcification and with a sharpening of the borders of the bony shadow. After several months the borders of the mass in myositis ossificans become sharply demarcated (Figure 9-11).

Treatment. Early treatment follows the principles we have discussed. Rest from function for the affected part may require the use of crutches when the lower extremity is involved and a sling when the upper extremity is involved. Maintenance of movement passively is important within the limit of pain. All other joints must be kept as functional as possible. Ultrasound and diathermy are modalities used to decrease pain and to improve range of motion. X-ray therapy is sometimes effective. If symptoms persist, and the callus matures, which may take a year or more, surgical excision may be necessary. Surgery before maturation of the callus results in recurrence of the mass.

KELOID SCARS

There is a painful condition of scar tissue in which keloid changes occur and which may produce a differential diagnostic problem. The changes may not be apparent superficially. Keloid scars are more common in the black race.

Treatment. X-ray therapy following the excision of the scar tissue is often effective. Ul-

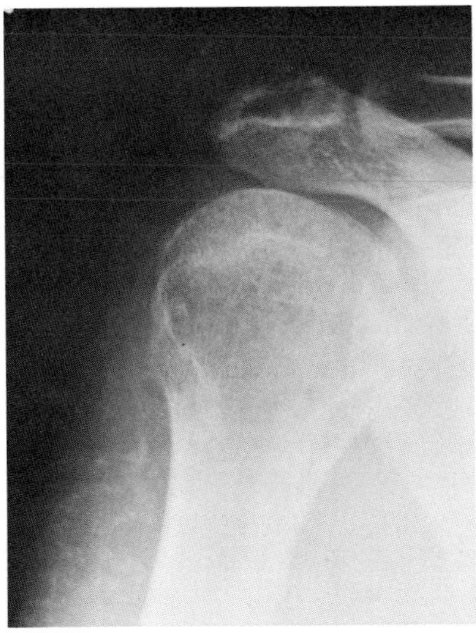

A

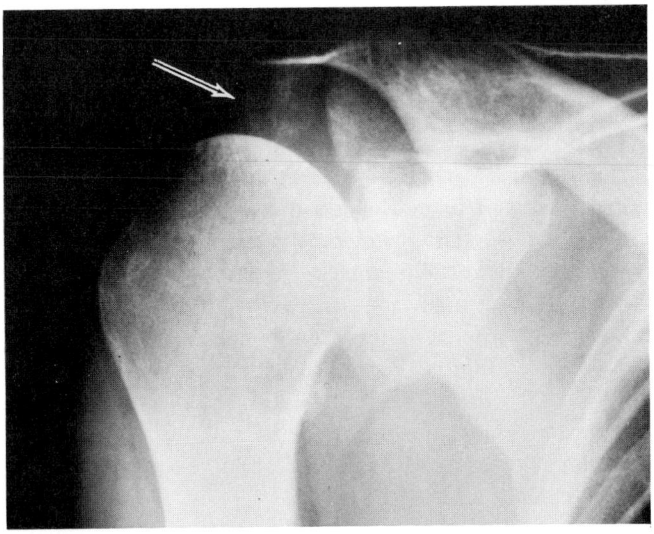

B

Figure 9-10. *A*, X-ray picture of early myositis ossificans showing poorly defined stippling throughout the deltoid muscle. *B*, Arrow shows incidental subacromial calcification in the same shoulder.

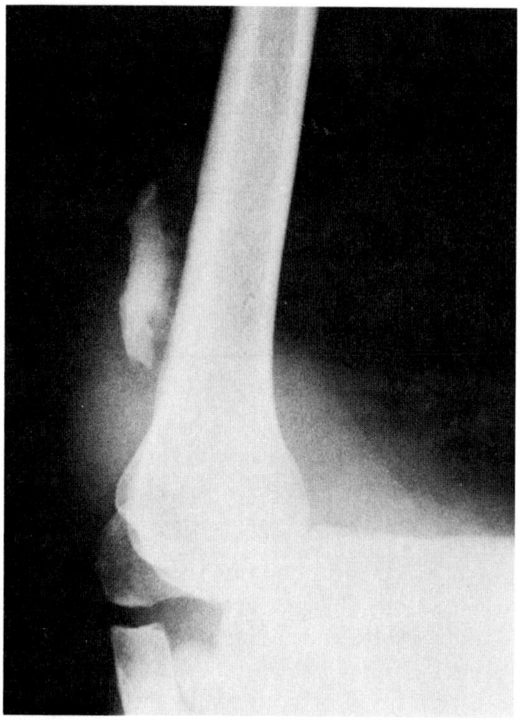

Figure 9-11. X-ray picture showing mature myositis ossificans.

trasound is also, at times, used successfully, as is the intradermal injection of steroids.

OSTEOPOROSIS SENILIS

Because osteoporosis senilis is common, we feel that its treatment should be a matter of common knowledge. Since this is a disease of protein metabolism and protein intake is often inadequate, the patient should be on a high protein diet. Protein concentrate added to milk is suggested as an easy and inexpensive way of correcting the diet. There is some clinical doubt as to the place of calcium in treatment of the disease, but the intake of milk with the protein takes care of this. Vitamin C and D supplements are beneficial.

It is generally accepted that hormonal deficiency plays a role in the etiology of the condition; hormonal replacement therefore forms part of the management of it. Recovery from osteoporosis senilis is a slow, drawn-out process, and for psychological reasons we recommend a 30-day course of estrogen once a day, in a dosage of 1.25 mg, followed by a 30-day course of testosterone once a day, in a dosage of 5 mg, followed by rest from hormones for 30 days. Very small doses are required. Ninety-day courses of treatment cannot allow a patient to expect any quick response "from taking medication," and lack of early relief of pain may be extremely disheartening. The criticisms leveled against hormonal treatment are rather spurious. At the age at which osteoporosis occurs, it is of little consequence if gynecomastia occurs in men. The possible carcinogenic factors seem to us of scant concern too, for the same reason. Parenthetically, the reader is reminded that a very early symptom of hormonal (especially estrogen) overdosage is pruritus.

Many patients suffering from osteoporosis senilis, because of their age and concomitant carelessness of themselves, are anemic and appropriate medical evaluation is necessary. If no other cause is demonstrated, then the patient should receive iron medication. Such patients also tend to avoid exposure to sunlight because they either stay indoors or overdress when they do go out. We have found that courses of tonic ultraviolet light seem to help them.

Finally, one of the most efficacious means of stimulating osteoblastic activity is the pull of muscle on bone. The pull must be active and associated with function as passive exercise is of little if any benefit in this problem. There is usually weakness of the extensor muscles of the spine as progressive disc degeneration increases the thoracic kyphosis and it must be corrected. However, as a pathological fracture most often draws attention to this condition, especially the crushing of a vertebral body, healing of the fracture (which occurs by the formation of osteoporotic new bone) and support of the fracture site by bracing (usually with the Taylor back brace) are equally important, as is treatment of muscle spasm by local injection, relaxation therapy, massage, and moist heat. Diathermy is contraindicated because it tends to increase the osteoporosis. A fine therapeutic line must be drawn between the benefits of bed rest and the necessity for exercise.

All this treatment regimen is necessary to achieve the main goal of relieving pain.

FIBROMYOSITIS AND TRIGGER POINTS

The diagnosis of fibrositis is synonymous clinically with the diagnosis of fibromyositis and myositis. It should be used to describe the condition of diffuse infiltration of the muscles with areas that are painful on palpation. Sometimes localized painful nodules are felt; hence the expression "painful deposits." These have to be differentiated from rheumatoid nodules, trigger points, and fatty tumors or herniations. No one, we believe, has ever demonstrated a specific pathological entity of a nodule; the only histological change demonstrated has been nonspecific local areas of round cell infiltration. For this reason many authorities refuse to recognize the condition as a pathological entity and ascribe fibrositic symptoms to a psychosomatic upset.

Clinically, however, there can be no doubt that such a condition exists. The symptoms and signs with which it is associated are remarkably constant. They can be relieved equally by suitable physical therapy without the addition of any overt psychotherapy to the treatment program.

Etiology. It is probable that the modern way of life causes people many times a day to lay undue stress or strain on their muscles sufficient to produce minute muscle fiber tears. These are most likely to occur in the extensor muscles of the back, which are the least trained and no doubt the most abused muscles in the body. The tears have to involve macroscopic areas of muscle before they can give rise to symptoms; nevertheless, when muscle fibers are torn, their healing is accomplished by the laying down of fibrous scar tissue. Probably the scarring is microscopic and insufficient to cause inefficiency in muscle action, and no symptoms ensue. But if joint dysfunction or any other cause of muscle spasm is superimposed, muscle relaxation becomes impaired.

Anything that interferes with physiological relaxation of muscle starts a vicious circle of events: Lack of relaxation produces fatigue, which produces muscle spasm, which produces further fatigue, further impairment of relaxation, more spasm, and so on. The flow of venous blood and lymph away from the muscle depends on perfect relaxation and contraction of muscle. Without relaxation, there are accumulations of lymph and an abnormal venous stasis in the muscle. With accumulation of tissue fluid and catabolites, which are noxious to the tissue, an inflammatory reaction develops. Pain ensues in scattered places, especially in preexisting traumatized areas. The addition of pain to the vicious cycle produces more spasm and more stasis in the affected muscles.

This is our concept of fibrositis, and this mechanism goes on throughout the hours of the patient's active day. The process can be reversed only by promoting normal flushing of the involved muscle and producing relaxation in it. Even when the cause of the impaired relaxation is psychological tension, the resulting spasm brings local pathological changes that are associated with physical manifestations, and the latter cannot be treated by psychological methods alone.

Infective Fibrositis

If to all the foregoing is added transient bacteremia or toxemia from any focus of infection in the body such as sore throat, an abscess, an infected tooth, a diseased viscus, or upper respiratory tract infection, infection may be set up and smolder in the ripe medium of the damaged muscle. This is the start of infective fibrositis. Treatment is ineffectual unless the focus of the infection is first identified and eradicated.

Cold Fibrositis

Local chilling by a draft may produce an acute attack of fibrositis. In this instance, besides the cycle of changes described above, there is some local vasoconstriction of the blood vessels, causing local ischemia that may give rise to severe pain. Treatment must include removal of the cause. Cold is one of the few causes of primary fibrositis.

Panniculofibrositis

Another form of fibrositis, found particularly in the back, is associated with any loss of joint movement from any cause. Panniculofibrositis is manifested by the adherence of the skin and superficial fascia to the deep fascia and even through the interstitial tissue throughout the muscles. It is detected by skin rolling directly over the spine, and when diffuse fibrositis is present in the muscles, skin rolling is tight over them too. It is not peculiar to the back, being also found over a pathologically tight iliotibial band and over diseased muscle in, for instance, patients with rheumatoid arthritis.

If in a well-localized area skin rolling is tight and tender in the back and there is clearly no pathological cause of pain in any structure in the back, the tight and tender skin rolling is very suggestive that something is wrong in the viscus beneath the area being skin-rolled. Skin rolling thus becomes a useful though nonspecific sign of the existence of pathology.

Pathological Fibrosis

When elastic fibers in soft tissue are lost, they do not regenerate but are replaced by fibrous tissue. The result is loss of elasticity, resilience, and length of the affected soft tissue, be it a tendon, a ligament, a capsule of a joint, the intermuscular septa, interstitial tissue, an iliotibial band, or plantar fascia. Painful fibrosis must not be mistaken for painful fibrositis; painful muscle spasm must not be mistaken for painful contractures.

Treatment. If cold is the cause of primary fibrositis, the treatment is to use local heat followed by kneading massage to help limber up the muscles. Movement, using the involved muscle(s), is encouraged within the limit of pain. A recalcitrant fibrositic nodule may require special attention. Pressure over it and friction massage around it may be sufficient to disperse it. If this fails, ultrasound over the nodule for 30 seconds to 2 minutes may be used. Locally applied anodal galvanism for 5 to 10 minutes is effective also. Sometimes the nodule has to be infiltrated with local anesthetic. Anodal galvanism enhances the effect of the injection. Treatment sessions should end with stroking massage to promote relaxation.

If fibrositis is secondary to a focus of infection, treatment fails unless the focus is found and eradicated. If the fibrositis is localized, the same treatment routine used in treating primary fibrositis is followed.

If the fibrositis is diffuse, a most effective precursor to the use of more specific modalities is the use of the moist-air baker. Sweating helps the excretion of muscle catabolites. The patient's temperature should not be allowed to rise above 102° F. This is followed by the application of ultrasound with faradic muscle stimulation to the involved muscles. The pumping action of muscle helps to flush the muscles and restore them to their normal physiological state. Any particularly recalcitrant nodule(s) is treated in the manner already described. Stroking massage completes each treatment session to promote relaxation.

The Trigger Point

The irritable trigger point in muscle is a manifestation of abnormal functioning of the muscle spindle apparatus resulting in painful muscle spasm, loss of the muscle's ability to assume its normal resting length, and pain, which is referred to places quite distant from it but occurs in predictable patterns. Further, the location of a trigger point is predictable, given a pattern of referred pain (Figures 9-12 to 9-15). Figure 9-12 illustrates the more common trigger points in the head and neck. Figure 9-13 illustrates the more common trigger points in the upper extremities. Figure 9-14 illustrates the more common trigger points in the torso. Figure 9-15 illustrates the more common trigger points in the lower extremities. In contradistinction, a fibrositic nodule has no predictable location, and pain from it has no

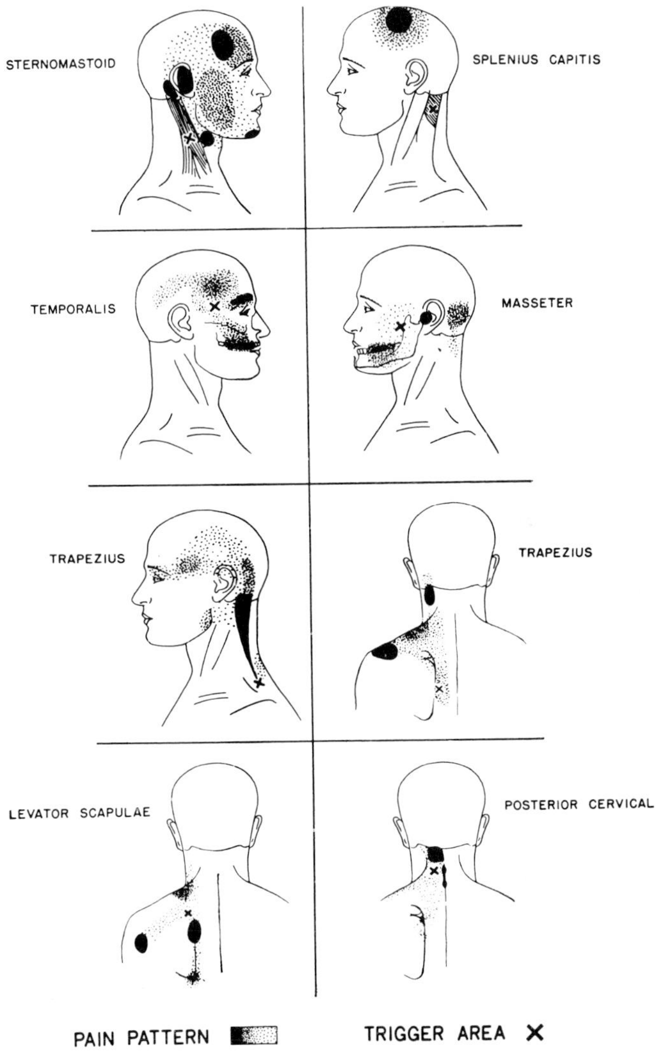

Figure 9-12. The main sites of predictable trigger points in the head, neck, and posterior forequarter areas and predictable patterns of referred pain from them. In each picture X indicates the trigger point. The black areas indicate the most common sites of referred pain. The heavily speckled areas indicate less common sites, and the lightly speckled areas are occasional sites of referred pain. (Figures 9-12 through 9-15 are from J. G. Travell, in *Postgraduate Medicine* 11: 425, 1952. Copyright by McGraw-Hill, Inc.)

predictable pattern. Fibrositis may be diffuse; the trigger point is localized, though there may be more than one trigger point.

There appear to be analogues of the muscle trigger points in tendons and joint capsules and their ligaments. But, whereas it is fairly certain that the muscle trigger point is an irritated spindle mechanism, to suggest that an irritable Golgi apparatus is the trigger point in a tendon and that an irritable Ruffini apparatus is the trigger point in joint capsules is more conjectural, though it seems logical and probable.

In competition with the pain receptor apparatuses mentioned above, which act through the spinal cord reflex arc, are the sensory nerve endings of the spinal nerves in the skin. These set off their reactions not only by way of the spinal cord reflex arc but also through their central nervous system connections via the spinothalamic tracts, the posterior columns of Goll and Burdach, and their connections with the sensory cortex of the brain. Reflex response to pain in this system is via the long motor tracts and their relays with the anterior horn cells and the peripheral nerves.

Noxious stimuli from the pain receptor organs appear to set up noxious conditioned reflex patterns in muscle, and successful treatment depends on the reestablishment of nor-

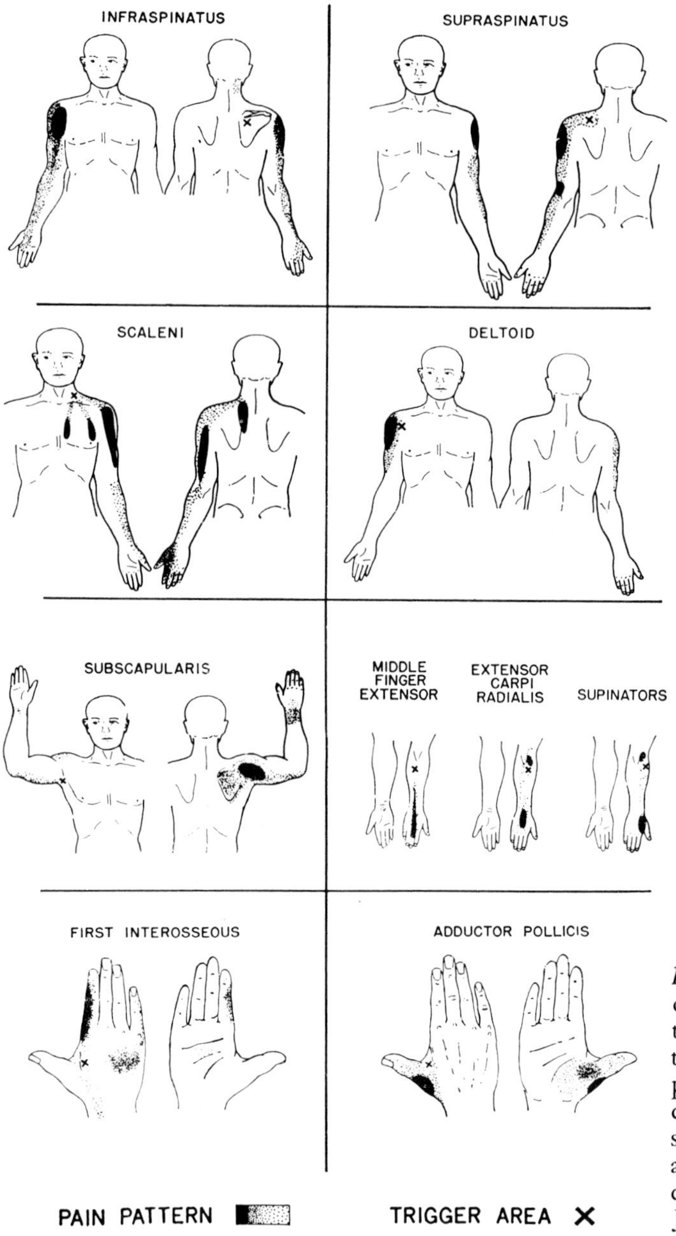

Figure 9-13. The main sites of predictable trigger points (marked X) in the muscles of upper extremities and their predictable patterns of referred pain. The black areas indicate the most common sites of pain. The heavily speckled areas are less common sites, and the lightly speckled areas are occasional sites of referred pain. (From Janet G. Travell.)

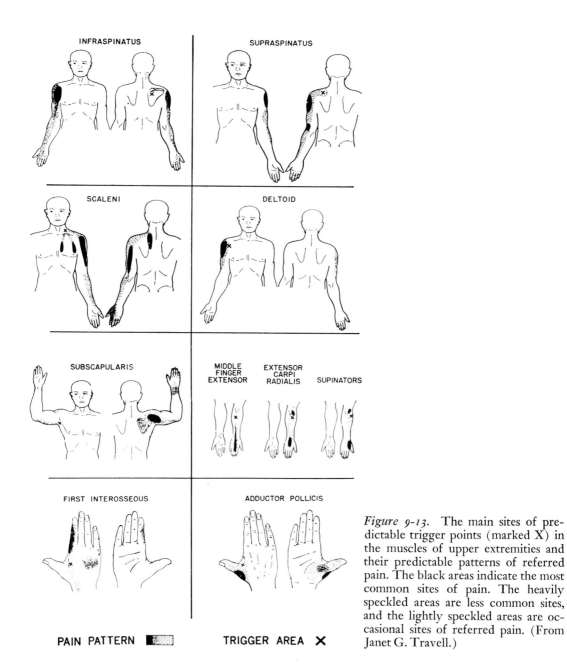

mal reflex patterns in the muscle, which is largely dependent on restoration of a muscle in spasm to its normal resting length. This would seem to be the basis for the use of so many varied modalities of physical treatment for the relief of pain. Particularly it explains the very rapid response of muscle pain and spasm in patients treated by the vapocoolant spray.

Treatment. Treatment of an irritable trigger point producing pain is by the use of the vapocoolant spray with stretch techniques or by injection therapy and stretch techniques or both. The proper use of the vapocoolant spray has been described in detail in Chapter 7, p. 126. The more common trigger points giving rise to predictable patterns of pain are illustrated on pp. 191–194, and these should be

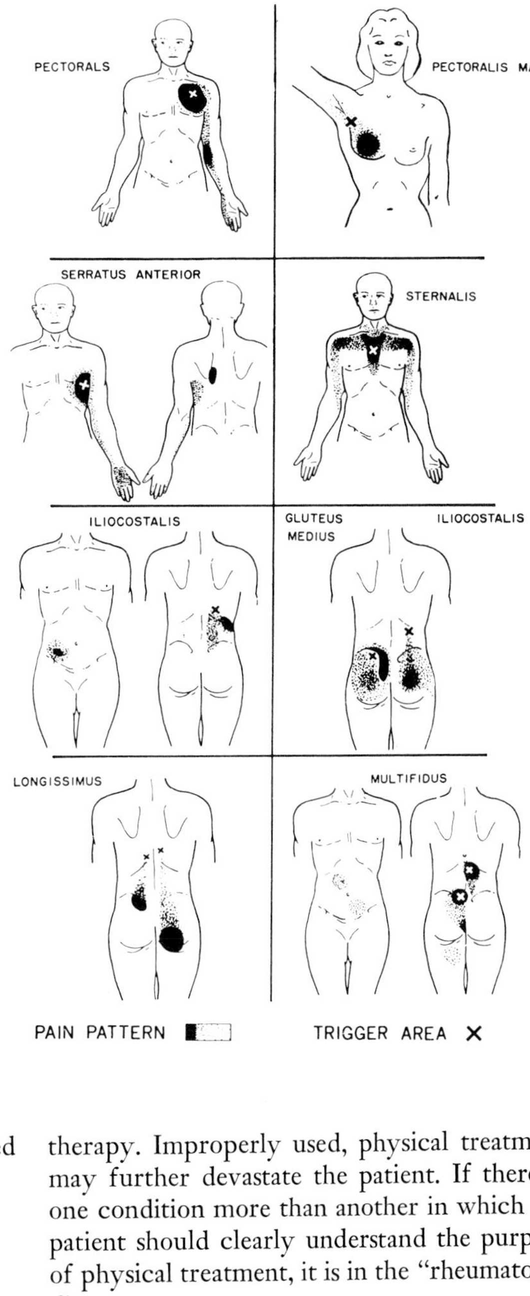

Figure 9-14. The main sites of predictable trigger points (marked X) in the muscles of the torso—particularly the back—and their predictable patterns of referred pain. The black areas indicate the most common sites of pain. The heavily speckled areas are less common sites, and the lightly speckled areas are occasional sites of referred pain. (From Janet G. Travell.)

consulted when therapy is being contemplated and undertaken.

COLLAGEN VASCULAR DISEASES

We wish to stress the importance of observing the principles of physical treatment outlined in Chapters 1 and 7, when one is devising a plan of physical and occupational therapy for the "rheumatoid" patient, no matter what the underlying disease entity may be. Properly used, physical treatment is an essential part of management of such a patient and must have its place in conjunction with appropriate drug therapy. Improperly used, physical treatment may further devastate the patient. If there is one condition more than another in which the patient should clearly understand the purpose of physical treatment, it is in the "rheumatoid" diseases.

Pain is an inevitable concomitant of the collagen vascular diseases. Too often patients are encouraged to believe that physical therapy is a treatment of the disease itself. If it is denied them, they feel deprived and neglected. Physical therapy easily becomes an addiction for these patients. Many "rheumatoid" patients, however, do not have physical therapy pre-

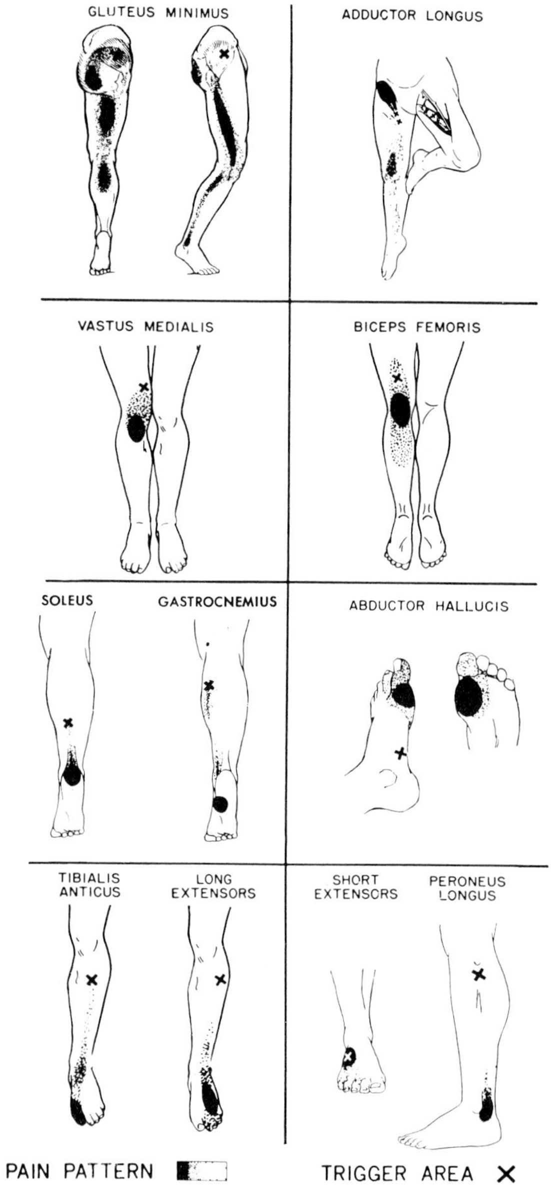

Figure 9-15. The main sites of predictable trigger points (marked X) in the muscles of the lower extremities and their predictable patterns of referred pain. The black areas indicate the most common sites of pain. The heavily speckled areas are less common sites, and the lightly speckled areas are occasional sites of referred pain. (From Janet G. Travell.)

scribed for them at all, and they certainly are deprived of useful adjunctive treatment which relieves pain and lessens disability.

The three phases of collagen vascular diseases are the acute phase, the chronic phase, and the phase of remission. Treatment should be tailored to the phase of the disease in which the patient is found. There are three types of programs of physical treatment. In the first, the modalities are chosen that relieve pain, produce relaxation, and prevent deformity. In the second, modalities are chosen to maintain as normal a state as possible. In the third, modalities are chosen to restore lost function and consolidate the patient's optimal functional state.

ACUTE PHASE OF THE DISEASE

Relief of Pain. The modalities used for the relief of pain are the same in each phase of the disease, but there are some painful conditions mistakenly attributed to the disease and these will be described later. Moist heat or cold packs are equally efficient in the relief of pain, but some patients react better to one or the other. When a number of joints are involved, the Hubbard tank* is effective. It not only supplies universal heat but produces relaxation, from the effects of weightlessness, and allows safe, painless movement of the involved joints. Jacobson's relaxation regimen helps muscles in protective spasm, which itself is a painful condition, to "let go." Sometimes, in muscles in spasm, there are irritable trigger points, the pain from which is readily relieved by the use of the vapocoolant spray or injection.

Prevention of Deformity. Prevention of deformity in the acute stage is achieved by the use of positioning splints. If the patient is in bed, preventing foot deformity and pain by not tucking in the bedclothes is very important. Shoulder pain can be relieved simply by propping the elbows up on pillows. Neck and shawl area pain can be relieved by the use of a soft cervical collar. Wrist pain can be relieved by the use of splints while knee pain can be reduced by the application of a posterior shell. In very "hot" joints, intraarticular injections of corticosteroids are most helpful in lessening the inflammatory reaction and pain.

* Available from Ille Electric Corp., 2245 Reach Rd., Williamsport, Pa.

CHRONIC PHASE

Relief of Pain. Often pain is still a feature in the chronic phase. The same modalities as in the acute phase may be used in preparation for the next step in treatment.

Maintenance. Maintenance of movement by means of movement therapy as opposed to exercise therapy is added to the treatment program. Splinting changes from the use of static to the use of dynamic splints, but prevention of deformities during the night remains an important goal, and the patient continues to use the static splint at night.

In ankylosing spondylitis postural exercises and breathing exercises are in order; even plaster casting may be essential to maintain normal bodily alignment and prevent back deformities. It should be remembered that in patients with ankylosing spondylitis who are intolerant of or refractory to medication properly applied x-ray therapy becomes an essential part of the treatment.

PHASE OF REMISSION

In this phase occupational therapy plays a major role in the restoration of function within the limitations of the patient's functional ability. Manipulation has a place in treatment but is absolutely contraindicated, especially manipulation of the cervical spine, in ankylosing spondylitis, in rheumatoid spondylitis, and, parenthetically, in people with neck pain associated with acute tonsillitis. Muscle reeducation in the form of a graduated exercise program is often advised. Bracing may be considered as a permanent adjunct to renewed activity, but splints may add to disability if they are thoughtlessly prescribed. In psoriatic arthritis, ultraviolet light may prove beneficial. The patient should be encouraged to participate actively in life within his or her tolerance.

Nonrheumatoid Pain in Rheumatoid Patients

Patients with collagen vascular disease often have pain that is attributed to the disease but in fact is arising from diagnosable and treatable conditions not primarily caused by the disease. Foot pain in a rheumatoid arthritic patient may well be arising from metatarsalgia, which can be relieved by mobilization of the metatarsophalangeal joints and the restoration of movement between the metatarsal heads. This, followed by faradic foot baths and redistribution of weight bearing by the use of a metatarsal head orthosis, is the proper treatment. Foot pain in a rheumatoid arthritic patient may also be caused by a dropped metatarsal head, which can be relieved by wearing a suitable orthotic device.

Ankle pain may arise from treatable joint dysfunction in the subtalar joint, and back pain may also fall into this category. Knee pain in a rheumatoid patient may be due to meniscal injury and can be relieved by meniscectomy.

Inability to supinate the forearm may be due to a meniscal injury in the ulnotriquetral joint and elbow pain may arise from the same cause. The various entrapment syndromes can occur in a rheumatoid patient and the symptoms can be relieved in the same way as in nonrheumatoid patients. This brief list of treatable conditions giving rise to pain not due to the disease condition is by no means complete but should alert the reader not to attribute all pain to arthritis.

COCCYGODYNIA

When coccyx pain follows direct trauma, it is regarded by many physicians as an almost insuperable pain problem, especially as removal of the coccyx, when all else has failed, often does not relieve pain or creates a chronic pain situation of its own. We believe that it is no more of a problem than any other musculoskeletal pain if the patient is subjected to a proper clinical examination following the system we advocate and if treatment for the cause of the pain adheres to the principles of physical therapy we advocate.

Examination of a patient with coccyx pain is incomplete without a rectal examination (an examination of coccyx movement with a finger in the rectum and the thumb on the surface posteriorly) and, in women, a pelvic examination. In the absence of a history of trauma, strain of one of the many ligaments that insert into the coccyx, especially in women, is a common cause of coccydynia. Early diagnosis, as with all other musculoskeletal pain, is a prerequisite to successful treatment. Once chronicity has settled in, the pain does tend to become intractable.

The joint structures of the coccyx are weight-bearing in that the coccyx plays a part in supporting the pelvis and its contents. Rest from function is an essential principle of treatment. This can be satisfactorily obtained by strapping the buttocks tightly together. The use of crutches for ambulation seems to add to the relief of pain. Sitting should be avoided because it increases pressure in the coccyx unless the patient is seated on an inflated ring. The patient should either stand or lie down.

Joint dysfunction is a very common cause of coccyx pain. It readily responds to joint manipulation. To achieve this, the involved junction, most often the joints at the sacrococcygeal junction, must be tilted open by means of the index finger, used as a fulcrum in the rectum, and the thumb on the surface, used as a mobilizing force. Prior infiltration of the junction with local anesthetic makes this a painless procedure. As a matter of fact, the simple injection therapy of local anesthetic with cortisone alone may produce relief.

When one suspects that synovitis is present, ultrasound applied under water is effective. We give this in a sitting whirlpool. The hot circulating water is also relaxing to the patient.

When ligament strain is suspected, infiltration with local anesthetic and the application of medical galvanism or ultrasound or phonophoresis with hydrocortisone are helpful. But rest from function between treatments is imperative.

Bruising of the periosteum responds well to ultrasound under water and rest from function.

Gynecological causes of coccyx pain must be corrected.

We feel that the most often overlooked forms of treatment are manipulation, local infiltration, and rest from function by strapping. We also feel that greater success would be achieved by surgery if patients were afforded good physical therapy postoperatively instead of being left to the healing powers of nature, which invariably in this location results in painful scarring. The coccyx is a functional part of the musculoskeletal system and adaptation of the associated structures must be achieved after its removal for surgery to be successful.

BONE PAIN

In our crossmatching of structures with possible pathological conditions as causes of musculoskeletal pain (see Chapter 2), principles rather than diagnostic entities were stressed. It happens that in all the sections, other than the one concerning bone, there are few if any possible omissions.

Recognizing that bone pain has many causes besides those already mentioned, even though the pain from them is not amenable to modalities of physical therapy for relief, we are deserting our format in this instance for a brief review of those causes of bone pain which may otherwise appear to have been overlooked.

Bone itself is an insensitive structure, and pain from it can arise only from pathological conditions that involve the periosteum or the endosteum. For this reason painful bone conditions are often quite far advanced before they manifest themselves clinically.

A group of systemic diseases that affect bone and often give rise to bone pain as a presenting symptom includes the leukemias, lymphomas, caisson disease, sickle cell disease, multiple myeloma, Gaucher's disease, and Letterer-Siwe disease. Pain from these conditions must stem from endosteal involvement and is characterized by symptoms of a deep, constant aching and throbbing.

When the periosteum is involved, pain becomes sharp, lancinating, and acute. Thus the character of bone pain during the course of a disease may change as the periosteum becomes secondarily involved.

Diseases involving the periosteum rather

than the endosteum that give rise to pain are Paget's disease, vitamin deficiency diseases, renal disease, thyroid disease, disease from disturbance of amino acid metabolism, osteogenesis imperfecta, and congenital fracture such as may occur in the tibia.

In considering conditions associated with fractures, there is a parenthetical observation to be made concerning vertebral body fractures. Involvement of the upper vertebral plate only usually reflects trauma as being the cause whereas involvement of both the upper and lower plates suggests an underlying pathological condition predisposing to fracture. Also, some back conditions characteristically involve either one vertebra, or two adjacent vertebrae with disc involvement, or two adjacent vertebrae sparing the disc, or the disc only. Tumors affect the vertebrae only. Paget's disease may be monostotic, and so may multiple myeloma and the blood dyscrasias already mentioned. Diseases associated with bone infection due to pyogenic organisms or tuberculosis or other granulomatous diseases involve two adjacent vertebrae and the intervertebral disc. A salmonella infection affecting the spine shows a predilection for the disc only. Infection such as occurs in heroin addicts also tends to select the disc for involvement. Ankylosing spondylitis spares the disc and does not deform the vertebrae, involving primarily the longitudinal ligament to produce the typical bamboo spine. Figure 5-16 illustrates this.

Two special types of chronic osteomyelitis cause bone pain: Brodie's abscess and the sclerosing osteomyelitis of Garré. Among the causes of pain from bone should be included causes of periostitis: syphilis, tuberculosis, sarcoidosis, and changes associated with some chronic lung diseases.

Sudeck's Bone Atrophy (Reflex Dystrophy)

An unusual, painful osteoporotic condition of bone occurs after trauma and immobilization. It is most commonly found in the foot and hand and is associated with an intense causalgia-like pain. The pain of Sudeck's bone atrophy may be checked by a stellate ganglion block or by a lumbar sympathetic block, depending on the site of the atrophy, following which physical therapy, including the use of manipulative techniques, readily relieves the disabilities arising from it.

We may have given the impression, by Figure 2-8, p. 22, that only a limited number of primary bone tumors give rise to pain. Obviously there are many other benign primary bone tumors and some other malignant primary bone tumors, but we were deliberately stressing the principles of the differential diagnosis of malignant tumors of bone. The same primary bone tumors may be found in the vertebrae as in the bones of the extremities, a fact that is sometimes forgotten.

Pain in any of the above conditions may result partly from reflex guarding muscle spasm. When this is the case, that part of the pain may indeed be relieved by physical therapy, if only temporarily, and particularly by the use of the vapocoolant spray and stretch techniques, by local injection therapy, by Hubbard tank, and by Jacobson's relaxation exercises. Bone pain may be relieved at least temporarily by the use of subtonal faradism or transcutaneous nerve stimulation. When bone pain is intractable there are neurosurgical procedures that may be successful.

We stress again the fact that just because pain is relieved by treatment it is by no means certain that the cause of that pain has been determined.

BIBLIOGRAPHY

Adams, R., Denny-Brown, D., and Pearson, C. M. *Diseases of Muscle* (2nd ed.). New York: Hoeber, 1962.

Ayres, S., Jr., and Mihan, R. Leg cramps (systremma and "Restless Legs" syndrome). Response to vitamin E (Tocopherol). *Calif. Med.* 111:87, 1969.

Bonica, J. J. *The Management of Pain.* Philadelphia: Lea & Febiger, 1953.

Boyd, H. B., and McCleod, A. C. Tennis elbow. *J. Bone Joint Surg.* (*Am.*) 55A:1183, 1973.

Conrad, R. W., and Hooper, W. R. Tennis elbow: Its course, natural history, conservative and surgical management. *J. Bone Joint Surg. (Am.)* 55A:1177, 1973.

Edwards, P. W. Peroneal compartment syndrome. *J. Bone Joint Surg. (Am.)* 51B:123, 1969.

Golding, D. N. Hypothyroidism presenting with musculoskeletal symptoms. *Ann. Rheum. Dis.* 29:10, 1970.

Gordon, E. E., Januszko, D. M., and Kaufman, L. A critical survey of stiff-man syndrome. *Am. J. Med.* 42:582, 1967.

Jackson, R. *The Cervical Syndrome* (3rd ed.). Springfield, Ill.: Thomas, 1971.

Kellgren, J. H. Observations on referred pain arising from muscle. *Clin. Sci.* 3:175, 1938.

Layzer, R. B., and Rowland, L. P. Cramps. *N. Engl. J. Med.* 285:31, 1971.

Leach, R. E., Zohn, D. A., and Stryker, W. S. Anterior tibial compartment syndrome: Clinical and electromyographic aspects. *Arch. Surg.* 88:187, 1964.

Leach, R. E., Zohn, D. A., and Stryker, W. S. Anterior tibial compartment syndrome due to strenuous exercise. *Mil. Med.* 129:610, 1964.

Mennell, J. M. Assessment of Residual Symptoms from a "Whiplash Injury." In Proceedings of the IVth International Congress of Physical Medicine, Paris, France, 1964. (Excerpted from Medical International Congress Series 107.) September 1964.

Mennell, J. M. *Foot Pain.* Boston: Little, Brown, 1969.

Millender, L. H., Nalebuff, E. A., and Holdsworth, D. E. Posterior interosseous nerve syndrome secondary to rheumatoid synovitis. *J. Bone Joint Surg. (Am.)* 55A:375, 1973.

Nirschl, R. P. Tennis elbow. *Orthop. Clin. North Am.* 4:787, 1973.

Schwarz, G. S., Berenyi, M. R., and Siegel, M. W. Atrophic arthropathy and diabetic neuritis. *Am. J. Roentgenol. Radium Ther. Nucl. Med.* 106:523, 1969.

Simpson, R. G. Nocturnal disorders of medical interest. *Practitioner* 202, 259, 1969.

Stevens, H., and Baxelon, M. Writer's cramp. *Trans. Am. Neurol. Assoc.* 91:342, 1966.

Swezey, R. L. Exercises with a beach ball for increasing range of joint motion. *Arch. Phys. Med. Rehabil.* 48:253, 1967.

Travell, J. G. Mechanical headache. *Headache* 7: 23, 1967.

Triger, D. R., and Joekes, A. M. Severe muscle cramp due to acute hypomagnesaemia in haemodialysis. *Br. Med. J.* 2:804, 1969.

Utz, J. P., Hart, A., and Johnson, E. W. A Study of Shin Splints. Paper presented to American Academy of Physical Medicine and Rehabilitation, Denver, Colorado, 1972.

Walton, J. N. *Disorders of Voluntary Muscle* (3rd ed.). London: Churchill, 1974.

Weeks, V. C., and Travell, J. G. Postural vertigo due to trigger areas in the sternocleidomastoid muscle. *J. Pediatr.* 47:315, 1955.

Zohn, D. A., and Leach, R. E. The role of the electromyogram in the diagnosis and management of the anterior tibial compartment syndrome. *Arch. Phys. Med. Rehabil.* 45:311, 1964.

10
Neurovascular Pain Conditions

NEURITIC CONDITIONS

This chapter deals with certain selected problems arising from the vascular or nervous system. We have included conditions only if they pose differential diagnostic considerations from those intrinsic to the musculoskeletal system (Chapter 9), if they have secondary manifestations in the musculoskeletal system, or if physical treatment plays a significant role in their management.

Carpal Tunnel Syndrome

Compression of the median nerve as it passes together with the flexor tendons and their synovial sheaths in a bony channel at the wrist produces the symptoms of the carpal tunnel syndrome. This channel is converted into a tunnel by the overlying transverse carpal ligament (Figure 10-1). Alteration of the ligament, of the contents of the tunnel, or of the bony structures forming the channel has as its end result entrapment of the median nerve.

ETIOLOGY

A wide variety of conditions may affect one or more of the structures involved, but the correct etiological factor often remains unidentified. Rheumatoid arthritis may produce both a thickening of the ligaments and tenosynovitis. Occupational stresses may directly damage the nerve. A fracture may alter the bony channel. Pregnancy with resultant edema causes compression of the contents, while numerous infectious and metabolic conditions such as tuberculosis, acromegaly, myxedema, and amyloid disease may produce alteration of the tunnel or its contents, leading to median nerve compression. Congenital anomalies of muscle, such as an aberrant abductor digiti quinti or aberrant lumbrical, a bifid palmaris longus, or a sublimis muscle belly extending into the carpal tunnel, are increasingly being recognized as causes of median nerve compression. New causes for the condition are constantly being acknowledged.

HISTORY

Paresthesia is the most common symptom. Its presence in the upper extremity at night should lead one to suspect either a carpal tunnel syndrome or a thoracic outlet syndrome. Numbness and pain are frequent complaints. More proximal referral of pain, usually to the elbow but occasionally to the shoulder, is a common manifestation. Weakness of the thenar muscles is usually a late occurrence. The condition is often bilateral.

PHYSICAL EXAMINATION

Because other pain syndromes may mimic the carpal tunnel syndrome, an examination of all cervicobrachial structures should be included.

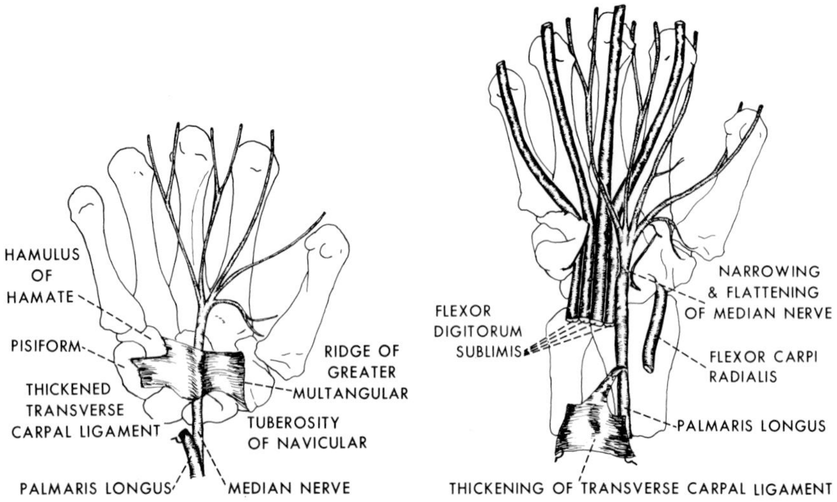

A *B*

Figure 10-1. *A*, Relationship of median nerve to structures making up carpal tunnel. *B*, Relationship of median nerve to other structures within the carpal tunnel.

The presence of a Tinel's sign at the wrist is the most common finding in the carpal tunnel syndrome. This is a painful tingling which is referred to the distal portion of the nerve on tapping over the site of damage. Techniques of examination such as direct compression of the median nerve in the carpal tunnel or forced flexion and extension of the wrist may reproduce or exacerbate the patient's symptoms. Atrophy of the thenar eminence may be observed but is not usually present. A neurological examination including motor and sensory testing of the median and ulnar nerves completes the examination.

ELECTRODIAGNOSIS

Electrodiagnostic testing is a valuable adjunct to clinical examination. Nerve conduction studies and electromyography should be combined, and comparison with the opposite side, if unaffected, should always be carried out. If symptoms are bilateral, comparison with the ulnar nerves should be made. Prolongation of the terminal latency is a common finding. Dispersion of the action potential (temporal dispersion) is also frequently found with this test. Sometimes there is a slowing of the conduction velocity of the segment of nerve from elbow to wrist, but this should not exceed 10 percent of normal values (see Table 6-4, p. 106). Greater slowing should make one suspect a process involving the nerve trunk itself.

Prolongation of terminal sensory latency, increased difficulty in obtaining a sensory latency, and disappearance of the sensory latency are all indications of abnormalities of sensory fibers at the wrist.

Electromyographic evidence of nerve degeneration of thenar eminence muscles, but not of the forearm muscles innervated by the median nerve, is evidence of entrapment at the carpal tunnel. This finding frequently appears later than abnormalities of nerve conduction, or appears only in more seriously injured nerves.

TREATMENT

Treatment may be nonsurgical. Splinting of the wrist by means of a cock-up splint (see Figure 7-8, p. 139) is both diagnostic and therapeutic. This splint is worn day and night and is designed so that the patient can perform pinch prehension but not grasp. If splinting proves successful as a treatment, the patient then discards the splint during the daytime but continues to wear it at night, maybe for months. If edema is suspected, as in pregnancy, a diuretic should be prescribed. Local injection of the tunnel with steroids is often effective but should not be done frequently, since crystalline deposits from the steroid may produce permanent nerve damage. Figure 10-2 shows a surgical dissection of the median nerve in a patient who has received steroid injection in the past. The crystalline deposits in the nerve are evident. We find no physical therapy modality to be beneficial.

Surgical division of the transverse carpal ligament is usually curative. Failure to completely divide the more distal portion of the ligament may lead to persistence of the patient's complaints.

Sometimes there is a variation in the course of ulnar nerve under the transverse carpal ligament, and the symptoms and signs of the carpal tunnel syndrome may be confusing.

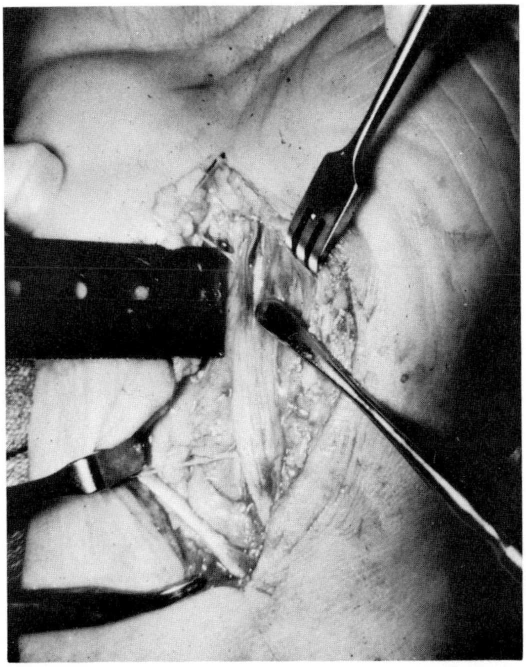

Figure 10-2. Crystalline deposit of steroid in the median nerve. (Courtesy of Bruce Butler, M.D.)

Other Entrapment Syndromes

Peripheral nerves are subject to entrapment from a wide variety of causes as they traverse soft tissues for long distances. Local, mechanical phenomena must be differentiated from disease processes involving the entire nerve, the *neuritides*. These other entrapment syndromes involve the ulnar nerve at the elbow and wrist, the radial nerve in the upper arm, the common peroneal nerve at the knee, the tibial nerve at the ankle, and the lateral femoral cutaneous nerve at the pelvic brim. Rarely, all of the other peripheral nerves may also be subject to entrapment. Parenthetically, the disc syndrome is really another entrapment syndrome.

ETIOLOGY

Causes of nerve entrapment are trauma, systemic disease, muscle spasm, space-occupying lesions, congenital narrowings and thickenings, and external compression.

Trauma may produce entrapment by direct compression of the nerve by hematoma or by causing it to be bound up in scar tissue or the callus if there is a fracture. Radial nerve entrapment in the callus formed by distal humeral fractures is not infrequently seen. Healing of the fracture may alter the angle of the bone and produce a gradual stretching of the nerve. This is seen in tardy ulnar palsy where the carrying angle of the elbow has been altered by previous trauma.

Systemic disease (rheumatoid arthritis, for example) causes synovitis, tenosynovitis, and ligamentous thickening with resultant mechanical compression of nerves. Of course, such a disease may also produce an inflammatory process involving the entire nerve trunk. The mechanical compression is ordinarily seen at the wrist, where compression of the median nerve produces the carpal tunnel syndrome, but it is also frequently seen at the elbow, where the ulnar nerve is compressed, or behind the medial malleolus, where the tibial nerve is compressed. Pressure on the tibial nerve from this or from other causes elicits symptoms similar to those seen with the carpal tunnel syndrome and results in what is known as the tarsal tunnel syndrome.

Muscle spasm may produce nerve entrapment although it frequently goes unrecognized as a cause. The complaint of posterior occipital headache, with or without antecedent trauma, may be at least partially explained on the basis of entrapment of the great occipital nerve where it lies free in the deep muscles of the neck. Similarly, neuritic-type pain radiating to the dorsum of the hand in tennis elbow may be explained by entrapment of the radial nerve by muscle spasm. Another cause of pain and weakness of the extensor muscles is entrapment of the posterior interosseous nerve by synovitis of the radiohumeral joint in rheumatoid arthritis. This has been mistakenly diagnosed as rupture of the extensor tendons at the wrist. As the median nerve passes between the heads of the pronator teres muscle in the forearm, it may be entrapped by muscle spasm. This condition is often mistaken for a carpal tunnel syndrome since the symptoms of paresthesia predominate in both conditions. Figure 10-3*A* illustrates the location of the pronator teres muscle by skin outline; *B* indicates the point in the muscle which, when palpated, reproduces the patient's pain and paresthesia. Entrapment of the anterior interosseous nerve in the arm produces a different picture since it is purely motor; here weakness of the long flexor of the thumb and index finger is present. We have even speculated that entrapment of the radial nerve as it dips into the brachioradialis muscle in the upper lateral one-third of the forearm may account for the ubiquitous flexed ulnar deviation deformity of the hand in rheumatoid arthritis.

Space-occupying lesions such as lipomas, ganglia, bony exostoses, and other tumors may produce entrapment of a nerve, particularly where a nerve passes through a fibro-osseous tunnel (for example, the ulnar nerve in the canal of Guyon at the wrist) and cannot accommodate to the new pressure.

Congenital narrowings and thickenings may in time result in nerve entrapment. These conditions are believed to be a cause of entrapment of the lateral femoral nerve of the thigh, producing meralgia paresthetica. Another common cause of this disorder is excessive weight gain.

External causes of nerve compression are many: tight-fitting clothing, ski boots, improperly applied splints, braces and casts, crutch palsy and Saturday night palsy, for example. The reader is referred to Chapters 3 and 6 for details of the common features in the history, physical examination, and laboratory aids in diagnosis.

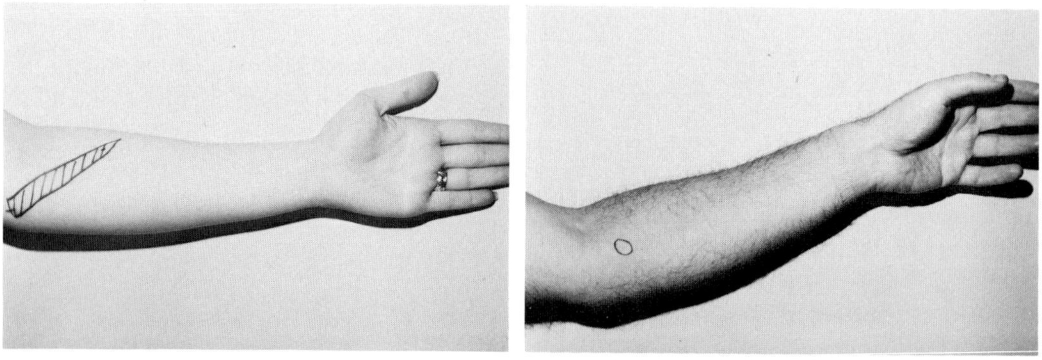

A *B*

Figure 10-3. *A*, Anatomical location of the pronator teres muscle. *B*, The point for deep palpation to reproduce the patient's symptoms. (Courtesy of Robert W. Downie, M.D.)

TREATMENT

Rest of the affected part by splinting and infiltration of steroid may prove helpful in mild cases. When muscle spasm is present, elimination of the spasm by injection of the muscle, by the use of ice massage or vapocoolant spray, and by stretching should be undertaken. Removal of the underlying cause by surgical means is frequently necessary.

Neuromas

Neuromas are painful enlargements of nerve that result from trauma. Most commonly the neuroma forms at the site of a severed nerve either in wounds or in amputation stumps, but trauma in the form of chronic pressure, irritation, or friction may produce a spindle-shaped neuroma on an intact nerve. An example of this is Morton's neuroma, which occurs at the bifurcation of the digital nerve to two toes, usually involving the nerve in the cleft between the third and fourth toes. Inadequate or unsuccessful surgical repair of severed nerves may produce a neuroma at the repair site.

HISTORY

Pain—sharp or burning (causalgic), intermittent or constant—is the most common symptom. At rest there is burning, and with the application of pressure this becomes a sharp acute pain. Vasomotor signs and alteration in the sweating pattern may be present. A neuroma may be the cause of the persistence of the symptom of phantom limb pain. Paresthesia, dysesthesias, and numbness may all be present.

PHYSICAL EXAMINATION

On palpation, localized tenderness to touch is invariably present. Gentle tapping over the site reproduces the patient's symptoms. Often, the enlargement of the nerve can be palpated.

ELECTRODIAGNOSIS

When a neuroma is suspected at the site of surgical repair, nerve conduction studies may be used to demonstrate an absence of impulses crossing the repair site.

TREATMENT

Physical therapeutic measures may be useful in treating the symptoms of pain resulting from a neuroma. Moist heat in the form of whirlpool baths reduces pain. Ultrasound decreases nerve conductivity, blocking the flow of painful impulses. Repeated tapping using a rubber mallet over the neuroma alters the sensitivity to pressure. In amputation cases, the patient can be instructed to tap the stump gently and repeatedly against a hard surface to get the same effect. Local injection of the neuroma with Novocain or a Novocain-cortisone combination may relieve the symptoms, as may the application of the vapocoolant spray over the neuroma site. Persistence of symptoms after trying the above methods may necessitate surgery.

Polyneuropathy

Polyneuropathy results from a wide variety of endogenous and exogenous causes and may produce problems in differential diagnosis from musculoskeletal pain. Involvement of multiple nerves is the usual picture, although we have seen single nerves clinically affected, and the extent of involvement varies from nerve to nerve. The distribution is usually symmetrical. Distal nerves are more commonly involved than proximal ones, but the infectious neuronitis of Guillain-Barré and the neuropathy of diabetic amyotrophy involve proximal nerves. Both motor and sensory fibers are involved as a rule, but some types of polyneuritis have a predilection for one or the other type of fiber. Facial nerves are spared except in the infectious neuronitis of Guillain-Barré.

ETIOLOGY

Since there are such varied causes of polyneuropathy, many different classifications are in use. A simplified list includes toxic, deficiency and metabolic, infectious and postinfectious, collagen disorders, and miscellaneous problems as causes.

Toxic causes include heavy metals (lead and arsenic), chemicals (tri-orthocresyl phosphate, various insecticides and fertilizers), and drugs (isoniazid, sulfanilamide).

Deficiency and metabolic causes include polyneuropathy secondary to chronic vitamin deficiencies, alcoholism, diabetes, uremia, and porphyria.

Infectious and postinfectious conditions include tetanus, diphtheria, and the neuronitis of Guillain-Barré.

Collagen vascular disorders such as polyarteritis, lupus erythematosus, and rheumatoid arthritis may produce polyneuropathy.

Miscellaneous problems include carcinoma, amyloid disease, and interstitial hypertrophic neuritis (Dejerine-Sottas).

HISTORY

Pain is the most common symptom. It is usually constant and described as burning or shooting in nature. Frequently the pain is inversely proportional to the degree of damage. Paresthesia or dysesthesia is also common, and sensory alterations in the form of either hypoesthesia or hyperesthesia are usually present. Patients often complain of weakness and difficulty with gait and with balance. Inquiries about recent infection, exposure to chemicals, and ingestion of drugs should be included in the taking of the history.

An unusual form of neuropathy is diabetic amyotrophy. Whereas the typical diabetic neuropathy is distal and primarily shows sensory involvement, this is proximal and primarily shows motor involvement. It occurs in elderly diabetics and is characterized by pain, weakness and wasting of the proximal muscles (usually of the lower extremities), and decreased reflexes. Sometimes there is involvement of the upper montoneuron, suggesting possible cord involvement as well. It is often reversible with good diabetic control.

PHYSICAL EXAMINATION

If lower extremity nerves are predominantly involved, the patient is observed to walk with a broad-based steppage gait. Muscle testing reveals weakness while sensory examination most commonly reveals alteration of pinprick and light touch and decrease in vibratory sensation. Deep tendon reflexes are decreased or absent, and if the process has been prolonged, muscle atrophy occurs. Tenderness on palpation of the muscles innervated by the affected nerve is a prominent sign. Tenderness on palpation of the affected nerve is always present.

LABORATORY TESTS

The laboratory is of value in differentiating systemic diseases associated with polyneuropathy. Cerebrospinal fluid examination may show an occasional pleocytosis and a mild increase in protein. In the neuronitis of Guillain-Barré, on the other hand, after an initial normal period a marked increase in protein unaccompanied by a rise in white blood cells occurs —the albuminocytological dissociation.

ELECTRODIAGNOSIS

Electromyographic examination reveals a variable amount of denervation potentials, not necessarily related to the degree of wallerian degeneration. Nerve conduction time studies show a decrease in the velocity, usually to below 40 meters per second. The decrease in velocity is most striking in the neuronitis of Guillain-Barré, in which velocities as low as 10 to 20 meters per second may be obtained. Terminal latencies remain normal or nearly so.

TREATMENT

Physical therapeutic measures consist in prevention of deformities, relief of pain, and muscle reeducation. Deformities occur either from soft tissue contractures or from overstretching of muscles. Proper bed positioning, appropriate splinting and bracing, and passive range of motion for the affected part help to prevent the occurrence of deformities. Moist heat and whirlpool relieve pain. Deep heat is contraindicated since it aggravates the pain. Interrupted galvanism is used during the period of paralysis, and muscle reeducation begins after the acute phase has passed. Electrical stimulation (faradism) is of value in the early phases of reeducation, particularly in severe involvement in which the patient has difficulty initiating movement. Strengthening of weakened muscles is accomplished by the use of graduated assistive and active exercises progressing into exercises with resistance. Ideally, function should not be resumed until a grade of "good" by manual muscle testing is achieved. Avoidance of fatigue is a cardinal rule here since fatigue increases weakness. Exercises performed on the mat, on the parallel bars, and standing free are prescribed to improve the patient's coordination and balance.

VASCULAR CONDITIONS

Peripheral Vascular Disease

This section deals with the differential diagnosis of musculoskeletal pain from peripheral vascular pain and the treatment by physical modalities of the various peripheral vascular disorders.

PAIN

Pain from involvement of the peripheral arteries is due to tissue ischemia. This may occur during activity or at rest.

Activity Pain. Intermittent claudication is a symptom due to ischemia secondary to peripheral vascular disease and is characterized by cramping pain. Rest promptly relieves it. The location of the pain is dependent upon the location of the vascular pathology.

Gradual obstruction of the aortic bifurcation (Leriche syndrome) produces bilateral buttock and leg pain, weakness and fatigue of the lower extremities, atrophy of the musculature of the leg, and difficulty in maintaining an erection. When the pathology is in the iliac artery, pain is present in the low back, buttock, and leg of the affected side. It may be accompanied by numbness. Involvement of the femoral artery, along its course or at the femoropopliteal junction, produces thigh and calf pain while obstruction of the popliteal artery or its branches produces pain in the calf, ankle, or foot. Coarctation of the aorta may give similar symptoms in the lower extremities.

These symptoms may be mistaken for those of a wide variety of musculoskeletal, neurological, and arthritic disorders. Figure 10-4 explains this diagrammatically. The vascular conditions are on the left and the conditions which might be confused with it are on the right. Conversely, the presence of vascular impairment of a minor degree may direct attention away from a primary disorder originating elsewhere. Such disorders include low back pain of musculoskeletal origin, nerve root

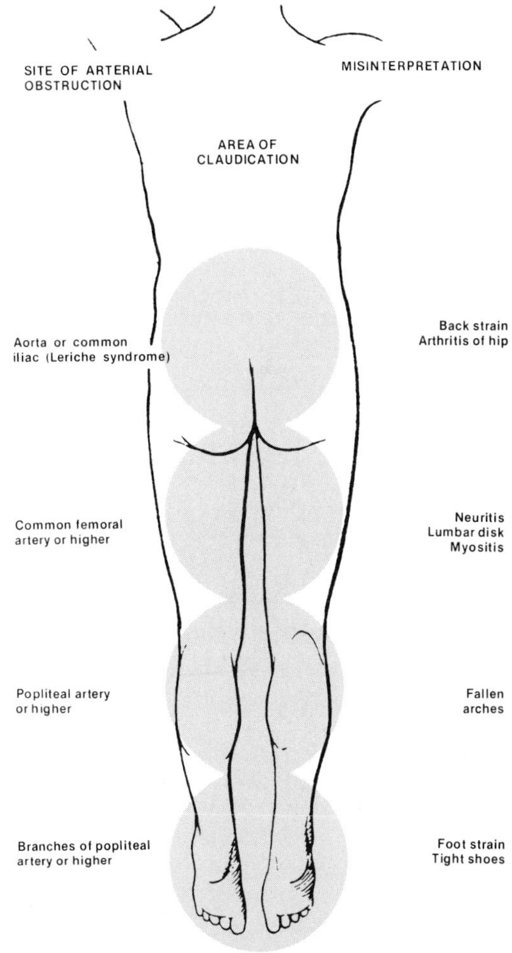

Figure 10-4. Conditions that might be mistaken for arterial obstruction at various sites. The reverse might also occur. (From R. W. Gifford and J. W. Hurst, *G.P.* 16:89, 1957.)

compression, spinal cord tumor, arthritis of the hip, peripheral neuritis, and a march fracture.

Rest Pain. Pain at rest arises from ischemia of a wide variety of tissues. The acute onset of severe unilateral extremity involvement accompanied by the "five Ps" of pain, pallor, pulselessness, paresthesia, and paralysis signifies the onset of acute arterial occlusion. Involvement of the vasa nervorum produces the symptoms of ischemic neuropathy. Pain in this instance is usually described by the patient as burning or shooting in nature and may be accompanied by paresthesia. Restless legs are probably an early manifestation of arterial insufficiency. Ischemia of the skin and subcutaneous tissues is known as pretrophic pain and is characterized by the patient as burning and boring. A more advanced state of this ischemia is frank ulceration and gangrene, which may also involve the underlying muscle and fascia, producing pain that is severe, steady, and boring. All of the above chronic causes of pain are usually worse at night and are relieved to some degree by the dangling of the leg over the side of the bed and by the frequent massaging of the extremity. The necessity of narcotics for pain relief is commonplace.

Venous Insufficiency Pain. Pain arising from varicosities, stasis, and ulceration is less severe than that from involvement of the arterial tree. It is usually described as a heaviness or an aching and is worse after prolonged periods of standing. Rest, elevation of the extremities, and elastic stockings provide some relief of symptoms.

PHYSICAL METHODS OF TREATMENT

Chronic Organic Occlusive Disease. Included in this category are arteriosclerosis obliterans, with or without associated diabetes, and thromboangiitis obliterans (Buerger's disease). General principles of treatment include prevention of complications by such means as relief of pain; proper foot care; elimination or reduction of adverse factors such as smoking and infection, control of cholesterol levels by diet and medication, and control of associated disease processes such as diabetes; use of vasodilator medications; and surgical treatment—sympathectomy, grafting, or amputation.

Relief of pain and promotion of circulation are the objectives of physical treatment. Whirlpool baths at body temperature usually reduce pain on a temporary basis. Improvement of the circulation can be aided by the application of heat. Since sensation may be altered, and since impairment of the circulation may prevent adequate cooling of the tissues, direct heating should be prescribed with great caution. Reflex heating, from a source of deep heat (as in diathermy) placed over the abdomen, is preferable. Coplanar heating with diathermy through the length of the extremity is helpful. Introduction of vasodilators such as histamine and carbachol through the skin by means of iontophoresis is also used. Perhaps the best way to improve circulation, or at least the ability of the patient to function, is by a series of graded active exercises. The stationary bicycle is fine, but just as effective is graded walking, both on the level and on an incline. Passive postural exercises (Buerger-Allen exercises) have little to offer except for the psychological benefit that accrues to the patient.

Massage, whether done by hand or by various pumping devices, would appear to be of minimal benefit in this problem. The same holds true for the devices that cause muscle contractions in an attempt to promote the pumping action of the muscles.

Ulcers. Ulcers may be due to pathology of the arterial or venous system. Most common are ulcers caused by chronic venous insufficiency: stasis ulcers. These develop rapidly and are characteristically located over the medial aspect of the lower half of the tibia, especially over the medial malleolus, where the saphenous veins of the leg originate. In chronic cases they are accompanied by hyperpigmentation of the skin.

Ischemic ulcers are most commonly caused by abnormality of the larger arteries secondary to arteriosclerosis obliterans or to thromboangiitis obliterans. They are usually located over the toes. Less common are ulcers secondary to involvement of the small arteries and arterioles, which occur secondary to hypertension. These are usually located over the lateral surface of the ankle or over the posterior and lateral surface of the leg.

Simple stasis ulcers can best be treated by continuous elevation and the application of warm saline soaks. An alternative method of treatment is the application of an Unna paste boot to provide firm pressure around the leg; the patient is then permitted to be up and walking. Figure 10-5 illustrates an Unna boot applied to a patient's lower leg and foot. The boot is applied only after the ulcer is clean and granulating. It is changed every week until healing has taken place. If the ulcer is extensive, and superimposed infection has occurred, whirlpool baths with a mild detergent added, manual debridement, and cold quartz ultraviolet light help to control the infection. Proteolytic enzymes are often added topically for purposes of further debridement. Zinc iontophoresis promotes healing by the production of a physiological dressing. Friction massage about the edges of the ulcer stimulates granulation tissue. Edema massage (see p. 137) as a treatment regimen for the whole leg should be used. When healing has commenced, an Unna paste boot may then be reapplied.

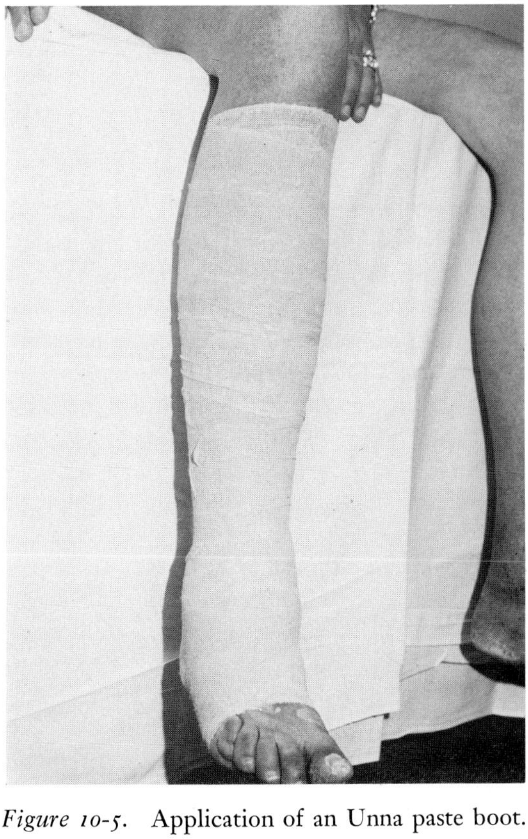

Figure 10-5. Application of an Unna paste boot.

In contradistinction to the treatment of venous ulcers, pain relief of ischemic ulcers is best accomplished by dependency of the extremities and avoidance of elevation. To help achieve this effect, the head of the bed may be raised. Whirlpool baths at body temperature are used for the relief of pain and debridement. Infection is controlled by manual debriding, application of topical debriding agents, and cold quartz ultraviolet light. Vascular dilation is assisted by distant reflex heating and by the local application of ultrasound to the circumference of the ulcer. The ulcer is treated locally in the same manner as the venous ulcer.

STROKE

It is the pain problems associated with stroke that concern us. Elderly people tend to have a wide variety of musculoskeletal problems. When a stroke occurs, subsequently there is interference with muscle action, loss of balance takes place, and abnormal mechanical stresses are superimposed. Many of these problems then become more acute, though they are not part of the results of the cerebrovascular accident.

A painful shoulder on the affected side is a common—and the most distressing—feature of the musculoskeletal involvement associated with stroke. Subluxation of the shoulder by the unsupported deadweight of the arm on the rotator cuff muscles is the usual explanation for shoulder pain. In fact, the dip apparent between the acromion and the greater tuberosity of the humerus is usually due to deltoid atrophy; the head of the humerus is not subluxated at all. This has to be differentiated from a true subluxation, an extreme example of which is illustrated in Figure 10-6.

When a true subluxation is present, a typical capsular pattern of pain is seen, with loss of passive range of motion in all planes. Palpation reveals a characteristic area of tenderness over the anterior portion of the greater tuberosity. Equally characteristic is an exquisitely tender trigger point in the belly of the infraspinatus muscle. There may be more than one trigger point. Firm pressure over these points produces local pain and often referred pain to the anterior aspect of the shoulder. Other trigger points are commonly noted in the middle and upper portions of the trapezius muscle, in the supraspinatus muscle, and in the deltoid.

When no true subluxation is present, the pain from disuse capsulitis is due to dysfunction and responds to manipulative therapy. There are often associated trigger points that must be treated as well. Maintenance of passive movement can be achieved by the patient with overhead pulleys or by a cane which he moves with the uninvolved hand, the involved hand being wrapped around it at the other end. Complementing this is a treatment program of heat, passive and active range of motion, resisted exercises, and gentle stretching.

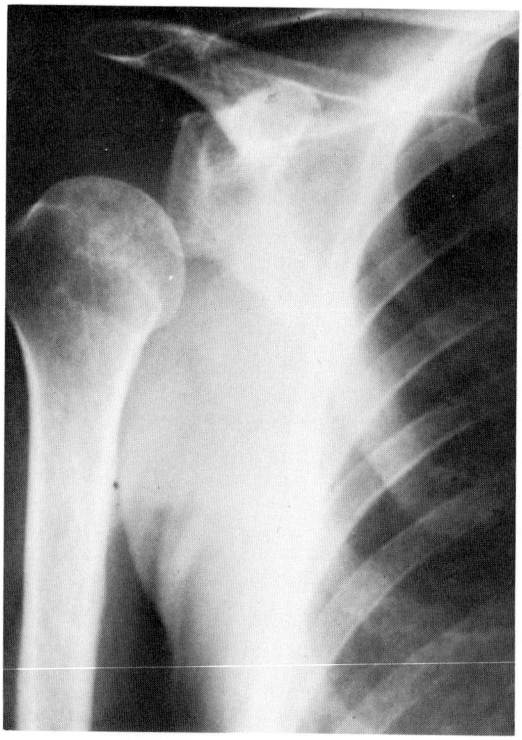

Figure 10-6. X-ray picture showing a true subluxation of the head of the humerus on the glenoid in a patient with stroke.

When subluxation is present, application of a sling in the recumbent position with the humerus elevated high in the glenoid fossa relieves the tension on the rotator cuff and deltoid muscles. Analgesics may have to be prescribed in the early stages of the condition. Spasticity can be relieved at times by injection into an involved muscle of 45 percent alcohol or 3 percent phenol. Unfortunately, the results are not usually permanent, and the injections have to be repeated at variable periods of time. A new drug that acts directly on the muscle to decrease spasticity is dantrolene sodium (Dantrium), given in doses from 75 mg per day to 300 mg per day. However, recent studies have demonstrated significant toxicity, limiting its usefulness.

Prophylactic treatment to prevent some of these morbid changes is achieved by good therapy in the first month of illness. It can be instituted even if the patient is unconscious. Maintenance of joint range of motion, actively and passively, and maintenance of muscle length by correct positioning of the patient in bed are simple maneuvers that can be performed by a therapist or nurse or by the family and can significantly reduce the incidence of these problems. The shoulder-hand syndrome, or reflex sympathetic dystrophy, is sometimes a complication of hemiplegia. When it is, the physical therapeutic measures described above may be sufficient to reverse the process. If not, one or more stellate ganglion blocks on the affected side will break the pain cycle and permit the use of a restorative program of therapy.

Prolonged immobilization, inadequate balance, loss of body image, loss of visual fields, the development of contractures, and muscle weakness are all factors leading to an increased incidence of disability. Many joints can have preexisting degenerative arthritic changes, and the further alteration of mechanics leads to an exacerbation of pain in them. Fracture of the hip occurs with much greater frequency on the hemiplegic side than on the uninvolved side.

Not all these problems are preventable, but awareness of some of the etiological factors may lessen the incidence of their occurrence. Pressure sores are almost uniformly preventable, and their presence indicates a lack of adequate nursing care.

COMBINED NEURITIC AND VASCULAR CONDITIONS

Thoracic Outlet Syndromes

A variety of mechanical deformities produce compression of the neurovascular structures as they traverse the thoracic outlet to the arm. These include cervical rib, scalene band, scalene muscle, pectoralis minor muscle, and reduced costoclavicular space brought on by poor posture. As other causes of cervicobrachial pain have become increasingly recognized, the diagnosis of pain attributable to one of the thoracic outlet syndromes has become less frequent. Among the important conditions to be differentiated from these syndromes are the various states that cause cervical root compression, and the peripheral pain syndromes, particularly those of nerve entrapment such as the carpal tunnel syndrome.

Other conditions may produce compression of the neurovascular structures in this area: subclavian or axillary artery aneurysm, apical lung cancer, pathological lymph nodes, enlarged thyroid, and fracture of the clavicle. Depending upon the type and location of decompressing factor, either neural or vascular symptoms will predominate.

ETIOLOGY

Four etiological categories for the thoracic outlet syndrome have been described. First is the cervical rib, which is present in one-half of one percent of the normal population. It may vary in size from a bossing of the transverse process to a completely formed rib. Attachment may be to the sternum, to the first costal cartilage, or to the first thoracic rib. When it is incompletely formed, attachment is by means of a fibrous band.

Second is the scalenus anterior muscle. It may produce compression by enlargement, by spasm, or by anatomical variations such as a broad insertion or insertion by means of two bands that surround the subclavian artery and lower trunk of the plexus. In Figure 9-13 (p. 192) the predictable pattern of pain from scalene spasm is illustrated.

Third, abnormalities of the costoclavicular space—the space between the clavicle and the first thoracic rib—either from congenital causes such as bony abnormalities or from acquired ones such as muscle weakness or fractures, may produce compression of the neurovascular structures. Fourth, hyperabduction of the arm, a maneuver that causes the pectoralis minor muscle to be drawn taut, compresses the neurovascular structures between the muscle and the coracoid process of the scapula.

HISTORY

Symptoms are usually unilateral but may be bilateral. They often appear after the carrying of a heavy object such as a suitcase. Paresthesia is the most common complaint, and its appearance at night during sleep is characteristic. Therefore, inquiries about the patient's sleeping position are important. Pain is usually along the ulnar border of the forearm and may enter the hand in either the ulnar nerve distribution or the median nerve distribution, since both are supplied by roots that originate from the lower trunk, namely, the eighth cervical and first thoracic. Weakness may be present, but it is likely to be described as a clumsiness. Numbness, or the sensation of numbness, is often noted. The hand may feel cold, and when compression is severe, Raynaud's phenomenon may be present. Wasting of the intrinsic muscles of the hand is rare.

PHYSICAL EXAMINATION

General features to be observed are abnormalities of posture such as severe sloping or rounding of the shoulders. Pendulous breasts may place an abnormal stress on the muscles that elevate the first rib. A leg length inequality may place an abnormal stress on one side or the other of the neck. Blood pressure readings of both arms should be carried out since occlusion of the subclavian or axillary arteries will lower the brachial artery pressure reading on the affected side. Observation of the extremity for color changes or atrophy and palpation for thyroid enlargement in the neck or other glands, tenderness of the scalenus anticus muscle (examined by palpation behind the lateral border of the sternocleidomastoid muscle), and temperature changes in the limb should be carried out.

Specific maneuvers should be performed to elicit reproduction of the patient's symptoms. Obliteration of the pulse without reproduction of the symptoms is seldom sufficient to make the diagnosis, since its occurrence is common in asymptomatic individuals. A cervical rib may be palpated as a prominence in the supraclavicular fossa. Deep pressure here may reproduce the symptoms. The Adson maneuver, performed with the arm on the lap, rotation and extension of the neck toward the involved side, and a deeply held breath, will bring about decrease in the pulse and reproduction of the symptoms in the cervical rib and scalenus anticus syndromes. Auscultation of the neck may yield a systolic bruit suggesting compression of the brachial artery. Local infiltration of the scalenus anterior muscle with Xylocaine, if it is felt to be in spasm, may temporarily abolish symptoms and is therefore an aid to diagnosis. The use of the vapocoolant spray and stretching has the same effect.

With the arms still in the lap, adopting the position of a military brace of the shoulders may elicit the same symptoms and signs in the costoclavicular syndrome. Elevation and extension of the arm over the head compresses the neurovascular structures between the pectoralis minor muscle and the coracoid process and reproduces the symptoms and signs of the pectoralis minor or hyperabduction syndrome. Care should be taken to flex the wrist during palpation of the pulse since extension eliminates the slight tortuosity of the radial artery which of itself may produce obliteration of the pulse.

Since the exact origin of the symptoms is often unclear, a complete examination of the neck and arm should be performed. Particular attention must be paid to detecting cervical radiculitis, ulnar nerve entrapment, and the carpal tunnel syndrome. Careful palpation for trigger points in the forearm and upper back should be carried out since these, when present, may mimic the neuritic symptoms.

LABORATORY TESTS

X-ray examination of the cervical spine may reveal the presence of a cervical rib. However, its causative role should not be presumed unless the clinical picture clearly eliminates other probable causes. Other abnormalities, such as a healed fracture of the clavicle with extensive callus formation, exostoses of the clavicle or first rib, or alteration in the bony dimensions of the outlet, may be demonstrated.

Electromyography and nerve conduction studies are usually unhelpful in these syndromes unless severe nerve damage has occurred. Slowing of motor conduction between Erb's point and the axilla, when present, is strong evidence for an outlet syndrome. Arteriography may be performed when the vascular component predominates and obstruction is suspected. Myelography should be performed if there is any suspicion that the cause of the symptoms is more proximal than the thoracic outlet.

TREATMENT

Nonsurgical treatment consists of medication to control the symptoms and physical therapeutic measures. In addition to the usual analgesic medications, the combination of Dilantin, 100 mg t.i.d., and Benadryl, 50 mg t.i.d., may produce adequate control of the symptoms. Physical therapeutic measures are directed at correction of postural abnormalities, faulty mechanics, secondary muscle spasms, extinguishing irritable trigger points, and joint dysfunction. If muscle weakness has resulted in inability to support the outlet adequately, cor-

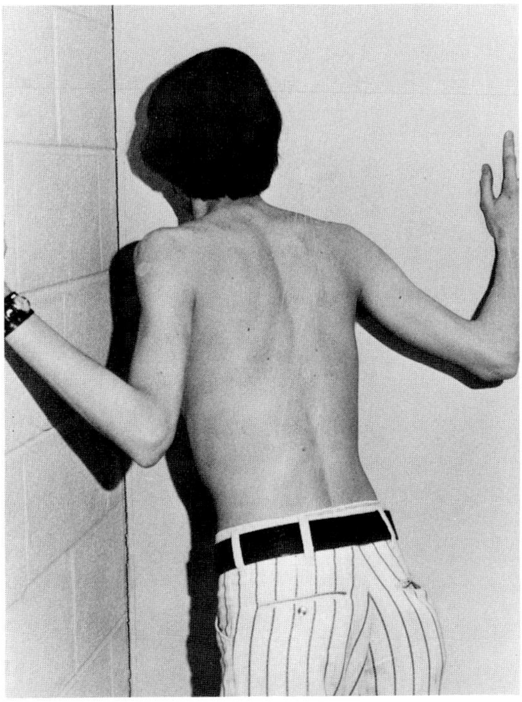

Figure 10-7. An exercise to stretch tight muscles and fascia in patients suffering from the thoracic outlet syndrome. The patient stands in the corner of the room and does pressing exercises.

rective exercises are applied. Tightness of the soft tissues or poor posture may be corrected by appropriate stretching exercise therapy. Figure 10-7 illustrates an exercise designed to open the thoracic outlets. The costovertebral joints of the first ribs may have to be mobilized by manipulation, which is conveniently performed by springing over the manubrium sterni. Compensation for inequality of leg length may be necessary to remove the results of abnormal stresses from a short leg. Wearing of a longline bra or the removal of bra straps for pendulous breasts may also reduce stress on the outlet. Poor sleeping habits such as hyperabduction of the arm may contribute to symptoms and should be corrected. Attempts at alteration of the carrying angle of the neck may be indicated. If trigger points, muscle spasm, and muscle shortening have developed in adjacent areas such as the upper back, efforts should be made to correct them. If all these measures have failed to produce relief of symptoms, surgical intervention may be necessary.

BIBLIOGRAPHY

Abramson, D. I. Physiologic basis for the use of physical agents in peripheral vascular disorders. *Arch. Phys. Med. Rehabil.* 46:216, 1965.

Fairbairn, J. F., Juergens, J. L., Spittell, J. A., and Allen-Barker-Hines. *Peripheral Vascular Disease* (4th ed.). Philadelphia: Saunders, 1972.

Garland, H. Diabetic amyotrophy. *Br. Med. J.* 2:1287, 1955.

Gifford, R. W., and Hurst, J. W. A note on the location of intermittent claudication. *G.P.* 16: 89, 1957.

Goodgold, J., Kopell, H. P., and Spielholz, N. I. The tarsal tunnel syndrome. *N. Engl. J. Med.* 273:742, 1965.

Hines, E. A. Some types of distress in the lower extremities simulating peripheral vascular disease. *Med. Clin. North Am.* 42:991, 1958.

Kopell, H. P., and Thompson, W. A. L. *Peripheral Entrapment Neuropathies.* Baltimore: Williams & Wilkins, 1963.

Nelson, P. A. Treatment of patients with cervicodorsal outlet syndrome. *J.A.M.A.* 163:1570, 1957.

Policoff, L. D. The management of the patient with arteriosclerosis obliterans. *Arch. Phys. Med. Rehabil.* 42:584, 1961.

Tardieu, G., Tardieu, L., Hariga, J., and Gagnard, L. Treatment of spasticity by injection of diluted alcohol at the motor point or by epidural route. *Dev. Med. Child Neurol.* 10:555, 1968.

Zohn, D. A., Hughes, A. S., and Haase, K. H. Carpal tunnel syndrome: A review. *Arch. Phys. Med. Rehabil.* 43:420, 1962.

Conclusion

11
Toward Improved Patient Care

It is to be anticipated that the problems of musculoskeletal pain will not be subject to any single medical breakthrough and vanish in the near future. They will continue to affect large numbers of people, who in turn will make up a high percentage of the patients seen in the physician's office. Principles of physical diagnosis and physical treatment of these problems have been emphasized in this text rather than discussions of specific treatment for each illness. These diagnostic and treatment methods have been based on physiological principles and have proved widely effective. Their limitations as well as their benefits have been stressed as well.

Persons in other specialties in medicine contemplate the use of computers in diagnosis and treatment. We believe that, for them, the day of the computer may be fast approaching. But we trust that no one imagines that day is anywhere close as regards dealing with diagnostic problems arising from the musculoskeletal system.

The medical education gap in this field has been alluded to, and it is our hope that this text will help narrow it. Failure of the medical profession to treat these problems adequately can only lead to the proliferation of unqualified cultists who will gladly step in where medicine fails to fulfill its role.

A further consideration is the extensive intrusion of medicolegal problems into the field of medicine. While this unfortunate development dictates in many cases what procedures will be carried out, whether they are indicated or not, we have tried to emphasize that reliance in the majority of these cases should be on history and physical examination and not on laboratory and x-ray studies.

Physicians from many specialties treat the problems of musculoskeletal pain. In reaching them with this text, we hope that a cross-fertilization of ideas will take place so that the patients with these problems will be the recipients of improved care.

Index

Index